Adelard of Bath, Conversations with his Nephew

On the Same and the Different, Questions on Natural Science, and *On Birds*

This book consists of editions and translations of the three known texts in which Adelard of Bath (*ca.* 1080–*ca.* 1152) addresses his 'nephew': an exhortation to the study of the liberal arts which constitute 'philosophy' (*On the Same and the Different*), a dialogue on the nature of things in which rational causes are sought (*Questions on Natural Science*), and a discussion concerning the upbringing and medication of hawks (*On Birds*). Adelard uses an urbane and lively style, and laces his texts with jokes, puns, and subtle echoes of classical Latin authors. The context is the Court schools of Norman bishops and dukes, and in particular, the education of King Henry II, with which Adelard seems to have been involved. A preface provides the results of the most recent research on the life and works of Adelard of Bath, and an index guides the reader through the diversity of the subject-matter in the texts.

Dr Charles Burnett is a Lecturer at the Warburg Institute, University of London. His books include editions and translations of Hermann of Carinthia's *De essentiis*, Abū Maʿshar's *Abbreviation of the Introduction to Astrology*, *Jesuit Plays on Japan and English Recusancy* (with Masahiro Takenaka), and a collection of articles in *Magic and Divination in the Middle Ages: Texts and Techniques in the Islamic and Christian Worlds*. In November 1996 he gave the Panizzi Lectures at the British Library, published as 'The Introduction of Arabic Learning into England'.

Cambridge Medieval Classics 9

General editor
PETER DRONKE, FBA
Professor of Medieval Latin Literature, University of Cambridge

This series is designed to provide bilingual editions of medieval Latin and Greek works of prose, poetry, and drama dating from the period *ca.* 350–*ca.* 1350. The original texts are offered on left-hand pages, with facing-page versions in lively modern English, newly translated for the series. There are introductions, and explanatory and textual notes.

The Cambridge Medieval Classics series allows access, often for the first time, to outstanding writing of the Middle Ages, with an emphasis on texts that are representative of key literary traditions and which offer penetrating insights into the culture of medieval Europe. Medieval politics, society, humour, and religion are all represented in the range of editions produced here. Students and scholars of the literature, thought, and history of the Middle Ages, as well as more general readers (including those with no knowledge of Latin or Greek), will be attracted by this unique opportunity to read vivid texts of wide interest from the years between the decline of the Roman empire and the rise of vernacular writing.

Already published

1 Nine Medieval Latin Plays, translated and edited by PETER DRONKE
2 Hugh Primas and the Archpoet, translated and edited by FLEUR ADCOCK
3 Johannes de Hauvilla, *Architrenius*, translated and edited by WINTHROP WETHERBEE
4 Dante, *Monarchia*, translated and edited by PRUE SHAW
5 Dante, *De Vulgari Eloquentia*, translated and edited by STEVEN BOTTERILL
6 Gregory of Nazianzus, Autobiographical Poems, translated and edited by CAROLINNE WHITE
7 *Digenis Akritis*, translated and edited by ELIZABETH JEFFREYS
8 Dhuoda, Handbook for her Warrior Son: *Liber Manualis*, translated and edited by MARCELLE THIÉBAUX

A scholar teaching a student, with four hawks on a perch above. The opening illustration to the 'Aviculaires des oisiaus de proie', an Anglo-Norman compilation of falconry based mainly on Adelard's De avibus tractatus.

Adelard of Bath, Conversations with his Nephew

On the Same and the Different, *Questions on Natural Science*, and *On Birds*

Edited and translated by

CHARLES BURNETT,

with the collaboration of Italo Ronca,

Pedro Mantas España and Baudouin van den Abeele

CAMBRIDGE
UNIVERSITY PRESS

PUBLISHED BY THE PRESS SYNDICATE OF THE UNIVERSITY OF CAMBRIDGE
The Pitt Building, Trumpington Street, Cambridge CB2 1RP

CAMBRIDGE UNIVERSITY PRESS
The Edinburgh Building, Cambridge CB2 2RU, United Kingdom
40 West 20th Street, New York, NY 10011–4211, USA
10 Stamford Road, Oakleigh, Melbourne 3166, Australia

First published 1998

Printed and bound in Great Britain by Biddles Ltd, Guildford and King's Lynn

Typeset in Photina 10/12pt [CE]

A catalogue record for this book is available from the British Library

ISBN 0 521 39471 6

Contents

Illustrations

Acknowledgements

Italo Ronca has shared the task of collating the manuscripts of the *Questiones naturales* and *De eodem et diverso* and has meticulously checked the editions and translations of both texts. Pedro Mantas España has worked closely with Charles Burnett on the *De eodem et diverso* as part of the subject of his doctoral dissertation ('El *De eodem et diverso* de Adelardo de Bath: la búsqueda de una "síntesis" en el renacimiento del siglo doce', Universidad Autónoma de Madrid, 1994); he has provided expert advice on computers, and has produced a concordance for the *De eodem et diverso*. Baudouin van den Abeele has written the introduction and commentary to, and prepared the text and translation of, the *De avibus tractatus*. The final versions of all three texts and translations are the responsibility of Charles Burnett, who has also written the introductory material and the commentary (except in the case of the *De avibus tractatus*) and the index.

We are indebted to Gerrit Bos, Luc Deitz, Michael Evans, Margareta Fredborg, Harry Hine, David Howlett, Yukio Iwakuma, Onno Kneepkens, Roland Mayer, Patrick McGurk, Constant Mews, Lodi Nauta, Michael Reeve, Andreas Speer, Patricia Stirnemann, Robert Sharples, David Vitali, Tessa Webber, Jeffrey West, and Irene Zwiep, for reading sections of the book, and for advice in their respective fields. Barrie Hall and Lucy McGuinness have generously checked the Latin texts and the English translations. Ron Thomson of Toronto kindly put at our disposal a draft translation of the first two thirds of the *Questiones naturales* that he had prepared. The editorial suggestions of Peter Dronke have helped to make the edition more reliable and the translation more readable. The Warburg Institute has provided the genial atmosphere in which most of the collaboration has taken place, and funding for microfilms. Finally, I am very grateful for the care with which the staff of Cambridge University Press has prepared this volume.

For Brian Lawn
senex prestigiosus

Introduction

I Adelard of Bath's Life and Works

At the beginning of his *Questiones naturales* Adelard says that a nephew of his urged him to reveal 'something new from the studies of the Arabs'.[1] It is novelty and inventiveness that characterize the works of Adelard of Bath (*ca.* 1080–*ca.* 1152), the twelfth-century English *philosophus* and translator of Arabic texts into Latin. And of all these works, those that feature his 'nephew', either as a silent auditor or as a spirited discussant, are the most finely wrought and the most entertaining and instructive. It is these works that are edited and translated in this volume.

In the Middle Ages Adelard was best known for his rediscovery and teaching of geometry. This well-deserved reputation rested on the fact that he made the first full translation of Euclid's *Elements* in the West – a translation of an Arabic version – and inaugurated the process of interpreting the text for a western audience.[2] The *Elements* were the opening item in an Arabic programme of study which culminated in mathematical astronomy and practical astrology, and Adelard also translated several texts relevant to this programme: the astronomical tables of al-Khwārizmī, the *Abbreviation of the Introduction to Astrology* of Abū Maʿshar, the *Centiloquium* (astrological apophthegms) attributed to Ptolemy, and the *Book of Talismans* (*Liber prestigiorum Thebidis secundum Hermetem et Ptholomeum*) of Thābit ibn Qurra.[3] These literal translations provided him with material for teaching the quadrivium (the mathematical sciences) as a whole, and evidence of this teaching is shown in his original works on the abacus (*Regule abaci*) and the astrolabe (*De opere astrolapsus*; this includes a summary of Ptolemaic

1 *QN*, Prologue: 'Nepos quidam meus . . . aliquid Arabicorum studiorum novum me proponere exhortatus est.'

2 See Clagett, 'Medieval Latin Translations' and Burnett, 'Latin and Arabic Influences'.

3 For these works, their manuscripts and editions, see Burnett (ed.), *Adelard of Bath: An English Scientist and Arabist of the Early Twelfth Century* and Bibliography below.

astronomy), and glosses to Boethius' *Arithmetic* and *Music*.[4] The *Regule abaci* is addressed to a certain 'H.', the *De opere astrolapsus* to 'Henry, grandson of the King';[5] in turn, a certain 'H. Ocreatus' addresses a work on doing arithmetic 'the Saracen way' to 'his master Adelard'.[6]

Besides these works on mathematics, however, Adelard wrote three texts which have the common feature of involving the *persona* of a nephew (*nepos*): the *De eodem et diverso* ('On the Same and the Different'; *DED*), the *Questiones naturales* ('Questions on Natural Science'; *QN*), and the *De avibus tractatus* ('Treatise on Birds'; *DAT*). These three works form a kind of literary triptych in which Adelard appears to be deliberately experimenting with different genres, and expressing his allegiance to different cultural traditions. In the first, Adelard takes up the style of the 'protrepticus' (i.e., exhortation to the study of philosophy) in the Greek tradition of Aristotle, Cicero and Boethius.[7] His avowed sources are Greek, and he gives prominence to his experiences in Sicily and Magna Graecia, i.e., the Greek-speaking part of Italy. In the second, Adelard adopts the genre of the 'questions on nature'. This also was an ancient form, with a pedigree going back to the *Problemata* of the Lyceum and extending through the *Problemata* attributed to Alexander of Aphrodisias to the 'natural questions' associated especially with the medical school of Salerno. This family, however, also had a distinguished Arabic branch,[8] and Adelard chooses to present his *QN* as a forum for Arabic learning and to refer in it to his experiences in the Principality of Antioch. For the third text, finally, Adelard takes over the genre of medical texts, addressing diseases arranged *a capite ad calcem* ('from head to toe'), and claims, this time, that his knowledge comes from an English source ('the books of King Harold'). This demarcation of three cultural traditions is a literary device of considerable sophistication. It remains to be seen to what extent it conforms to Adelard's true sources, and genuine experiences.

4 See Burnett, 'Adelard, Music and the Quadrivium'.

5 The designation of Henry II as 'nepos regis' (with 'nepos' in the classical sense of 'grandson' or 'descendant' rather than the specific medieval sense of 'nephew') occurs also in Henry of Huntingdon, *Historia Anglorum*, x.33, ed. D. Greenway, Oxford, 1996, p. 761: 'Dux Henrice, nepos Henrici maxime magni . . . '

6 Burnett, '*Algorismi vel helcep*'; this article includes an edition of Ocreatus' *Helcep Sarracenicum* ('Saracen calculation'), of which the prologue is headed: 'Prologus H. Ocreati in Helcep ad Aelardum Batensem magistrum suum'.

7 Cicero's *Hortensius* and its Greek model, the *Protrepticus* of Aristotle, survive only in fragments, but were the progenitors of the genre of 'exhortations to the study of philosophy'.

8 See Daiber, 'Masāʾil'.

Adelard's biography can be only partially reconstructed from his own testimonies and from independent references.[9] He is consistently referred to as being 'of Bath', and his connection with the city is assured by a self-reference in his *QN* (Q28, p. 150: 'non Stoicum me set Bathoniensem dico') and the use of Bath in practical examples in his works on the astrolabe and on talismans. The name 'Ađelhardus' occurs four times in the cartulary of Bath Priory, twice (in a document from William II's reign and another of 1100) as *dapifer* ('steward') to the bishop of Bath, John de Villula (1088–1122), and twice (in 1106 and 1122) among the 'milites et ministri' of the same bishop.[10] The document of 1106 describes Ađelardus as 'filius Fastradi'; Fastrad was one of the principal tenants of Giso, John of Villula's predecessor. On the presumption that this Ađelardus is Adelard of Bath the scholar, his birth must be placed in 1080 or before, but it is perhaps surprising that Adelard should have reached the high position of *dapifer* to the bishop by 1100, and the four Ađelarduses mentioned in the Bath cartulary may not be the same person. Two further charters mention an Adelard from Bath: 'Adelard(us) de Bada' appears in the Pipe Roll of 1130 among tenants in Wiltshire exempted from a murder fine,[11] and 'Adelard' de Bath'nian' was a witness to a charter of King Stephen drawn up in favour of the Abbey of Bury St Edmunds between 1135 and 1139.[12] That Adelard was teaching at Bath itself is suggested by the fact that his pupil Ocreatus probably belonged to the most prominent family in the city at that period – the Husseys.[13]

Most of the remaining details concerning his biography must be inferred from his writings, and due allowance must be made for the

[9] The pioneering study of the biography of Adelard is the chapter on 'Adelard of Bath' in Haskins, *Studies in the History of Mediaeval Science*, pp. 20–42; for further information see Bliemetzrieder, *Athelhard von Bath*, Thorndike, *History of Magic*, II, pp. 14–49, Lawn, *Salernitan Questions*, pp. 20–31, Burnett, 'Introduction' and M. Gibson, 'Adelard of Bath' in Burnett (ed.), *Adelard of Bath: An English Scientist and Arabist of the Early Twelfth Century*, pp. 1–6 and 7–16, and Southern, *Robert Grosseteste*, pp. xlv–xlvii and 85–8. A useful up-to-date guide to the historical background and to the various hypotheses concerning Adelard's biography is provided by Louise Cochrane in *Adelard of Bath: The First English Scientist*.

[10] W. Hunt, *Two Chartularies of the Priory of St Peter at Bath*, Somerset Record Society, 7, 1893, I, nos 34, 41, 53, 54.

[11] *Pipe Roll 31 Henry I*: 'Nova placita et novae conventiones', Pipe Roll Society, London, 1929, p. 72.

[12] *Regesta regum anglo-normannorum 1066–1154*, vol. 3, eds H. A. Cronne and R. H. C. Davis, Oxford, 1968, p. 282, no. 764. The editors consider 'Adelard' de Bath'nian' (apostrophes indicate abbreviation-marks) to be a corruption of 'apud Bath'nian' ('at Bath').

[13] See Burnett, 'Ocreatus'.

literary context of these details. Those in the *DED* suggest that he studied in Tours – this famous centre in the Loire valley is the setting of the work, and the place where he read music as part of his *Gallica studia*, met a famous, but unnamed, wise man who elucidated the science of astronomy for him, and (perhaps) played the cittern in front of a queen – and that, on travelling from Salerno, he met a Greek philosopher in 'Grecia Maior' who was an expert in medicine and the nature of things. Those in the *QN* state that Adelard dismissed his pupils outside Laon, in preparation for a seven-year period of travel devoted to the *Arabum studia* ('the studies of the Arabs'), and mention specifically his experiences in Tarsus and during an earthquake at Mamistra (Misis, ancient Mopsuestia) in Cilicia. Both these places were in the Norman Principality of Antioch, and possible confirmation of Adelard's visit there is provided by a reference in a fuller translation of the same *Book of Talismans* by Thābit ibn Qurra that Adelard had translated: the translation by John of Seville and Limia. In the preface John refers to a previous, partial, translation by an 'Antiochene'; the reference would seem to be to Adelard's translation.[14]

The *DED* is dedicated to William, bishop of Syracuse in Sicily; his bishopric must fall between the last mention of a predecessor (1105) and the first mention of a successor (1124). The *QN* are dedicated in most manuscripts to Richard, bishop of Bayeux (one family of manuscripts dedicates the work to 'presul G' who could again be the bishop of Syracuse).[15] This Richard was either Richard FitzSamson, bishop from 1107 to 1133 or Richard of Kent, bishop from 1135 to 1142. The prologue also mentions a recent return to England during the reign of Henry the son of William (= Henry I, 1100–35) after a long absence.[16]

This information suggests that Adelard grew up in Bath just at the

[14] See the prologue to Thābit ibn Qurra's *De imaginibus*, ed. Burnett in '*Magister Iohannes Hispalensis et Limiensis* and Qusṭā b. Lūqā's *De differentia spiritus et animae*: a Portuguese Contribution to the Arts Curriculum?' *Mediaevalia. Textos e estudos*, 7–8 (1995), pp. 221–67 (p. 253): 'Hunc ergo librum ab ipso, Dei iuvante, quem nullus Latinorum preter quendam Antiocenum, qui quondam eius partem habuit, adeptus fuerat'. The identification of this Antiochene with Adelard was first suggested by Richard Lemay in 'The True Place of Astrology in Medieval Science and Philosophy: Towards a Definition', in *Astrology, Science and Society*, ed. P. Curry, Woodbridge, 1987, pp. 57–73 (see p. 70).

[15] See below, p. xlvi. This 'presul G' could, however, also be William of Corbeil, archbishop of Canterbury, 1123–36 or Godfrey, bishop of Bath and Wells, 1123–35; see further Bliemetzrieder, *Athelhard von Bath*, pp. 78–9.

[16] For attempts at a more precise dating of the *DED* and *QN* see Haskins, *Studies*, pp. 21–22 and 26–7, Thorndike, *History of Magic*, II, pp. 44–9, Bliemetzrieder, *Athelhard von Bath*, pp. 25–6 and Lawn, *The Salernitan Questions*, pp. 28–30. No consensus has been reached.

time when Bishop John de Villula had chosen to move the seat of his bishopric from Wells to Bath, where he restored the Roman spa and, himself a doctor, attracted doctors and scholars to the city.[17] John may have encouraged Adelard both to go to Salerno and to study in Tours, his *natale solum*. The 'wise man of Tours', whose interest in astronomy inspired Adelard, is not identifiable, but Adelard's literary style and interests are remarkably close to another alumnus of Tours, Hildebert of Lavardin, who returned to Tours as bishop in 1125. On the other hand, the dialectical arguments he employs in his *DED* are in the tradition of Beringerius of Tours (d. 1088) and Roscelin of Compiègne who was at St Martin de Tours between 1100 and 1110 and perhaps for longer.[18] Adelard was with his 'nephew' in Tours, applying himself to *Gallica studia* which included music and astronomy. In the *QN* Adelard takes his nephew and other students to Laon, where his nephew promises to pursue *Gallica studia* while Adelard departs elsewhere to devote himself to *Arabum studia*. At the turn of the twelfth century, the school of Laon, under the direction of Anselm, was a favoured place of study for the administrators of the English Court. In Laon Adelard was probably doing the same as his contemporary William of Corbeil, the future archbishop of Canterbury, who was teaching the sons of Ranulf Lambard, Henry I's judiciary, whilst following Anselm's lectures.[19] Moreover, Laon also had a reputation for the study of the quadrivium, and the theologian and mathematician Ralph, Anselm's brother and successor as head of the school, would have been teaching there at the time of Adelard's visit.[20]

So far, Adelard's biography has followed that of the typical bright young scholar, aspiring to a career in the Church or royal administra-

17 William of Malmesbury, *Gesta pontificum*, ed. N. E. S. A. Hamilton, London, 1870, pp. 194–5. For evidence of Bishop John's erecting the King's bath and other buildings see B. Cunliffe and P. Davenport, *The Temple of Sulis Minerva at Bath*, I.1, Oxford University Committee for Archaeology, Monograph no. 7, 1985, pp. 79–82 and 186–8.

18 For the intellectual tradition at Tours in the eleventh to twelfth centuries see Peter Dronke's introduction to his edition of Bernardus Silvestris, *Cosmographia*, and Mews, 'Nominalism and Theology before Abaelard'.

19 T. A. Archer, 'Ranulf Flambard and his Sons', *English Historical Review*, 2, 1887, pp. 103–112, discussing the evidence of the *Liber de miraculis sanctae Mariae Laudunensis*, 2, 6, in PL, 161, col. 977B–C. The two nephews of Roger, bishop of Salisbury, the successor of Ranulf as royal justiciary, also went to Laon.

20 Ralph (d. 1133) and his brother Anselm (a pupil of Anselm of Canterbury; d. 1117) had been running the cathedral school since about 1080. In their joint work on biblical exegesis they were accompanied by William of Champeaux (d. 1122), a student of Anselm and later a teacher of Abelard, and perhaps also of Adelard (see pp. 52–3 below): see *Theologische Realenzyklopädie*, III, Berlin and New York, 1978, pp. 1–5.

tion. At this point, however, Adelard's life takes a novel turn: he visits southern Italy and Sicily, and may well have spent some time in the entourage of William, bishop of Syracuse, whom he praises in particular for his skills in the mathematical arts.[21] Syracuse, the city of Archimedes, was ruled by the Norman count, Tancred, and another Norman, also called Tancred and probably a cousin, was the prince of Antioch, Adelard's next port of call.[22] Adelard was presumably caught up in the enthusiasm for visiting the Middle East which followed the success of the First Crusade in 1098. The earthquake he experienced in Mamistra may have been the one that affected the area in 1114.[23] On his return to England, Adelard probably settled in Bath. This return occurred well within the reign of Henry I, for, in 1122, he attested a charter in the city. It has been suggested that he also served the king, perhaps as an adviser on the running of the Exchequer.[24] It was during this period that he wrote his version of the tables of al-Khwārizmī, which have a starting-point (*radix*) of 1126 (there are no dates connected with Adelard's other translations from the Arabic, but they may have been written around the same time).

What happened after the death of Henry I in 1135 and during the period of civil war between Henry's daughter, the Empress Matilda, and King Stephen is less clear. Doubt has been cast on the reading 'Adelard' de Bath'nian' in the charter of King Stephen from the late 1130s,[25] and the horoscopes drawn up by a partisan of the king in 1150 and 1151 are not necessarily the work of Adelard.[26] A different allegiance is suggested by the dedication of the *De opere astrolapsus* to 'Henrice, nepos regis'. For this can hardly be anyone other than the future Henry II, grandson of Henry I, especially since Adelard goes on to mention the dedicatee's reaching the 'age of discretion' and refers to the importance of rulers being philosophers. Henry Plantagenet, son of the Empress Matilda and of Geoffrey, Count of Anjou and (since

21 *DED*, p. 2: 'omnium mathematicarum artium eruditissime'.

22 R. L. Nicholson, *Tancred: A Study of His Career and Work*, Chicago, 1940.

23 See Cochrane, *Adelard of Bath*, p. 32, referring to J. Milne, 'Catalogue of Destructive Earthquakes' in *British Association for the Advancement of Science Report for 1911*, Portsmouth, 1912, p. 665.

24 This is the argument of R. L. Poole (*The Exchequer in the Twelfth Century*, Oxford, 1912, pp. 56–7) who suggests that his exemption from a murder fine in the *Pipe Roll 31 Henry I* implies that he was in the king's service.

25 See n. 12 above.

26 These horoscopes are discussed in North, 'Some Norman Horoscopes'. North makes a case for Adelard as the astrologer, but the claims of Robert of Chester, who wrote astronomical texts in London, Stephen's capital, in 1147 and 1150, are perhaps stronger.

1144) Duke of Normandy, was formally declared to have come of age by his father on his return from a triumphal tour of England, early in 1150. At the same time he was invested with the duchy of Normandy and his succession to the English throne looked probable. Hence the appropriateness of addressing him as 'grandson of the king'.[27] The *De opere astrolapsus* accompanies in the manuscripts another text which was intended for the sons of Geoffrey, Duke of Normandy: the *Dragmaticon* of William of Conches. William is the earliest scholar to know Adelard's *QN*, which he uses extensively, without acknowledgement, in his *Dragmaticon*.[28] The young Duke Henry provides the link between Adelard and William. Adelard's presence in Normandy is also suggested by the fact that the earliest copies of his translations of the works of Thābit ibn Qurra and Abū Maʿshar were in a manuscript from the Benedictine abbey of Mont St-Michel.[29] It was probably in Normandy, too, that he dedicated the *QN* to Richard, bishop of Bayeux.[30] Whether he returned to England after Henry's accession to the throne in 1154 is unknown; the *De opere astrolapsus* is the last testimony to his existence.

It appears that Adelard never held an ecclesiastical post. As a layman he was free to travel widely, and it is perhaps rash to try to confine him to the few places that are directly mentioned by him, or where he is mentioned in external sources. Aside from the parallels between his work and that of William of Conches, who had been nurtured at Chartres, there are other factors that suggest that Adelard had some connections with the cathedral school there. His version of the tables of al-Khwārizmī and the *commentum*-version of his translation of Euclid's *Elements* were known in the city by the

27 These arguments were first set out by North, 'Some Norman Horoscopes', pp. 155–61.

28 Lawn, *Salernitan Questions*, p. 52, n. 3, D. Elford, 'Developments in the Natural Philosophy of William of Conches: A Study of his *Dragmaticon* and a Consideration of its Relationship to the *Philosophia*', unpublished PhD dissertation, Cambridge, 1983, pp. 233–7, and Burnett, 'Adelard, Music and the Quadrivium', p. 80. The involvement of Adelard and William with the education of Henry II is developed in greater detail in Burnett, *The Introduction of Arabic Learning into England* (Panizzi Lectures, 1996), London, 1997, lecture 2.

29 Now Avranches, Bibliothèque municipale, 235.

30 It is likely that this was Richard FitzSamson, since William of Conches already appears to know some of the subject-matter of the *QN* in his *Philosophia* (written in the 1120s; see p. xxiii below): this Richard was the son of Samson, bishop of Worcester, and therefore familiar in the area of Adelard's earlier activity; he is known to have had intellectual interests. However, it is possible that Adelard dedicated a composition that he had started to work on many years previously, to the second bishop called Richard; for this Richard of Kent was the son of the Duke of Gloucester, the Empress Matilda's principal supporter.

1140s, since they were included in the *Heptateuchon* (written in the early 1140s) of Thierry, the chancellor of Chartres.[31] On the other hand, Adelard's interests and scientific knowledge correspond to those of scholars associated with the cathedrals of Hereford and Worcester. The earliest copy of his tables of al-Khwārizmī is in a manuscript written in the cathedral priory of Worcester before 1140, where it accompanies a work on the Church calendar (*computus*) by Robert, bishop of Hereford (1079–95) and two astronomical texts written by Walcher, the prior of Great Malvern Abbey (just outside Worcester), who died in 1135. One of the latter texts reports the opinions on the lunar nodes of Petrus Alfonsi, the converted Jew from Huesca in Aragon, who had also made a version of the tables of al-Khwārizmī, apparently from the same Arabic manuscript as that used by Adelard.[32]

Finally, it may be significant that the only medieval reference to the existence of Adelard's *DED* is in a catalogue of the books of Waltham Abbey of *ca.* 1200, where a '*summa*' of Pseudo-Quintilian's *Declamationes maiores* is also attributed to Adelard.[33] This *summa* is probably the dialogue-version of the *Declamationes*, edited by Lehnert under the title 'Excerpta Parisina',[34] which, although anonymous in the manuscripts, has some of the hallmarks of a work by Adelard.[35] The *Excerpta Parisina*, the *QN* and *DED* occur together in the manuscript which gives what appears to be Adelard's authorial revision of the text of the *Questiones*.[36] This manuscript, described below as 'L/P', is

31 For the evidence that the *Heptateuchon* was written at Chartres see P. Stirnemann, 'Où ont été fabriqués les livres de la glose ordinaire dans la première moitié du xiie siècle?' in *Le xiie siècle: mutations et renouveau en France dans la première moitié du xiie siècle*, ed. F. Gasparri, Paris, 1994, pp. 257–302 (see pp. 275–6). A second copy of the tables of al-Khwārizmī once existed in Chartres, Bibliothèque municipale, 214: this appears to have been the manuscript from which the text in the *Heptateuchon* was copied. For the *commentum*-version of Euclid's *Elements*, which I believe is a direct result of Adelard's teaching, see Burnett, 'Latin and Arabic Influences'.

32 The relationship between Walcher, Petrus and Adelard is explored in Burnett, 'Works of Petrus Alfonsi', pp. 45–7 and 52–4.

33 The Waltham Abbey Bible (Passmore Edwards Museum, Stratford, East London), fol. 158r, item 39: 'Quintilianus de causis et summa magistri Adelardi de eodem'; ibid., item 46: 'Sic faciunt cause rerum. De eodem et diverso. Moralium dogma'; see M. R. James, 'MSS from Essex Monastic Libraries', *Transactions of the Essex Archaeological Society*, new series, 21, 1937, pp. 34–46 (42–3).

34 G. Lehnert, in *Quintiliani quae feruntur Declamationes XIX maiores*, Leipzig, 1905, pp. 389–431; see also H. Dessauer, *Die handschriftliche Grundlage der neunzehn grösseren pseudo-quintilianischen Declamationen*, Leipzig, 1898, pp. 62–5.

35 This question is more fully explored in Burnett and Ronca, 'Transmission of Adelard's *Questiones naturales*'.

36 See below, p. xlv.

written in an English hand of the late twelfth century which resembles hands in the few surviving Waltham manuscripts of the time.[37] The privileged position of Adelard's works at Waltham requires some explanation.

The evidence suggests, then, that Adelard spent his life as a tutor to the sons of the nobility and perhaps, even, as a *philosophus* in the court.[38] His subject was the whole range of the secular sciences, which he would have described as *philosophia*. His original writings demonstrate that he not only promoted the seven liberal arts which formed the backbone of the traditional Latin curriculum of *philosophia*, but was also aware of a more widely encompassing range for the subject: for, while the topic of the *DED* is this traditional curriculum (with a particular emphasis on logic), the *QN* deal with the new subject of physics or natural science, and end with the promise of a book which could only be described as being on metaphysics.[39] Whether Adelard owed this new conception of philosophy at all to the *Arabum studia* remains to be investigated.

II The Texts

De eodem et diverso

The *DED* takes the form of a dramatic dialogue between Philocosmia, who advocates worldly pleasures, and Philosophia, whose defence of scholarship leads into a summary of the contents of each of the seven liberal arts.[40] Philocosmia's claim that philosophers cannot be trusted because they never agree with each other leads Philosophia to attempt to prove the compatibility of the philosophies of Plato and Aristotle. This gives Adelard the opportunity to advance a well-argued theory of knowledge, which was the first element of Adelard's work to attract the attention of modern scholars.[41] The contrast between *res*, the perceptible realities with which Philocosmia deals, and *verba*, the mental concepts of Philosophia, runs through the whole work.

Adelard's principal classical model is Boethius' *Consolation of Philo-*

37 E.g., London, British Library, Add. 34749.

38 For the role of the professional 'philosophus' in the twelfth and thirteenth centuries see Burnett, 'Master Theodore', pp. 248–52.

39 See pp. 226–7 below.

40 For other discussions of *De eodem et diverso*, see Jolivet, 'Adélard de Bath et l'amour des choses', Drew, '*De eodem et diverso*', Pabst, *Prosimetrum*, I, pp. 389–407 and Speer, *Die entdeckte Natur*, pp. 19–27.

41 Hauréau, *Histoire de la philosophie scolastique*, chapter 14, pp. 345–61.

sophy, which provides him with much of his vocabulary and phraseology,[42] with his broader philosophical concepts,[43] and with the form of the work, in which arguments in prose are concluded with passages of verse.[44] The sections on the liberal arts, on the other hand, reflect the textbooks on each of these arts which were currently being used in Adelard's circle.[45] Each section is divided into two parts: first there is a description of the allegorical figure representing the art, in which the range and importance of that art is indicated; then there is a summary of the doctrines of the art, put into the mouth of the allegorical figure or of the man she has chosen to be its founder or foremost exponent. These *personae* are reminiscent not only of the allegorical figures in Martianus Capella's *Marriage of Mercury and Philology* but also of the allegorical women and practical men of the medieval iconography of the arts, and in particular of the sculptures on the Portail Royal of Chartres Cathedral.[46] The practical men (not always named by Adelard),[47] are presumably Donatus for grammar, Cicero and Aristotle for rhetoric, Aristotle for logic, Pythagoras for music and presumably Euclid for geometry.

In the field of the trivium Adelard relies on Cicero's *De inventione* and the logical works of Boethius, of which he refers specifically to the *Introductio ad syllogismos categoricos*.[48] In his account of the quadrivium, Adelard is more expansive. For arithmetic he summarizes definitions in Boethius' *Arithmetic* and arranges the material in a symmetrical way, so that the 'beautiful diagram of Nicomachus' for the even numbers matches 'the sieve of Eratosthenes' for the odd numbers. The doctrine in the section on music is also dependent on Boethius – this time his *Music* – , but Adelard adds items from his own experience and that of the English.[49] In geometry he does not yet

42 Courcelle, 'Adelard de Bath et la *Consolation* de Boèce', and Bliemetzrieder, *Athelhard von Bath*, pp. 377–83.

43 Burnett, 'Adelard of Bath's Doctrine on Universals'.

44 See P. Dronke, *Verse with Prose from Petronius to Dante*, Cambridge, Mass., 1994 (for Adelard see pp. 46–7); Bliemetzrieder, *Athelhard von Bath*, p. 31, sees a deliberate contrast between the poetic forms used by Philocosmia and Philosophia.

45 For details on sources, see commentary to the English translation, below, pp. 74–9.

46 See M. Evans, 'Allegorical Women and Practical Men: The Iconography of the *Artes* Reconsidered', in *Studies in Church History, Subsidia 1: Mediaeval Women*, ed. D. Baker, Oxford, 1978, pp. 305–29 and E. Jeauneau, *L'Age d'or des écoles de Chartres*, Chartres, 1995.

47 It is possible that the names of Donatus and Euclid were originally written in the margin, as in the case of other authorities not named in the text of the *DED*, Plato (p. 20), Aristotle (p. 20) and Pythagoras (p. 12).

48 For the context, see K. M. Fredborg, 'The Unity of the Trivium', *Geschichte der Sprachtheorie*, 3, ed. P. Schmitter, Tübingen, 1995, pp. 325–38.

49 This section is discussed in Burnett, 'Adelard, Music and the Quadrivium'.

show himself to be familiar with Euclid's *Elements*, but rather depends on a text of the tradition of the Roman land-surveyors (*agrimensores*): the work known in modern scholarship as the *Geometria incerti auctoris*, which was already in existence in the early eleventh century.[50] In the case of astronomy, Adelard includes mention of its usefulness in prediction, but forbears from discussing it at length because of the difficulty of its subject-matter.

The classical sources in the *DED* are interpreted in the light of the teaching that Adelard had received, presumably mainly at Tours, but possibly also in Laon. It is these contemporary sources that can be glimpsed in the way Adelard attempts to show how the epistemology of Plato and Aristotle can be reconciled. For he propounds a theory of universals which is very similar to what was later called the 'indifference' theory: i.e., that the terms (*voces*) 'individual', 'species' and 'genus' can be applied to the one thing (*res*) and differ only according to the different levels at which that thing is considered. Adelard's intention is to rescue such *voces*, as reaching 'the causes and beginnings (*initia*) of the causes of things', in the face of an Epicurean attitude in which only the sensible 'things' (*res*) are regarded as of any relevance.

The status of *res* and *voces* was very much the concern of the early 'vocales' (later known as 'nominalists') of which Roscelin was one of the most important. Adelard's arguments are especially close to those of the Pseudo-Rhabanus' commentary on the *Isagoge* of Porphyry and of an anonymous commentary on the same text in MS Munich, Bayerische Staatsbibliothek, Clm 14458.[51] In Adelard's summary of rhetoric there are parallels with theories of William of Champeaux, who may have been in Laon when Adelard was there. Adelard seems also to have been influenced by the anonymous *Glosule ad Priscianum*. This late eleventh-century work was present in a manuscript in Chartres; it has recently been suggested that it was written by John of Rheims, the teacher of Roscelin.[52]

50 See L. Toneatto, *Codices artis mensoriae*, 3 vols, Spoleto, 1995, pp. 1015–1110. Chapters from this work are also found in the Mont St-Michel manuscript that included the earliest copies of Adelard's translations of Arabic texts; see n. 29 above. For other sources and parallels to the doctrine in this section see Burnett, 'Latin and Arabic Influences'.

51 Y. Iwakuma, 'The Realism of Anselm and his Contemporaries', in *Anselm: Aosta, Bec and Canterbury*, eds. D. E. Luscombe and G. R. Evans, Sheffield, 1996, pp. 120–35 (for Adelard, see pp. 134–5).

52 See Mews, 'Nominalism and Theology before Abaelard', and 'Philosophy and Theology 1100–1150: The Search for Harmony', in *Le xiie siècle: mutations et renouveau en France dans la première moitié du xiie siècle*, ed. F. Gasparri, Paris, 1994, pp. 159–203. For the

The *DED* concludes with the advocacy of travelling to seek for knowledge in other cultures, and a metaphor from human physiology. This nicely prepares the way for the *QN*, one of whose questions (Q49) is included in the anecdote, concerning a journey from Salerno through Magna Graecia, with which Adelard ends the *DED*.

Questiones naturales

The *QN* is the most substantial of the three works included in this volume.[53] It consists of a set of seventy-six questions arranged in three parts of almost equal length: on plants and brute animals (Q1–14), on man (Q15–47), and on earth, water, air and fire (meteorology and celestial physics) (Q48–76). Adelard has adopted the standard genre of natural questions[54] as a convenient framework for a literary masterpiece with his own particular stamp. The evidence for this is both the use of leitmotifs to bind the work together – in particular *ratio* and *opinio* – and the structure of the work. It develops an argument of rational causes in nature from the lowest creatures physically – i.e., from the very roots of the plants – to the heavens. Of the three parts, that on man discusses his corporeal aspects only, but both the others end with questions concerning the incorporeal soul, which are discussed more extensively than most of the other natural questions. The first part ends with the question of whether brute animals have souls; the whole work ends with the problem of whether the heavens are ensouled.

That Adelard was using a pre-existent set of natural questions is indicated by the facts that some questions are raised but not discussed because, according to the protagonists, they have been dealt with elsewhere (e.g., 'Why do men grow grey in old age?', 'Why do we gulp down hot foods?', 'Why do some waters taste of soda?', 'Why are some waters hot in themselves?', and 'Why does the Moon suffer eclipse?');[55] others are said to be answered in the wrong order (e.g.,

attribution of the *Glosule* to William of Champeaux himself, see I. Rosier, 'Le Commentaire des *Glosulae* et des *Glosae* de Guillaume de Conches sur le chapitre *De voce* des *Institutiones Grammaticae* de Priscien', *Cahiers de l'Institut du moyen-âge grec et latin*, 63, 1993, pp. 115–44 (see p. 118).

53 For other discussions of the work see Jolivet, 'Les "Quaestiones naturales" d'Adélard de Bath', Maurach, 'Adelhard von Bath' and Speer, *Die entdeckte Natur*, pp. 27–75.

54 The whole tradition is set out expertly by Lawn, *Salernitan Questions*: see pp. 20–31 for a discussion of the sources and parallels to Adelard's questions.

55 These questions are mentioned in Q20, Q34, Q59 and Q70 respectively.

the question of the cause of the obliquity of the ecliptic, which should precede that of the contrary movement of the planets).[56] The most telling (but perplexing) reference appears only in MS L/P:[57] 'Hic deest cur in occipitio careatur oculis' ('Here is lacking the question on why one lacks eyes in the back of the head'). There clearly was a list of questions in which this one occurred in this position.[58] The much greater correspondence between the questions addressed in William of Conches' *Philosophia* and *Dragmaticon* and those of the *QN*, in comparison to that between their answers,[59] also suggests that Adelard and William were both using a similar set of pre-existent questions.[60]

Since at the end of the *DED* Adelard specifically and pointedly mentions a discussion with a Greek philosopher in Magna Graecia in which one of these questions was asked, one would expect his own questions to come from the same source as the collections of Salernitan questions which were popular in the Middle Ages.[61] The problem is that there is no evidence that Adelard was able to read or translate Greek, and there is no earlier Latin list of questions that is nearly as complete as Adelard's *QN*.[62] Nevertheless, it is probable that Adelard's inspiration came from Southern Italy. One of the few works which Adelard quotes directly in both *DED* and *QN* is Nemesius' *Premnon phisicon*, which also happens to be found in the earliest

56 The beginning of Q71.
57 See p. 152 below. For the explanation of the manuscript sigla see pp. xliii–xliv.
58 I.e., just before Q30. Note that this question and its answer are included in one of the Hebrew versions of the *QN* after the equivalent to Adelard's Q30: see Gollancz, *Dodi ve-Nechdi*, pp. 36–7. Moreover, this Hebrew translation also gives an answer to the problem of gulping hot foods, which *QN* passes over, allegedly because Macrobius had already provided the answer; ibid., p. 41, see pp. 162–3 below.
59 The following questions are shared by Adelard and William: Q1–7, Q18, Q20–21, Q23, Q37, Q41–2, Q51–2, Q54–7, Q59, Q63–5, Q69–72. Of these, in only Q18, Q21, Q23, Q54 and Q57 are there resemblances between the answers in Adelard and in William's *Philosophia*. By the time William writes the *Dragmaticon* he clearly uses Adelard's answers in Q1–5, Q12, Q23, Q30 and Q37.
60 Another hint at such a set of questions can be found in Baudry de Bourgueil's poem addressed to Adela, daughter of William the Conqueror (Baldricus Burgulianus, *Carmina*, ed. K. Hilbert, Heidelberg, 1979, p. 183: Poem no. 135, lines 1287–94): 'Cur homo sit calvus, cur non sit femina calva? / Cur quoque gignat homo, femina concipiat? / Cur homo barbatus, imberbis femina cur sit? . . . '. In Q52 the nephew refers to a question as being 'well known'.
61 Eleven collections of these questions have been edited by Brian Lawn in *The Prose Salernitan Questions*.
62 The extant earlier sets of questions in Latin available to, and apparently used by, Adelard are the *Vetustissima translatio* of the combined pseudo-Aristotelian and pseudo-Alexandrian *Problemata*, and the short list of questions in Statius' *Thebaid*, VI, vv. 360–4, explicitly cited by Adelard. There is no evidence that Adelard knew Seneca's *Natural Questions* or Priscianus Lydus' *Solutiones ad Chosroem*.

manuscript of Adelard's *QN*.[63] The *Premnon phisicon* is a translation from Greek of Nemesius' *On the Nature of Man*, made by Alfano, the learned archbishop of Salerno (d. 1085), who had visited Constantinople and wrote both Latin poetry and medical texts based on Greek sources. It is conceivable that a set of Greek natural questions accompanied Alfano's Greek copy of Nemesius' *On the Nature of Man*. For, when Burgundio of Pisa comes to make a second translation of the same work in 1165, he promises Frederick I Barbarossa, the dedicatee, solutions to a set of natural questions as a sequel to his translation; most of the specific examples he gives of natural questions can be found in Adelard's *QN*.[64] Alfano was obviously interested in the natural-questions genre, since a work entitled 'Alfani Salernitanensis de quibusdam questionibus medicinalibus' was formerly in the library of Christchurch, Canterbury.[65] It is possible, then, that Alfano provided the conduit through which a Greek set of natural questions became available to Adelard.[66]

As for the answers, there is, again, very little correspondence between Adelard's *QN* and the extant versions of the Salernitan

63 MS G below. This manuscript includes a second, anonymous, translation of Nemesius' chapter on the elements, edited by R. C. Dales, 'An Unnoticed Translation of the Chapter *De elementis* from Nemesius's *De natura hominis*', *Medievalia et humanistica*, 17, 1966, pp. 3–19; since Dales' article another manuscript of this text has been found: British Library, Add. 22719, fol. 161v–162r.

64 Ed. G. Verbeke and J. R. Moncho, Leiden, 1975, pp. 2–3: 'Si in his vos exercitari perpensero, et altiora Vobis transferre curabo: de corpore caeli, de forma et motu eius et de omnibus passionibus quae sunt a caelo deorsum, ut de lacteo circulo et cometis et ventis et coruscationibus et tonitruis et iride et pluviis et grandine et pruina, et cur mare salsum est et cur tot fluminibus influentibus nec augetur nec dulcoratur, et de terrae motu qualiter fiat'; cf. Adelard's Q50, Q51 and Q53. G. Vuillemin-Diem and M. Rashed ('Burgundio de Pise et ses manuscrits grecs d'Aristote: Laur. 87.7. et Laur. 81.18', *Recherches de théologie et philosophie médiévales*, 64, 1997, pp. 136–98), argue persuasively that Burgundio is referring to Aristotle's *Meteorologica*, which includes discussion of all these subjects.

65 M. R. James, *The Ancient Libraries of Canterbury and Dover*, p. 59.

66 This is also the conclusion of Lawn, *Salernitan Questions*, p. 37, n. 3 and p. 54. On pp. 24–5 Lawn points out that three of Adelard's questions (Q12, 33 and 47) are the same as the Greek questions which occur in parts of the pseudo-Aristotelian and pseudo-Alexandrian *Problemata* of which we have no Latin translation predating Adelard. The question allegedly posed by the Greek philosopher in Magna Graecia (*DED* and Q49) is not attested elsewhere. Although the *Problemata* of Pseudo-Aristotle was translated into Arabic (*Problemata physica arabica*, ed. L. S. Filius), and comparable sets of natural questions can be found in Arabic authors (e.g., in Pseudo-Apollonius, *On the Secrets of Nature*, ed. U. Weisser, *Buch über das Geheimnis der Schöpfung und die Darstellung der Natur von Pseudo-Apollonios von Tyana*, Aleppo, 1979 which includes some Pseudo-Aristotelian problems no longer extant in Greek), it seems unlikely that an Arabic set would have been Adelard's immediate source.

questions.[67] Nor does Adelard appear to know the medical texts and translations from Arabic of Constantine the African who had passed through Salerno on his way to Monte Cassino where he died before the end of the eleventh century. Here, too, the comparison to William of Conches is instructive, because, already in the *Philosophia*, William includes some of the Salernitan solutions,[68] alongside quotations from Constantine the African.[69] Some of Adelard's answers may reflect an earlier stage in the genre of natural questions: a stage that predates the impingement of the translations of Constantine the African on Salernitan learning, and may be representative of Archbishop Alfano's circle.[70] However, what is striking is that most of the arguments in the solutions, and some of the questions to which Adelard gives most attention, bear no relationship at all to the natural-questions genre.

Sometimes Adelard uses a pre-existent question simply as a prop on which to hang a topic in physics that concerns him more deeply: e.g., the cooling nature of fans (Q34) leads him to investigate essential and accidental qualities; the movement of winds (Q59–61) leads to the problem of infinite movement.[71] In regard to the soul, however, Adelard appears to introduce completely new questions. The topic of the soul is specifically excluded from the Salernitan questions in a statement near the beginning of the fullest set of those questions:

> The discussion of the soul, on the other hand, has little or no relevance to theoreticians or philosophers, since its creation and infusion belong to the divine power in such a way that human reason can have no sensual experience of its disposition and only faith can judge its worth.[72]

Given this statement, it is not surprising that we do not find the soul discussed in any of the extant sets of *Problemata* or Salernitan questions.

On the other hand the soul was a very hot topic of discussion at the time that Adelard was writing the *QN*. There is a proliferation of writings amongst Latin theologians and philosophers; not only William of Conches' works with their controversial doctrine of the *anima*

67 The only similarities are in the answers to Q12, Q23, Q29, Q34, Q51 and Q58.

68 This is the case for William's answers to Adelard's Q6, Q37, Q18, and Q73.

69 Ronca, 'The Influence of the *Pantegni* on William of Conches's *Dragmaticon*', in *Constantine the African and ʿAlī ibn al-ʿAbbās al-Maǧūsī: The* Pantegni *and Related Texts*, eds. C. Burnett and D. Jacquart, Leiden, 1994, pp. 266–85.

70 Lawn (*Salernitan Questions*, pp. 21–3) shows how the Galenic theories as given in Constantine the African's *Pantegni* are absent in Adelard's work.

71 Thus Speer writes (*Die entdeckte Natur*, p. 75): 'In dem Bestreben, die "*causae rerum*" zu erforschen, versichert sich Adelard der großen Themen der Physik: Materie, Bewegung, Elementenlehre und schließlich Kosmologie.'

72 MS Bodleian, Auct. F.3.10, fol. 118, *The Prose Salernitan Questions*, ed. Lawn, p. 2.

mundi, which derived from a new interest in Plato's *Timaeus*, but also Hugh of St Victor's *De unione spiritus et corporis*, Isaac of Stella's *De anima*, Alcher's *De anima et spiritu* and other texts.[73] The urgency of the debate was caused by the threat to orthodox Christian discourse on the salvation of an incorporeal soul through good works, from Platonic and Aristotelian conceptions of the soul on the one hand, and from medical works which discussed the activity of corporeal spirits in the body on the other. Adelard was responding to a contemporary interest, as he himself states, in the *persona* of his nephew:

> Nephew: Now that we have discussed everything else about animals, we should tackle the last and most urgent question: 'Do brute animals have souls?' (p. 111 below)

If Adelard did not draw on Salernitan solutions, where did Adelard find the answers to his questions? He claims that they derive from *Arabum studia*. As we have seen, his nephew had asked for 'something new from the studies of the Arabs',[74] and in the introduction to the questions Adelard asserts that he had undertaken to 'scrutinize the studies of the Arabs as much as he could',[75] with the result that, in the *QN*, he was putting forward the views of the Arabs, not his own.[76] A short way into the text, Adelard includes his well-known defence of following reason rather than authority, which he prefaces with the words: 'I have learnt one thing from my Arab masters, with reason as guide',[77] and the nephew, on two occasions, pours scorn on Adelard's addiction to the views of the 'Saracens'.[78] And yet, up to now, only Greek and Latin sources have been identified.[79] A correspondence has been recognised between the answer of Q12 and a Greek pseudo-Alexandrian problem that is not known to have been translated into Latin in Adelard's time;[80] several of Adelard's solutions resemble those

73 See V. Liccaro, *Ugo di S. Vittore, I tre giorni dell'invisibile luce e l'Unione del corpo e dello spirito*, Florence, 1974, and B. McGinn, *Three Treatises on Man*, Kalamazoo, Michigan, 1977.

74 Note 1 above.

75 *QN*, introduction to the questions: 'Arabum studia ego pro posse meo scrutarer.'

76 Ibid.: 'hoc tantum vitato incommodo, ne quis me ignota proferentem ex mea id sententia facere, verum Arabicorum studiorum sensa putet proponere . . . causam Arabum, non meam, agam.'

77 *QN*, Q6: 'Ego enim aliud a magistris Arabicis ratione duce didici.'

78 *QN*, introduction to the questions: 'cum Sarracenorum sententias te sepe exponentem auditor tantum notaverim'; Q5 'Itane igitur illi Sarraceni subtilibus te nugis illaqueabant'. It is not clear whether the nephew's use of 'Sarraceni' rather than 'Arabes' has a derogatory tone.

79 This enigma is well discussed in Lawn, *Salernitan Questions*, pp. 21–7.

80 Ibid., pp. 24–5; the same question and answer also occur in the Salernitan Questions.

of Nemesius' *Premnon phisicon* in Alfano's translation.[81] The classical Latin sources for some of the arguments include Boethius' *Consolation of Philosophy*[82] and Macrobius' *Saturnalia.*[83] There are, however, two classical works which appear to provide exactly the arguments and methodology which Adelard claims to be 'Arabic'.

The first of these is Plato's *Timaeus*. The nephew regards Adelard as being a faithful follower of Plato in every branch of philosophy,[84] and Adelard himself characterizes Plato's doctrines as 'divine'.[85] This statement is applied to Plato's doctrine of vision, on which Adelard quotes a substantial passage from Calcidius' translation of Plato's *Timaeus*. Several further passages from the *Timaeus* are quoted under the name of 'the Philosopher' (*Philosophus*). For Adelard, the Philosopher is Plato, and he himself is one of the 'Academics' who follow him. After the initial statements concerning Adelard's Arabic studies and Arabic masters, the role of rational guide is taken over by Plato.[86]

The other text is the *De natura deorum* of Adelard's fellow 'Academic', Cicero. Adelard's very claim (following his 'Arabic masters') to use rational arguments rather than those of authority recalls the argument at the beginning of Cicero's *De natura deorum*, in which Cicero agrees to use rational arguments rather than tradition, to discuss the nature of the gods.[87] That Adelard used this work is clear from his arguments for the different kinds of fire and of movements in the universe, which are taken directly from the *De natura deorum*,[88] although Adelard refers them to Cicero's authorities,

81 Ibid., pp. 22–3, concerning Q1–5, Q18.
82 Boethius is cited by name in Q20–21 and Q46.
83 Macrobius' *Saturnalia* provided two questions that Adelard answered – Q20 = *Sat.* 7.10 and Q34 = *Sat.* 7.8.14 – and one question that he deliberately did not answer, between Q34 and Q35. Adelard's answers to Q20 and Q34 are partially indebted to Macrobius. For the importance of the *Saturnalia* as a source of scientific information (especially its seventh book which is also found separately in the manuscripts), see Lawn, *Salernitan Questions*, pp. 14–15.
84 Q24 'Et in philosophicis contemplationibus et in phisicis causarum effectibus, ethicisque etiam consultibus Platoni te penitus consentire percepi'; Q27 'dum Platonem sequeris'; Q29 'Plato tuus'.
85 Q23 'Achademicorum divina quam sanxit Plato sententia'; ibid., 'auctor huius divine rationis Plato'.
86 Neither the Arabs nor the Saracens are mentioned after Q6.
87 See p. 103 and n. 14 below.
88 See pp. 220–1 below and *De natura deorum*, II.xv (40–41) and II.xvi (44). This source was identified by Silverstein in 'Adelard, Aristotle and the *De natura deorum*'. A copy of *De natura deorum* once belonged to Philip of Harcourt, bishop of Bayeux and was bequeathed to Bec by him by 1163; it therefore followed the same itinerary as an early manuscript of Adelard's *QN* and Seneca's *Natural Questions* (see p. xxxii below); see Reynolds, *Texts and Transmission*, pp. 126–7 (article by R. H. Rouse).

Cleanthes and Aristotle, rather than to Cicero himself. The whole question of the nature of God, addressed at the end of the *QN*, is couched in the same form as in Cicero's treatise.

Some of these classical sources were mediated through the filter of works closer to Adelard's own time. The application of reason to theological questions in the works of Beringerius of Tours and Anselm of Canterbury savours of the same intellectual climate as the *QN*. Specific examples of borrowings cannot be pointed to, but it can be shown that Adelard read Boethius' theory of sound propagation through the medium of the *Glosule ad Priscianum* or a related work.[89] The only classical texts quoted *in extenso* and, presumably, directly, remain Calcidius' translation of Plato's *Timaeus* and Boethius' *Consolation of Philosophy*.

There seems, therefore, to be good reason for regarding the learned Arab in the *QN* as 'an expository device . . . introduced expressly for the presentation of difficult or hazardous views'.[90] Such a device could, for example, protect Adelard from accusations of heresy in his discussion of animals' souls. For here, he sets up a contrast between the views of the nephew, which are the orthodox ones – that animals have sensation, not reason, and this is a bodily function which perishes with the animal – to those of the 'Adelard' of the dialogue, by which animals not only have discernment and judgement and can understand the spoken word, but also have incorporeal souls which survive the body. For this, Adelard gives no authority besides the persuasiveness of reason.[91] It is no wonder that Adelard had to protect himself from such radical views by saying 'Don't blame me! Blame the Arabs!'

There is still the probability, however, that, beyond its function as a literary device, the allegation of *Arabum studia* has some justification. We need not doubt Adelard's claims to have visited Tarsus and Mamistra,[92] especially in view of the likely identification of Adelard with the 'Antiochene' mentioned by John of Seville and Limia.[93] The anecdotes that Adelard relates concerning these visits are of the same

89 See Burnett, 'Adelard, Music and the Quadrivium', pp. 79–80 and p. 231, n. 37.

90 This is the formulation of Margaret Gibson, in 'Adelard of Bath', p. 11.

91 There is, however, a striking resemblance between the tenor of Adelard's arguments for the continuing existence of animals' souls after the death of their bodies and that of Johannes Scotus Eriugena's arguments in *Periphyseon*; see p. 230, n. 25.

92 As Margaret Gibson has done, 'Adelard of Bath', p. 10, n. 25: 'The reference to Tarsus is so close to Acts 21.39 that it need not be a literal statement. In short, Adelard had heard of the earthquake and he had read Acts; I hesitate to construct a visit to the Holy Land on that evidence, unsupported.'

93 See p. xiv above.

tone as those referring to episodes nearer to home – his experience of Mt Etna in the *QN* and his playing of the cittern in front of the Queen in the *DED*. They involve meetings and discussions with wise men, not encounters with books, and Adelard is quite careful not to imply that he read any Arabic books: he had searched out the intellectual occupations of Arabs and he had learnt his scientific methodology from 'Arabic masters' (*magistri Arabici*). These terms recall those that Adelard was to use later, in his *De opere astrolapsus*, in which Arabic learning is undoubtedly present, and where he explicitly undertakes to 'write down in Latin what I have learnt in Arabic from my masters'.[94]

It is quite likely that Adelard actually studied with masters whose language of scientific discourse (whether they were Jews, Christians or Muslims) was Arabic, both in the Principality of Antioch and when he was back in the West Midlands. The direct results of these Arabic studies were his translations of Euclid's *Elements* and the astronomical, astrological and magical texts referred to at the beginning of this introduction. These studies in mathematics, too, are hinted at in the *QN*, when he promises to write about astronomy[95] and admits that he was once keen on finding out more about magic.[96] The absence of any clear influence of Arabic sources in the *QN* may be partly due to the fact that their subject-matter – physics – hardly overlaps with the Arabic mathematical texts (in the broadest sense) which principally engaged the attention both of Adelard and of his contemporary translators. Some of the arguments concerning questions of natural science could well have been picked up in conversations with Arabic scholars, especially those that have a distinctly Aristotelian ring to them.[97] But because they have been written down in a Latin which is replete with Classical formulations and vocabulary, and transmuted

94 This is a combination of two phrases: *De opere astrolapsus*, ed. Dickey, p. 148, line 10: 'quod Arabice didici, Latine subscribam', and ibid., p. 172, lines 11–12: 'De hoc itaque quod a magistris acceperim . . . non minus subscribendum est'. That these masters are, indeed, masters of Arabic learning, if not Arabic speakers themselves, is indicated by several statements in the *De opere astrolapsus*, e.g., ibid., p. 216, lines 7–8: 'Nunc de casarum equatione, quam magistri 'tezweietelbuiut' [i.e., Arabic 'taswiyat al-buyūt'] vocant, dicendum est'. For the details of this argument see Burnett, 'Adelard of Bath and the Arabs'.

95 See below, p. 147.

96 See below, p. 193.

97 For example, the problem of infinite movement in Q61 which recalls Aristotle, *Physics*, Book 3, 2 (202a3), and the reason why winds move around the earth and not upwards, which recalls Aristotle, *Meteorologica*, 2, 4 (361a23). Adelard's explicit reference to 'Aristoteles in physicis' is a misremembering of a misleading citation in Nemesius; see p. 230, n. 31 below.

by the alchemy of Adelard's strongly individual style, Arabic counterparts are no longer easy to detect.[98]

The previous paragraphs have suggested that different sources have to be sought for the list of questions themselves, and the answers to those questions. In the manuscript tradition of the *QN*, the questions continue to have a loose relationship with the solutions. Sometimes they precede the text (MSS GL/PBeSWBCcNJ), sometimes they are inserted into the text just before their solutions (MSS QOx), sometimes they appear in both positions (MSS VXYRiEsEF); on one occasion the questions are listed without the text (MS Qu).

There remains the problem of the title. In the case of the *DED*, the title is unambiguously given to the text by Adelard himself in the prologue: 'Hanc autem epistolam "de eodem et diverso" intitulavi' ('I have entitled this letter *De eodem et diverso*'). For the *QN*, on the other hand, no such authorial confirmation of the title exists. The manuscripts themselves provide many different titles (before more or less settling for 'De naturalibus quaestionibus'). Adelard himself refers to the work in a descriptive manner (rather than with a title) at the beginning of his *DAT*: 'Quoniam *in rerum causis* disserendis animus noster admodum est fastiditus . . .' ('Since our mind has become quite sated with discussing *the causes of things*'). In MSS GL/P and their family the only title is 'Ricardo Baiocensi Episcopo Adelardus Bathoniensis sal(utem)', which is parallel to the *actual* heading of the *DED* in the only manuscript that contains that work (MS P): 'Siracusio episcopo Willelmo Adelardus sal(utem)'. Thus both works take the form of letters ('epistolae'), which is the genre to which *DED* is explicitly assigned ('in epistola hac . . .'; 'hanc autem epistolam'). As a letter, no further title would be required. In the same position in the prologue (equivalent to a protocol in a letter) in which, in the *DED*, Adelard explains the title (i.e., just after the mention of the name of the dedicatee), in the *QN* Adelard writes that he will reply to his nephew 'super rerum causis' ('on the causes of things'). This agrees with the subject-matter as referred to in *DAT*, and may explain the origin of the title in MS W: 'De rerum causis', a descriptive title.

Several manuscripts (MSS GFBSLWVXNEs) add the motto at the

98 A similar problem confronts anyone seeking the sources of Petrus Alfonsi's *Dialogi*. This work is full of philosophical and cosmological speculation, which Petrus without doubt acquired from his Arabic learning in the Muslim kingdom of Saragossa before his conversion to Christianity in 1106, but his sources are very difficult to trace to specific Arabic texts; see Burnett, 'Works of Petrus Alfonsi'.

end of the list of *capitula* 'Sic faciunt cause rerum'. In MS G, and G only, this is preceded by 'Incipit liber'. The full phrase 'Incipit liber sic faciunt cause rerum' would suggest that the true title of *QN* was 'sic faciunt cause rerum' which would be no longer a descriptive title, but an emblematic one on a par with 'de eodem et diverso'. The Waltham Abbey catalogue understands this to be the title of the work.[99] However, there is at least one problem with this interpretation. The words 'Incipit liber' in MS G are on a separate line from 'Sic faciunt cause rerum' and are followed by a punctus. They would rather go with what precedes: 'Expliciunt capitula' ('The chapter-headings end; the book begins'), and 'Sic faciunt cause rerum' would remain a 'motto' as in the other manuscripts that give it. That the scribe of MS G did not consider this as the title is indicated by the fact that for the running titles he uses a shortened version of the epistolary heading: 'Baiocensi episcopo Adelardus'. The compiler of the Waltham catalogue might have understood the motto to be a title either because, in the manuscript in Waltham (which cannot be identified with any of the extant manuscripts) there was no punctuation between 'Incipit liber' and 'Sic faciunt cause rerum', or because the prologue and the list of questions was missing and 'Sic faciunt cause rerum' had become the heading. Nevertheless, this phrase, combined with Adelard's own references to 'rerum cause' as the subject-matter of the *QN*, might suggest that *De causis rerum* is a more authentic title than *Questiones naturales*. In this case Adelard may be deliberately recalling Virgil's well-known line 'felix qui potuit rerum cognoscere causas',[100] which, ironically, he puts into the mouths of the ridiculed philosophers in *DED*.[101] However, since the work has become widely known as the '*Questiones naturales*', and this title has some manuscript authority, it is retained here.

The *QN* appear to have had immediate success. They were copied on both sides of the English Channel. They were soon in the library of the Augustinians at St Victor in Paris,[102] and may have been in Chartres.[103] Altogether, thirteen manuscripts from the twelfth century have survived. Some of them (MSS BeEJW) are in a 'pocket-book' format, suggesting that they were meant to be carried around for

99 See no. 33 above.
100 Virgil, *Georgics*, 2.490; compare also Ovid, *Metamorphoses*, 15.68 (referring to Pythagoras): 'et rerum causas et quid natura, docebat'.
101 p. 8 below.
102 MS V, copied in 1140–50.
103 Patricia Stirnemann (private communication) detects a Chartrian hand in MS Q, another mid-twelfth-century manuscript.

entertainment or teaching. In one (MS J) the *QN* are surrounded by notes, apparently taken from lectures by Peter Abelard or one of his followers in Paris. Sometimes Adelard's work was accompanied by the recently recovered *Natural Questions* of Seneca.[104] This combination was known to William of Conches, who used both works extensively in his *Dragmaticon*. With the entry of Aristotle's natural science from the late twelfth century onwards, the *QN* become progressively less popular. Ten of our extant manuscripts date from the thirteenth century, three from the fourteenth, and two from the fifteenth. Nevertheless the *QN* are a source for Vincent of Beauvais' *Speculum naturale* (1253),[105] and Roger Bacon quotes with approval Adelard's advocacy of reason over authority.[106]

In academic circles the doctrine of the *QN* was replaced by that of Aristotle's natural science, and as a literary work it was abandoned in favour of classical texts.[107] However, it continued to be popular in vernacular literature. It was translated twice into Hebrew, once by Berachya Ha-Nakdan, perhaps in Oxford in the late twelfth century, and a second time anonymously in an abbreviated form.[108] Several scholars have suggested that, in view of the number of French words used in the Hebrew versions, an intermediate French translation of Adelard's work once existed.[109] No such text has survived, but a large proportion of the early fourteenth-century Tuscan *Questioni filosofiche* has been shown to be a direct translation of the *QN*.[110]

With the rise of humanism in the Renaissance interest in the *QN* revived. Giovanni Pico della Mirandola quoted the work.[111] The text was printed three times, and the well-known humanist and book-collector, Raphael de Mercatellis (1437–1508), had a deluxe version

104 This combination can be seen already in the manuscript of the *QN* bequeathed by Philip of Harcourt to the abbey of Bec in 1163, and in MSS VEsY and Z.

105 Vincent of Beauvais cites 'Adelardus' (twice 'Adelardus ad nepotem', once 'Adelardus Bathoniensis philosophus') in 8 chapters in his *Speculum Naturale*. I am grateful to Monique Paulmier-Foucart for sending me the references.

106 *Opus Maius*, I, pp. 5–6.

107 In the case of one of the fourteenth-century manuscripts (MS Z), the classical work of Seneca has been retained, but the scribe has stopped copying Adelard's work after writing the first page, and subsequently erases what he has copied.

108 Both versions are edited and translated in Gollancz, *Dodi ve-Nechdi*.

109 See discussion in Gollancz, *Dodi ve-Nechdi*, p. ii.

110 See Geymonat, *Questioni filosofiche*.

111 *Adversus astrologos*, III, c. xv, ed. E. Garin, 2 vols, Florence, 1946, II, p. 308: 'Adelandus (*sic*) ex opinione Saracennorum causam accessus maris atque recessus hanc et narrat et probat. Luna vero in causa non est' (cf. Q52). It does not seem to be the case, however, that Adelard is the same as 'Adelandus Arabs' whose conclusions are listed among the *Conclusiones nongentae* of Pico (ed. A. Biondi, Città di Castello, 1995, pp. 36–8).

(MS Se) copied from one of these printings. This revival of Adelard's work, however, was short-lived, and thereafter, apparently, it was only in England that sporadic interest was shown.[112] The first versions to be published in modern times were the Hebrew versions, to which Gollancz added an English translation of the Renaissance printing of the *QN*.[113] Martin Müller provided the first critical edition of the Latin text, but he was unaware of some of the best manuscripts. The present edition, it is hoped, has benefited from the knowledge of the whole manuscript tradition behind it and the application of the exigencies of present-day scholarship in the edition of texts.

De avibus tractatus

The treatise on hawking by Adelard of Bath (*DAT*) is not the oldest manual written on this subject in western Europe, but it is the first that can be attributed to an identifiable author. The practice of hunting with birds of prey had been introduced into the West some centuries earlier, in the aftermath of Antiquity. Roman civilisation had been ignorant of it, but in the course of the fourth century A.D. it became known by way of the Germans.[114] Hawking, in a broad sense, is the technique of training captive birds of prey, in order to hunt with their help. According to the main categories of birds used, one speaks of falconry for hunting with falcons (peregrine falcon, gerfalcon, etc.), and of hawking where the sparrow-hawk or the goshawk are used. Manuals to help falconers and hawkers were written at first in Latin, and the oldest ones that have been preserved date back to the middle of the tenth century, namely two collections of recipes for curing illnesses of birds of prey.[115] The twelfth century saw a sudden proliferation of such texts, of which no fewer than eight have survived, mainly concerned with remedies.[116] Some were written in the Norman kingdom of Sicily, such as those attributed to Dancus and

112 For further details see Burnett and Ronca, 'The Transmission of Adelard's *Questiones naturales*'.

113 Gollancz, *Dodi ve-Nechdi*, pp. 85–161.

114 On these questions, see K. Lindner, art. 'Beizjagd', in *Reallexikon der germanischen Altertumskunde*, 2nd edn, II, 1976, pp.163–73.

115 One was edited recently by B. Bischoff, 'Die älteste europäische Falkenmedizin (Mitte des zehnten Jhs.)', in id., *Anecdota novissima*, Stuttgart, 1984, pp.171–82. The second, ascribed to a certain Grimaldus, has been studied and edited in the D.E.A. dissertation of An Smets, 'Le traité d'autourserie de Grimaldus Ms 184 (288) de la Bibliothèque municipale de Poitiers (fol. 70r°–73v°)', Poitiers, 1995–6.

116 For references and a *status quaestionis*, see Van den Abeele, 'Les Traités de fauconnerie latins du xiie s.'

Guillelmus, and perhaps also that of Gerardus, while the treatise by Adelard of Bath originates in England; others however cannot be related to any precise context: this is the case with the apocryphal Letter on the care of hawks, allegedly sent to king Ptolemy by Aquila, Symachus and Theodotion.[117] Comprehensive and detailed works appear in the thirteenth century, the masterpiece being the *De arte venandi cum avibus* of the Emperor Frederick II of Hohenstaufen.[118] At the same time a translation of an Arabic work on falconry was produced at the imperial court: it is known under the name of *Liber Moamin falconarii.*[119] Most of the later works were written in the vernacular, with a strongly individual literature in fourteenth-century France and Spain, and significant contributions in Italy, Germany and England.[120]

The *DAT* forms part of the renewal of the genre in the twelfth century. The attribution to Adelard of Bath is supported by various arguments. The manuscripts mention *Adelardus* or *Alardus* by name in the text, and the very choice of the form, a dialogue between a master and his nephew, is paralleled in the two other works edited in this volume. English origin is shown by the allusions to English practices and the inclusion of English plant-names. The mention of a previous discussion concerning the *causae rerum* must refer, as Charles Haskins first noticed, to the *QN*, whose paternity is unchallenged.[121] Moreover, the style is Adelard's, as becomes apparent when comparing the phrasing of the nephew's questions in the *QN* and in the *DAT*. Furthermore, Adelard alludes to his own experience of hawking in the section on music in *DED*.[122]

The *DAT* comprises 29 sections or chapters. Compared to the other

117 These are the names of three of the translators involved in producing the Septuagint in the first and second century A.D.

118 *Friderici Romanorum Imperatoris Secundi De arte venandi cum avibus*, ed. C. A. Willemsen, Leipzig, 1942.

119 The Franco-Italian translation has been edited by H. Tjerneld in *Moamin et Ghatrif*, Studia Romanica Holmiensia 1, Lund, 1945. Since then the Latin text and its Italian translations have been edited by M. D. Glessgen, *Die Falkenheilkunde des "Moamin" im Spiegel ihrer volgarizzamenti: Studien zur Romania Arabica*, 2 vols, Beihefte zur Zeitschrift für romanische Philologie, 269–70, Tübingen, 1996.

120 An overview of the complete Latin tradition is given in Van den Abeele, *La Fauconnerie au moyen âge.* Still useful for vernacular treatises in spite of its age is H. Werth, 'Altfranzösische Jagdlehrbücher nebst Handschriftenbibliographie der abendländischen Jagdliteratur überhaupt', *Zeitschrift für Romanische Philologie*, 12, 1888, pp. 146–91 and 381–415, 13, 1889, pp. 1–34.

121 C. H. Haskins, 'Some Early Treatises on Falconry', *Romanic Review*, 13, 1922, pp. 18–27, reprinted in his *Studies in the History of Mediaeval Science*, pp. 346–55.

122 See below, pp. 54–5; cf. the hawking metaphor in the same text, p. 69 and n. 59.

treatises of the twelfth century, it stands out for its detailed attention to the processes of 'manning' – the word for taming in the language of the falconers – and training the hawks. After a lively prologue, the first sections explain the qualities required from those who want to devote themselves to this activity, the way to capture young hawks and to give them adequate care, and the first steps of their training. Since Adelard speaks only of *accipitres*, the *DAT* is strictly speaking a manual of hawking, and not of falconry. In medieval glossaries written in England, *accipiter* is translated by 'goshawk',[123] and the surviving Anglo-Norman versification of part of Adelard's treatise[124] mentions *osturs*, which was the specific word for a goshawk. Hunting with the indigenous goshawk seems to have been the main, if not the only, form practised in early German times, and it remained predominant in the northern parts of Europe.[125]

Sections seven to twenty-nine of *DAT* are of a therapeutic nature.[126] Adelard explains to his nephew which remedies to apply if the hawk gets various diseases, proceeding in the classical medical order from the head to the feet. The symptoms of the illnesses are generally left out, which is a common feature in twelfth-century treatises on bird-medicine. For each illness, one or several remedies are listed in simple terms, with some allusions to vernacular English names.

According to the prologue, Adelard has gathered his knowledge from older sources – the 'books of King Harold' – and from direct contact with practitioners. What is meant by King Harold's books remains a puzzle, since no older text of English origin has come down to us.[127] This Harold is presumably the unfortunate king Harold Godwinson who died at the battle of Hastings and whose keenness on falconry is illustrated in the Bayeux Tapestry. Some of his books may have survived in a royal library under the Norman kings, where

123 This is what one finds in thirteen out of sixteen cases recorded by Hunt, *Teaching and Learning Latin* (see index).

124 See below, p. xxxvi.

125 On the various medieval senses of *accipiter*, see Van den Abeele, *La Fauconnerie au moyen âge*, pp. 75–9.

126 One should note that the bird diseases in the *DAT*, as in other Latin treatises, cannot always be properly identified, because of a lack of descriptive elements. Present-day veterinary knowledge has lost the link with the traditional terminology.

127 The first scholar to draw attention to this was Haskins, 'King Harold's Books'. There might have existed an older text on hawking, according to a mention in a catalogue of Christ Church Cathedral Priory, Canterbury, written about 1300: 'Liber Alvredi Regis De custodiendis accipitribus'; see M. R. James, *The Ancient Libraries of Canterbury and Dover*, Cambridge, 1902, p. 60, no. 496. It does not seem to have survived.

Adelard could have consulted them.[128] At any rate, the remedies given by Adelard cannot be traced back to any existing source: they are found neither in the two tenth-century collections, nor in the other twelfth-century treatises. Nor do they show traces of Oriental influence, as could be expected if the work had been written after Adelard's journeys in the Mediterranean. Rather, genuinely English practice seems to be reflected in his little work.

The work has survived in two forms, a long version (α-version) and a short version (β-version). The latter is one of the texts included in the *Practica avium*, an anonymous collection of five abridged twelfth-century treatises.[129] Here it is headed: 'quarta <particula> quid Alardus anglicus nepoti suo interroganti responderit'. As for the title of the long version (which is presumably closer to Adelard's original text), Swaen, its previous editor, opted for *De curis accipitrum*, which he adapted from the last sentence of the Vienna copy: 'Hec habui de cura accipitrum que dicerem'. However, as has been noticed by Dafydd Evans, this is never a title in the manuscripts, and gives only a partial account of the contents.[130] In the same Vienna manuscript the heading is 'Hic est de avibus tractatus', and *De avibus tractatus* proves to be more appropriate to the range of the subject-matter.

In 1922 Charles Haskins wrote that 'Adelard's little work does not seem to have been widely used'.[131] This opinion must now be corrected. With at least six manuscripts, its direct tradition is not negligible, and an investigation of later Latin and French treatises reveals an abundance of excerpts from it. A partial Anglo-Norman versified translation was edited in 1943 by the principal scholar of medieval cynegetic texts, Gunnar Tilander.[132] However, he failed to recognise the relationship to Adelard's work, which was pointed out quite recently by Evans, who signalled some resemblances to three further Anglo-Norman treatises.[133] There exists also a hitherto unknown Old French translation of the treatise, in MS Lyons, Bibliothèque Municipale, 765, fols 74v–81r. Although it does not preserve the dialogue form, it follows the α-version quite faithfully, except that

128 This is a hypothesis of Evans, 'Adelard on Falconry'. For Adelard's possible connections with the royal court, see pp. xvi–xvii above.

129 This collection was previously thought to have been made for Enzio, the son of Frederick II Hohenstaufen, but this seems unlikely in the light of the new manuscripts discovered; see Van den Abeele, 'Les traités de fauconnerie latins du xiie s.', p. 282.

130 Evans, 'Adelard on Falconry', pp. 25–6.

131 Haskins, 'Some Early Treatises', p. 347.

132 Tilander, 'Fragment d'un traité de fauconnerie anglo-normand en vers'.

133 Evans, 'Adelard on Falconry', p. 26.

it omits the prologue and a few chapters at the beginning and in the middle of the text.[134] Substantial excerpts of the α-version occur also in various anonymous Latin treatises and collections of recipes: the *Physica avium* of MS Oxford, Bodleian Library, Lat.Misc. c.73, the *Liber de universis passionibus falconum, accipitrum, austurum, spervariorum, et qualiter eos curare poteris et qualiter eos nutrire et mudare debeas*, of Cambridge, University Library, Ff.6.13, and the *Doctrina nutritionis accipitrum et rapacium avium*, of Milan, Biblioteca Ambrosiana, Z 175 Sup. The compilation of Simon de Hembrad, entitled *Practica avium de raptu viventium* in Vienna, Österreichische Nationalbibliothek 2414 and Toulouse, Bibliothèque municipale (not catalogued), reproduces 21 of the 29 chapters of the *DAT*. A single Latin compilation draws on the β-version: the *Liber de ancipitribus et falconibus* of Oxford, Corpus Christi College, 287. Finally, one must mention the extensive use of Adelard's *De avibus* in the only known Provençal treatise on falconry, the *Dels auzels cassadors*, written by the troubadour Daude de Pradas around the middle of the thirteenth century.[135] Evans has drawn attention to a couple of related chapters,[136] but, in fact, one can recover the majority of Adelard's counsels and recipes in the Provençal text.

Of all these works, only the Anglo-Norman versified translation and the work of Daude de Pradas have been edited.[137] This may be a reason why the influence of Adelard's work has been underestimated. A new edition of the *DAT* should provide an impulse and a basis for further study of a whole range of medieval texts on falconry.

III Adelard's Language and Style

All the texts edited here betray some of the idiosyncrasies by which Adelard's language and style can be recognized. On the one hand he likes variety of expression;[138] on the other hand, he has his own

134 Chapters 2, 3, 13, 14, 15, 18 and 21 are omitted.

135 Schutz, *The Romance of Daude de Pradas*, Columbus, Ohio, 1945.

136 D. Evans, 'Le traité de fauconnerie en vers provençaux *Dels Auzels Cassadors*, son intérêt culturel', in *La Chasse au moyen âge. Actes du Colloque de Nice (22–24 juin 1979)*, Publications de la Faculté des Lettres et des Sciences Humaines de Nice, 20, Paris, 1980, pp. 9–17 ; see also id., 'Difficulties in the Botanical Lexis'.

137 A critical analysis of all the Latin treatises on falconry is included in the second volume of B. Van den Abeele, *Les traités de fauconnerie latins du moyen âge*, Diss., 4 vols, Louvain-la-Neuve, 1991, which is being prepared for publication.

138 E.g., in *QN*, Q40 he uses 'volunt, postulat, affectat, petit' and 'deposcit' all to mean the same thing ('desire, seek').

favourite phrases.[139] He uses asyndeton frequently,[140] but also adds 'et' for emphasis; in both cases probably as much for considerations of prose rhythm as for sense. One conspicuous feature is his use of double negatives.[141] While this is quite common in classical prose,[142] it becomes almost an affectation in Adelard's works, and as such has been preserved as far as possible in the English translations here. Some non-classical meanings of classical words are found.[143] Non-classical forms also occur in the manuscripts, but it is difficult to know whether Adelard is responsible for them, rather than a later copyist.[144] Here the judgement has to be made by the editor.

For the three works edited here Adelard deliberately employs three different styles. The *DED* is primarily didactic, the first part employing the language of the lawcourt. The arguments are set forth in balanced phrases, usually each of two parts, and commonly with rhyme or assonance at the end of each part.[145] The language is often exhortatory,[146] and the text is interlarded with proverbs or proverb-like statements, sometimes enhanced by alliteration.[147] The whole work is

[139] 'Unde fit ut' occurs seven times in the *DED*, eleven times in *QN*; 'si nescis', five times in the *QN*, 'ades animo', once in the *DED* and three times in the *QN*; 'de his que circa . . . ', four times in the *QN*; 'cedo' ('I agree'), three times in the *QN*, once in *DAT*. The invocation to let other scholars 'look to what they have said' occurs in similar form three times: 'videant igitur illi quid dixerint' (Q52); 'videant itaque quidam quid de motu dixerint' (Q60): 'viderit ille quid dixerit' (Q74).

[140] E.g., *DED*, p. 2: 'reprehensionis metum patienter ferrem, accusationi iniuste pro posse meo responderem'.

[141] *DED*: 'haut iniuste', 'haut obscure', 'haut imprudenter', etc.; *QN*: 'non dissimiliter', 'non dissimilis', 'non improbabiliter', etc.; *DAT*: 'non invideo'. The first two texts give examples of a pile-up of these double negatives: *DED*, p. 28: 'tu quantum supra ipsam possis eadem victoria *haut obscure* ostendetur. *Neque* enim michi quin superior discedam *diffido*, cum et ea que nuper in familiaris tui consolatione *haut imprudenter* disseruisti mente teneam, et ex illis alia *non minus probabilia* ratione duce argumenta contexere *non torpescam*'; *QN*, Q31: 'ex *non dissimili* causa procedant, eandemque explicationem *non fugiant*.'

[142] E.g., Macrobius, *Saturnalia*, 1.2.1 (introducing the work): 'tibi ipsi, quantum arbitror, non iniucundum'.

[143] 'Fatiscere' is commonly used for a failure in mental activity, and perhaps might have been regarded as an iterative of 'fatigo' (*DED*, pp. 26 and 44; *QN*, Q17 and Q38); 'resipere' is used in the sense of 'resipiscere' (*DED*, pp. 8 and 32).

[144] E.g. *QN*, Q9: 'virosus' for 'viriosus'; Q23 'protundetur' for 'protendetur'. On the two occasions in the *DED* on which the perfect of '-pingo' is used, the manuscript gives '-pixi' rather than '-pinxi', and '-pinxi' is never found in the works of Adelard edited here; the form 'depixit' is also found in Gregory of Tours, according to *TLL*., s.v.

[145] This is especially marked in the prologue: ' . . . et illos facundos iudico, et hos taciturnos appello. Quippe nec illi omnia noverant, nec isti omnia ignorant . . . ne si invidie attrectationes metuant, inscitie accusationem incurrant.'

[146] There are frequent imperative forms and 'o!' is used four times.

[147] p. 8 '[Philosophi] pannosi per plateas sibilaverant', ibid. 'quot panes diviseris, tot

written in an elevated style following the best classical *auctores*; words or phrases influenced by colloquial or vernacular speech are avoided, but Greek words are inserted to lend *cachet* to the discourse. In keeping with this literary style, on two occasions Adelard inserts passages of elegiac couplets, which conform to the strict rules of classical Latin and do not show the rhymes at the *caesurae* or the tendency for rhythmic symmetry that medieval Latin verse of a lower register is wont to do. One conspicuous feature is his use of *versus rapportati*, in which parallel phrases are interwoven to form a pattern a, b, c, aa, bb, cc, aaa, bbb, ccc, etc.[148] Adelard's principal model, both for form and for language, is Boethius' *Consolation of Philosophy*; amongst the *moderni* who used this model was Hildebert of Tours, who also had a predilection for *versus rapportati*.[149]

The *QN*, on the other hand, is written in a freer style of Latin. Adelard continues to use alliteration[150] and proverbial phrases;[151] he quotes the classical poets by name or tacitly.[152] But most conspicuous is his use of paronomasia or punning (puns are not obvious in the *DED*).[153] The puns either play on the ambiguity arising from a single word with two different meanings,[154] or change a word slightly to produce its 'true' meaning, in the manner beloved by the ancient etymologists.[155] Classical words are used with non-classical mean-

philosophos habebis', p. 10 'estque in summa verbositate summa eorum fiducia'; for alliteration, see the verse p. 12 'In tenebris tenebrosa docens tenebrosus Appollo'.

148 p. 28 below: 'Ver, autumnus, hiemps, cur pingat, compleat, artet,/Prata, domos, latices, gramine, farre, gelu'. For this form see E. R. Curtius, *European Literature and the Latin Middle Ages*, trans. W. R. Trask, Princeton, 1953, pp. 286–7.

149 Cf. Hildebertus, *Carmina minora*, no. 30, vv. 3–4, ed. A. B. Scott, Leipzig, 1969, p. 19: 'rebus, avis, specie, locuples, excelsa, celebris, / abstinuit luxu, fastibus, opprobriis'.

150 Cf. the *annominatio* of 'violentos principes, vinolentos presules' in the third sentence of the *QN*, and Q23 'quod michi obscuris omnibus obscurius est'.

151 E.g., Q19 'Ita percutit qui non videt', and the proverbs in Q34 and Q45.

152 Q22 Virgil (not named); Q34 Martial (not named); Q34 Terence; Q49, Q53, Q59 and Q65 Statius; Q55 Horace (the 'Satirist').

153 Subtle puns may be intended in *DED*, p. 28: '*visibus* ignoret nostris cur sidera quedam *invideant*' (the verb 'invidere' means 'begrudge', but is put together from the words meaning 'not' and 'seen') and ibid., p. 50: 'earumque ipsarum partium descriptiones haut *imparatas* habet' ('imparatus' means 'not well-thought-out' but is used in the context of the discussion of even and odd numbers – 'par' and 'impar').

154 E.g., Q71: 'NEPOS: Qua excessione caloris ac frigoris eliminata vigeret mortalis res publica mansueta mediocritate *exhilarata*. ADELARDUS: *Exhilarata* certe foret; neque enim hilaris esse posset' ('exhilarari' = (1) to be joyful, (2) to be 'out of joy'); Q76: 'Quare que harum diffinitionum tibi *sedeat*, *sedens* expecto' ('sedere' = (1) 'to be settled', (2) 'to sit').

155 E.g., Q25: '*Metitur* enim, immo ut verius dicam *mentitur*' ('For the common person measures – I should rather say, mis-shares'); Q58: 'anilibus imbuti *sententiis*, nescio an *sentibus*' ('immersed in the thoughts – not to say thorns – of the old woman').

ings.[156] A phrase from the Vulgate is recalled.[157] There are playful diminutives[158] and technical terms from late classical and more recent medical texts.[159] In two cases the meaning of the word can be shown to be influenced by its vernacular Romance cognate;[160] in another, this cognate is specifically referred to.[161] 'Haut' ('not'), which is non-productive in the Romance languages, is dropped almost altogether in favour of 'non'.[162]

This style is in keeping with the fact that the *QN* is a dramatic work recreating a real conversation (if not actually performed as a public disputation).[163] There are echoes of the plays of Terence.[164] But the spirited nature of the dialogue is above all in the style of Plato, Adelard's 'friend' and 'familiar'. Calcidius' translation of the first two thirds of Plato's *Timaeus* is the text most frequently and extensively quoted, both in the *DED* and in the *QN*. It is difficult, however, to see much in common between the turgid prose of Calcidius' translation (even in the sections of rapid exchange) and Adelard's lively language.[165] Adelard is unlikely to have known any other dialogue of Plato directly.[166] Rather, he follows in the tradition of classical Latin

156 Adelard appears to use 'molestia' as a derivative of the root 'moles' and having the meaning 'bulkiness' (Q15, Q38 and Q67). In the case of 'succurrere' Adelard appears to use the original and literal meaning 'run (out) under' rather than the derived meaning which had become the common meaning: 'run to help' (Q58).

157 Q64: 'Et tu ex illis es' (Matthew 26, 73).

158 E.g., Q31 'mammulis suspendulis' – the organs of smell.

159 Q18 'occipitium' for the front part of the head (a technical term in medical texts and in Macrobius, *Saturnalia*, 7.10. 2 and 7.10.5; cf. the more poetic 'prora' used in the *DED*, p. 70); Q16 'aperitivus', of the pores in the skin; Q12 'albugineus', one of the humours of the eye (earliest use attested in MlatWb, s.v., is in Pseudo-Copho, *Anatomia* of 1110–1120); Q23 'visibilis' for the visual spirit (used in Medieval medical sources: e.g., by Constantine the African); 'insita' as a noun for graft (Q6; in an equivalent passage William of Conches uses the Classical 'surculus'); 'sensile' as a noun for the sensible universe (Q61 and Q69).

160 For 'iecur' in the sense of 'gizzard' rather than 'liver' see p. 229 below; 'unctum' for grease (Q7 'quod communi usu unctum'; cf. v. Wartburg, XIV, pp. 28–9: French 'oint' = 'graisse de porc'); cf. also 'foliata', Q64.

161 Q20: 'poros ipsius cranei – id est teste'; cf. v. Wartburg, XIII, p. 281.

162 There are fourteen examples of 'haut' in the *DED*, all but three of them before an adjective or adverb with a negative connotation; in the *QN*, which is more than twice the length of the *DED*, there are only three examples of 'haut', and adjectives and adverbs are regularly negated with 'non'.

163 The language of disputation is used in Q28: 'oppositio . . . responsa'.

164 E.g., the Terentian phrases 'sic sic' (Q25) and 'scrupulum inicere' (Q6) and the uses of asides (Q35 'Cum protervo michi quantum intelligo sermo susceptus est').

165 That is not to say that Adelard does not use Calcidius' words: he employs 'Equum postulas' (cf. *Timaeus*, 17B) four times, and follows Calcidius' usage of the terms 'sensilis mundus' and 'irrationabilis' (rather than 'irrationalis').

166 The *Meno*, whose style recalls that of the *QN* more closely than that of the *Timaeus*, was

authors who admired Plato's dialogues – Cicero, and, above all, Macrobius in his *Saturnalia*. The form of the *Saturnalia* is deliberately modelled on that of a Platonic dialogue.[167] Adelard follows Macrobius in referring to the conversation between 'Adelard' and the nephew as a 'tractatus',[168] which takes place during a holiday.[169] The participants provoke each other by taking opposite sides in an argument,[170] and relate personal experiences.[171] Both authors use the metaphor of knots or snares for getting into tricky situations.[172] Macrobius, however, does not appear to use puns. This tendency is found, rather, in one of Adelard's near contemporaries, Honorius Augustodunensis (*ca.* 1070–*ca.* 1140).[173] Another contemporary note is struck by Adelard's use of the language of deductive argument of Euclid's *Elements*, especially that found in the *commentum*-version of the *Elements*, which belongs to Adelard's intellectual circle.[174]

The *DAT* takes a further step towards a colloquial and 'low' style. The urbane and conversational tone allows for punning[175] and irony.[176] Only in the prologue is an attempt at literary elegance made;

translated into Latin by Henricus Aristippus who praises the library of Syracuse in Sicily a generation after Adelard dedicated his *DED* to the bishop of Syracuse.

167 Macrobius specifically invokes the example of Plato in *Saturnalia*, 1.1.3 and 1.1.5, and presents the whole work as a conversation between cultivated citizens sharing a meal during a festival and recalled some time later, as in Plato's *Symposium*.

168 E.g., *QN*, Prologue: 'ego tractatum subscriptum recepi', Q23 'nichil hoc tractatu iocundius', Q37 'quod quia ad compositionem hominis attinet, in philosophico tractatu tangi non pudebit' (also Q49 and Q69); compare *Saturnalia*, 7.1.1 'philosophicis . . . tractatibus'.

169 Compare *QN*, Q47: 'Quoniam has ferias his gignasiis deputavimus . . . ' ('Since we have dedicated this holiday to these exercises . . . ') with *Saturnalia*, 1.1.2: 'Nam per omne spatium feriarum meliorem diei partem seriis disputationibus occupantes . . . ' (' . . . occupying the whole period of the holiday [i.e., the New Year's festival of the *Saturnalia*] with serious disputations').

170 E.g., *Saturnalia*, 7.5.1, 7.7.8 and 7.13.21.

171 E.g., *Saturnalia*, 7.7.14 and 7.13.18.

172 *Saturnalia*, 7.8.8–10: 'Inretitum te iam, Disari, teneo et si vere opinor, nusquam hodie effugies . . . ' to which the reply is 'retia quibus me involutum tenes nimis rara sunt, nimis patula: ecce me, Euangele, sine nisu inde exemptum videbis'; compare *QN*, Q27: 'sequitur tamen aliquid unde tibi nodum innectam'.

173 Honorius, *Elucidarium*, III.30, p. 452: 'si rogare auderem, libenter de Antichristo audirem.' For the connections of Honorius with Canterbury and (possibly) the diocese of Worcester, see V. I. J. Flint, 'Honorius Augustodunensis', in *Authors of the Middle Ages*, II, ed. P. J. Geary, Aldershot, 1995, pp. 100–107.

174 Compare the phrases in the *QN* 'Hoc enim ad superiora sequitur', 'ex praedictis ea . . . procedunt' with the injunctions such as 'premissa proposicione argumentum elicito' in the *commentum*-version: see *Robert of Chester's (?) Redaction*, p. 120 *et passim*, and Burnett, 'The Latin and Arabic Influences'.

175 *DAT*, c. 16: 'fellera vero, nisi fallor, potest curari.'

176 Ibid., c. 16: 'credo quod non vituperabis me'; ibid., explicit: 'ceterum tibi vel alicui si suam placet addere sententiam non invideo.'

the rest of the text is written in what might be called 'utilitarian' Latin: its aim is simply to impart information. Several of the Latin words used are attested elsewhere only in vernacular garb;[177] they could be back-formations from the Anglo-Norman vocabulary with which Adelard was familiar. In other cases Anglo-Norman or Anglo-Saxon equivalents of Latin terms are specifically referred to.[178]

IV Editorial Principles

De eodem et diverso

The *DED* survives in only one manuscript: Paris, Bibliothèque nationale, lat. 2389, s.xii^{med}, fols 81v–91v (old foliation 82v–92v = P), where it immediately follows a copy of the *QN*. The work was edited from this manuscript by Hans Willner in Beiträge zur Geschichte der Philosophie des Mittelalters, IV.1, Münster, 1903. Our edition is based on a completely fresh reading of the manuscript. Those emendations to the manuscript text made by Willner that seem necessary and sensible have been adopted, but in certain cases different emendations have been proposed, and in other cases the manuscript reading has been retained. Willner restored classical Latin orthography to the text and introduced German conventions of punctuation. Here the medieval spelling of the manuscript is kept, with occasional adjustments for internal consistency; while adopting modern English punctuation, we have been sensitive to the punctuation given in the manuscript.

Questiones naturales

The *QN* appear in the following manuscripts. The sigla used in this volume are listed, and the headings in the manuscripts are included.[179]

177 E.g., the word 'patellata' (*DAT*, c. 4), a cake produced in a 'patella' ('small pan') is not found in classical Latin literature nor in the *Novum Glossarium*, but is clearly cognate with Old French 'paelee', and Old Provençal 'padelada'; see v. Wartburg, VIII, p. 2; of similar formation are 'budellus' ('bowel'; c. 18) and 'mutatio' ('mewing'; c. 24).

178 E.g., c. 3 'quam vulgo firmam dicunt' [Old French 'ferme']; c. 15 'quod escinum vocant' [Anglo-Norman 'eskignon']; c. 16 'Angli suo ydiomate 'nigram herbam' vocant' [presumably Anglo-Saxon 'brūnwyrt']; c. 27 'que apud veteres Anglos vocatur 'milda', alia purgativa que apud eosdem 'strica' dicitur' [Anglo-Saxon 'milde' and 'stræc/strec'].

179 The folio numbers are recto unless otherwise stated. For further details see Burnett and Ronca, 'The Transmission of Adelard's *Questiones naturales*'.

B	Praha, Národní kníhovna, VIII.H.32, s. xii–xiii, fols 54v–69, Ricardo baiocensi episcopo adelardus batonensis salutem.
Be	Bergamo, Biblioteca civica, a VII 40, s. xii, fols 1–54v. *No title.*
C	Paris, Bibliothèque nationale, lat. 6415, s xii², fols 23v–39v. Adelardi Bathensis quaestiones physicae (early modern hand).
Cc	Oxford, Corpus Christi College, 86, s. xiv, fols 163–87. Incipit liber Adelardi de causis naturalium compositorum.
D	Oxford, Bodleian Library, Digby 11, s. xiii, fols 97–103. Incipit prologus super dialogum adelardi et nepotis.
E	Eton College, 161, s. xii^{med}, fols 1–37 (entire MS). Incipit liber Adelardi Batoniensis ad nepotem suum de solucione questionum naturalium.
Es	Escorial, Real Biblioteca, O III 2, s. xiv, fols 1–72. Incipit liber adelardi batensis de quibusdam naturalibus questionibus.
F	Firenze, Biblioteca Medicea Laurenziana, Gadd. Rel. 74, s. xv, fols 3–34. Liber probleumatum seu naturalium questionum per Adelardum bathoniensem philosophum Ad Ricardum Baiocensem Episcopum Incipit. Ricardo baiocensi episcopo Adelardus Bathoniensis salutem.
G	London, British Library, Cotton Galba E IV, s. xii^{med}, fols 214–28. RICARDO BAIOCENSI EPISCOPO ADELARDUS BATHONIENSIS. SAL(UTEM).
J	Salzburg, Stiftsbibliothek St Peter, a.V.2, s. xii, fols 35v–63v. Title cut off at top of page, but last two words appear to be 'naturis rerum'.
L/P	Leiden, Bibliotheek der Rijksuniversiteit, Voss. Lat. Q. 84, fols 12v–16v + Paris, Bibliothèque nationale, lat. 2389, fols 65–81v, s. xii^{med}. RICARDO BAIOCENSI EPISCOPO ADELARDUS BATHONICENSIS (sic!) SALUTEM.
M	Montpellier, École de médecine, 145, s. xiii, fols 63–90v. *No title.*
N	Paris, Bibliothèque nationale, lat. 18081, s. xiii, fols 195–210v. *No title.*
Or	Oxford, Oriel College, 7, s. xiii^{1}–xiii^{med}, fols 192–4v. *No title.*
Ox	Oxford, Bodleian Library, Bodley 679, s. xiii^{ex}, fols 108–27. Hunc librum compilavit adelardus bathoniensis ad nepotem suum de solucione questionum naturalium. Incipit prologus. (*lower margin*) Incipit tractatus Adelardi ad nepotem.
P	*See* L/P
Q	Paris, Bibliothèque nationale, lat. 6739, s. xii^{med}, fols 1–71. Incipit Prologus ADELARDI bathoniensis.

Qu Paris, Bibliothèque nationale, n.a.l., 372, s. xiii^{in}, fol. 67. *Table of questions only.*

R Reims, Bibliothèque municipale, 877, s. xii, fols 1–27. Ricardo baiocensi epo. adelardus batoniensis sal.

Ri Reims, Bibliothèque municipale, 872, s. xiii–xiv, fols. 109–11. *No title.*

S Stuttgart, Württembergische Landesbibliothek, Phil.Theol. Q 159, s. xiii, fols 12v–22. Ricardo Baiocensi episco. Adelardus Bathoniensis Salutem.

Se Sevilla, Biblioteca universitaria, 332/155, A.D. 1491, fols 135–52v. Incipit prologus Adelardi Bathoniensis in suas questiones naturales perdifficiles.

V Paris, Bibliothèque nationale, lat. 6628, A.D. 1140–50, fols 80v–114v. Incipit liber adelardi bathensis de quibusdam naturalibus questionibus.

W London, Wellcome Foundation, 4, s. xii^{med}, fols 1–24v (entire MS). Adelardi angli de rerum causis. Dialogus inter ipsum et nepotem.

Wi Wien, Österreichische Nationalbibliothek, 2351, s. xiii, fols 1–17v. Physica anshelmi (*sic*) anglici.

X Paris, Bibliothèque nationale, lat. 14700, s. xiii, fols 273–96. Incipit liber adelardi bathensis de quibusdam naturalibus questionibus.

Y Paris, Bibliothèque nationale, lat. 6286, s. xiii, fol. 55v (opening only). Incipit liber adelardi bathensis de quibusdam naturalibus questionibus.

Z Paris, Bibliothèque nationale, lat. BN 6385, s. xiv, fol. 51v (opening only). Incipit liber adelardi batensi (*sic*) de quibusdam naturalibus questionibus. Rubrica.

The text was printed three times in the Renaissance, twice on its own in Louvain in *ca.* 1475, and once with Pseudo-Albertus Magnus' *Liber aggregationis seu liber secretorum de virtutibus herbarum et de virtutibus animalium quorundam*, in Antwerp in *ca.* 1488–91. The heading in all three printings is 'Incipit prologus Adelardi Bathoniensis in suas questiones naturales perdifficiles'.

For the present edition a collation of all the manuscripts and the Renaissance printings was made for samples of the text taken from the beginning, the middle and the end. On the basis of this collation the relationships among the manuscripts were established, and five manuscripts were chosen on the grounds either of their giving the best

text or their being at the head of subfamilies of manuscripts. These manuscripts are G, L/P, M, Q and V, and they have been collated for the whole text. It is evident that none of these manuscripts is free from mistakes, or clearly exhibits the original text of Adelard. Moreover, they each have peculiarities which preclude them from being simply copies (with progressive corruptions) of an authorial 'published' text. These peculiarities may derive either from revisions by Adelard himself, or from those of later 'editors', and are as follows:

MS G includes extensive intelligent corrections to a hastily written text; these corrections have been added from at least two different manuscripts and are contemporary with the original copying of the manuscript (as can be seen from the fact that they have been made before the bifolia were stitched into gatherings); some of the corrections may be authorial. MS L/P is a copy of a text which includes two passages (in Q26) and two questions (one in the *capitula*, one between Q29 and Q30) not found in any other manuscript, and may be additions of Adelard himself. MS M gives a text which is full of rhetorical changes (mostly of word order) in respect to the other manuscripts. These would appear to be due to a later 'editor' who, perhaps, worked in the environment of a Cistercian monastery (the manuscript comes from Pontigny). Occasionally, however, it preserves a correct reading lost in all the other manuscripts. MS Q gives a text dedicated to a different bishop from the other manuscripts and with the question-headings (*capitula*) distributed through the text rather than grouped at the beginning. It also includes a fuller and more accurate transcription of a passage from Calcidius' translation of Plato's *Timaeus* in Q23 and generally has a very good text. It is clearly a variant edition of the text, made at an early date, and it is possible that Adelard himself was responsible for this edition. MS V is important because it stands at the head of a family of manuscripts which were copied in Normandy and at St Victor in Paris. Again, an attempt at organising the material in a logical way (with numbered chapter-headings at the beginning, and chapter-headings distributed throughout the manuscript) can be seen, but the exemplar of MS V was already seriously corrupt, having at least two columns of text in the wrong place, and missing out several passages throughout the text. Some passages towards the end of the text give the appearance of being completely rewritten.

Another manuscript should be mentioned in this context: MS J, which has the distinction of including Adelard's *De opere astrolapsus* as well. This text gives several unique readings, which may be authorial

(including the title 'de naturis rerum', and an additional passage in Q34), but is so full of mistakes apparently due to hasty copying (the whole manuscript being rather a 'note-book' than a library copy) that it cannot be used as a basis for a critical edition.

MSS M and V must be discounted as providing 'Adelard's' text, on the grounds that they are later 'editions'. We are left with the choice of giving one of what we presume to be Adelard's own 'editions', those of MSS G and L/P which were dedicated to Richard, bishop of Bayeux, or that of MS Q, which was dedicated to 'bishop G'. There are several reasons for choosing the text of MS L/P, corrected by MS G:

1) When MSS L/P and G are taken together they can, in most instances, provide readings which seem to be authorial, especially when one takes into account the corrections and readings preceded by 'al.' (= *alibi* or *aliter*) in MS G.

2) The variants between MSS L/P and G (especially the 'al.'-readings in G) are often shared by MS V and MSS MQ respectively; these manuscripts, therefore, take us back close to the beginnings of the split between these two textual traditions.

3) The *QN* and *DED* form a unit in MS L/P, and in L/P alone. Our edition reproduces the same grouping, while swapping their order.

4) The text of MS Q is available (though not always reliably so) in the Italian edition and translation of Balossi, Di Giovanni and Ferrari.

In the following edition, then, MS L/P is privileged. However, whenever MS G gives a plausible alternative reading, especially when this is shared by other manuscripts in the small group, this is reported. When MS L/P uniquely gives a plausible reading, this is also reported, because it may be part of the 'revision' suggested for MS L/P. However, changes of word order occurring only in MS L/P, small variants in spelling, obvious wrong transpositions of letters, faulty word-divisions and such like, are not mentioned. The alternate readings of MSS MQV have in every instance been considered, and occasionally adopted, when it appears that both MS L/P and MS G have been miscopied. Also, the presence of the same reading in MSS MQV often corroborates whether MS L/P or MS G is correct; in cases of word order the decision of the majority of all these manuscripts is usually followed. Wherever variant readings are reported, the readings of five manuscripts – MSS GL/PMQV – are given, unless the variant reading is unique to one manuscript. The punctuation and orthography follow MS G whenever possible, with the exception that the *e-caudata*, used inconsistently in this manuscript, has been replaced by a simple 'e'.

The aim, then, is not to provide, on the one hand, a diplomatic

transcription of any one manuscript, nor, on the other, to reconstruct a text which one can, with confidence, claim to be Adelard's *ipsissima verba*. Rather, it is to give an impression of what could have been the common archetype of the edition of the *QN* (presumably Adelard's own edition) represented by MSS L/P and G, while at the same time introducing modern conventions of punctuation, consistency of spelling and (occasionally) emendations of obvious errors in the text. Following the example of MSS L/P and G the questions have been listed at the beginning of the text and not inserted as headings to their solutions. However, the numbering of the questions (which is confused in both MSS L/P and G) has been rationalized.[180]

The following edition can be regarded, additionally, as complementary to that of Martin Müller in Beiträge zur Geschichte der Philosophie und Theologie des Mittelalters, XXXI.2, Münster, 1934. Müller reports the readings of MSS PBCEJS but did not use MSS G or Q, and did not know MS L. Moreover, Müller provides classical spelling throughout, so his text may be consulted by anyone who may be confused by the Medieval spellings retained in this edition.

De avibus tractatus

The text of the *DAT* is in a much worse condition than those of the *DED* and the *QN*. What appears to have happened is that Adelard wrote a work in a similar style to that of the *QN*. This text proved quite popular and was progressively modified in the direction of practical use, and vernacular language. These modifications obviously led most frequently to abbreviation, but it is possible that further recipes were also added. It is possible, too, that Adelard himself sketched out two or more versions of the text; the absence of a dedication might indicate that he never officially 'published' his text.

The version which appears closest to Adelard's original is the long version (α-version). This is found in three manuscripts:

A Wolfenbüttel, Herzog August Bibliothek, 47.13.1 Aug. 4to, s. xiv–xv, fols 25–31.

R Bruxelles, Bibliothèque royale, 8486–91, s. xiii, fols 25v–26v.

180 The numbering of the questions in this edition is the same as that in Müller's edition (which follows the Renaissance editions) with the exception of our Q35–7 = Müller Q36–8. Müller's Q35 ('Quare quaedam calidissima levius transglutiendo quam manu tenendo feramus') has no manuscript authority, whereas our Q38 ('Quare homines tam tenera habeant membra'; omitted by Müller) does.

W Wien, Österreichische Nationalbibliothek, 2504, *ca.* AD 1200, fols 49–51.

Adelard's hand is clear in this version above all because of the presence of a prologue, which occurs uniquely here: it refers to the *QN* and its language and style matches that of the *DED* and *QN*.[181] The original text of Adelard, however, is no longer recoverable. Probably at a very early stage in the diffusion of the work a split occurred which separated off MS A from the rest of the tradition. MSS RW on the whole share readings against MS A, and often the readings of MSS RW are reflected in the abbreviation known as the 'short version' (or β-version). MS A shows more vernacular influence than the common readings of MSS RW, in particular using *ipse* as the demonstrative pronoun (with no emphatic force) where MSS RW use the more classical *is*. But non-classical Latin words (or classical Latin words with non-classical meanings) occur throughout the text in all three manuscripts.[182] MS A also adds chapter-headings which are different from the isolated headings in MS R (MS W lacks chapter-headings), and identifies some of the vernacular terms used in the *DAT* with the Latin technical medical terms (e.g., c. 16 *ictericia* and c. 20 *tenasmon*). These headings would not be original, and hence are included in square brackets in this edition.

MS R has a different prologue from MS W and omits several sentences, especially towards the end; on the other hand, it includes chapter 29, which is missing in MS W.[183] When MS A agrees with MS R against MS W or with MS W against MS R we are likely to have an early reading. Otherwise, it is impossible to go further back in the tradition than the variants between MS A and MSS RW. Thus, all three manuscripts have been taken into account in preparing this edition, and their readings, wherever they are plausible, have been mentioned. Certain readings have been adopted because they conform to Adelard's vocabulary and phraseology. Trivial variations and obvious copying errors[184] have not been recorded, and when the

181 E.g., *DAT*, c. 1 'elegantius' ('more finely expressed') appears in a similar context at the end of *DED*, p. 68: 'opere precium erit diversarum gentium doctores adire, quodque in singulis elegantius reperies, memorie mandare'; *DAT*, c. 1 'Sit sane' is used frequently in the *QN* (Introduction to the questions, Q7, Q34).

182 See pp. xli–xlii above. Sometimes RW have the non-classical word where A has the classical one (e.g. c. 21 'frisso' R, 'frisco' W, for 'recenti' A).

183 R is the only manuscript of the three which appears to have been used by a practitioner. It is in a small format, easy to carry around, and includes several works on human medicine. On the pages which contain the *DAT* there are distinct signs of blood-stains.

184 One cause of error is that the copyist of MS A (or one of his predecessors) has apparently

readings of MSS A, R and W are equally plausible, the readings of MS A have been adopted.

As has been noted above,[185] several Latin, Anglo-Norman and French texts derive from the *DAT* (α-version). All these have been compared with the text presented by MSS ARW, and the testimony of those that have been found useful in determining a possible reading in the *DAT* has been reported. The following texts have proved helpful:

DN = *Doctrina nutritionis accipitrum et rapacium avium*
LUP = *Liber de universis passionibus falconum*
SdH = Simon de Hembrad, *Practica avium de raptu viventium*[186]

It is noticeable that some of the Adelardian features of the *DAT* (α-version) are omitted in these versions. The dialogue form disappears in all of them. In *DN* the pun 'fellera vero nisi fallor' (c. 21) is changed into 'fellera vero, ut mihi videtur'[187] and the distinctive description of the 'seat of the bile' (ibid.) is dropped.

The β-version also occasionally corroborates a reading or provides a solution to a problem in the α-version. It has been fully collated from the three known complete manuscripts:

H Cambridge, Clare College, 15, s. xiii, fol. 186r–v.
L Manchester, Chetham's College, 11380, s.xiii–xiv, fol. 147r–v.
Q Vatican, Biblioteca Apostolica Vaticana, Pal.lat. 1253. s. xiii, fols 256r–257v.

The previous edition of the text by A.E.H. Swaen (*De cura accipitrum: A Mediaeval Latin Treatise by Adelard of Bath*, Groningen, 1937) was based on W (Swaen did not know A and R). He prints the edition alongside the text of one manuscript of the β-version (Cambridge, Clare College 15; a further two manuscripts of this version were unknown to him). Swaen forbears from introducing emendations in the text, but suggests some corrections in the footnotes and commentary. He regards the manuscripts as being too faulty for a critical edition to be made. With the benefit of two further manuscripts (MSS AR), and of knowing Adelard's style and vocabulary from his other works, one may be more optimistic about providing a critical text.

expanded abbreviations in the wrong way; e.g. in c. 4 'abstrahere' has become 'aliqua de causa'.

185 See pp. xxxvi–xxxvii above.

186 The Vienna manuscript is used. This text is particularly close to the *DAT* (α-version), almost serving as a fourth manuscript of the work. Its text is closest to that of MSS RW, but it includes chapter 29.

187 The pun is also omitted in R.

However, the text offered must still be regarded as provisional and hypothetical, and could be radically altered if a better manuscript is found. The aim has been to provide a text which makes sense, and which conforms as far as possible to what is known about Adelard's style and vocabulary.[188]

Punctuation

In collating the manuscripts for the texts edited here the punctuation of the scribes has been rigorously recorded, and the use of the *punctus elevatus*, semicolon, colon and other signs observed. In the case of the *QN* the scribes of MS G invested great care in punctuating the text, often correcting one punctuation to another, or moving the punctuation. This helps us to understand which words go with which (as does modern punctuation). To a large extent the function of the punctuation of MS G can be reproduced by modern punctuation. What is lacking in modern punctuation, however, is the *punctus elevatus*, indicating a rise in the voice. Its use in MS G may be illustrated in the following phrases, in which the *punctus elevatus* is represented as '´':

ut que calida est ´ calidum, que frigida est ´ frigidum, sicca ´ siccum, humida ´ humidum affectet (Q3).

Modern punctuation, on the other hand, has the advantage of being able to indicate asides, inserted remarks, and direct speech, by means of brackets, long dashes and inverted commas. Proper names have been capitalized.

Orthography

In general the orthography of the manuscripts has been preserved: that of MS G (usually corroborated by MS P) for the *QN*, that of MS P for the *DED*, and that of MSS W and R for the *DAT*. An internal consistency has been imposed (e.g., 'dextra' is written where MS P has both 'dextera' and 'dextra'; 'set' where MS P has 'sed' and 'set', 'repperi' where MS P has 'reperi' and 'repperi'). The reader used to the conventions of classical Latin must be aware that an 'h' may be redundant or missing; 'ph' and 'f' may be interchanged; 'e' stands for classical 'ae' (this digraph is represented by 'ę' in MS G of *QN*); 'ti'

188 A full critical edition, giving the α- and β-versions in parallel columns and full manuscript variants, has been prepared by Baudouin van den Abeele, and should be published at a later date.

and 'ci' before a vowel may be interchanged; 'i' and 'y' may be interchanged. In addition, the following 'medieval' spellings should be noted: ancipiter (*for* accipiter), archanum, Appollo, coctani (*for* cottani), efficatia, gignasiis, mercennarios, Occeanus, pedissequus, perhennis, sepum (*for* sebum).

A positive apparatus criticus is provided; i.e., first the adopted reading and the manuscript(s) in which it occurs (or the scholar responsible for the emendation) are given, then the variant readings in other manuscripts are mentioned.[189] In cases where only one of the key manuscripts gives a variant reading, only this reading and its manuscript are mentioned: it is to be presumed that *all* the other manuscripts have the reading in the text. The following abbreviations are used in the apparatus criticus and notes:

a.c.	before correction
p.c.	after correction
al.	a reading prefixed in MS G by 'al.' (= *aliter*, or *alibi*)
vl	a reading prefixed in MS G by 'vel'
i.m.	a reading in the margin without prefix
supra	a reading written above the word in the text
om.	omits
add.	adds
sc.	'scilicet'; prefixes the 'correct' reading after citing the 'incorrect' but apparently authorial reading of the manuscripts.
transp.	transposes
post/ante	places before/after in the case of a more distant displacement than transposition
scripsimus	a reading proposed by the editors, but not attested by the MSS
addidimus	a reading added by the editors
seclusimus	a reading rejected by the editors
( )	indication of, or realisation of, an abbreviation
† †	a corrupt or unintelligible word or phrase
[ ]	word or words which the editors consider not to be authorial
< >	word or words which have been added by the editors or translators
Willner	An emendation by Hans Willner

[189] Errors in previous editions, or emendations that have not been adopted, are not generally mentioned.

Müller	An emendation by Martin Müller
LUP	*Liber de universis passionibus*
DN	*Doctrina nutritionis*
SdH	Simon de Hembrad, *Practica avium*
α-version	The long version of the *DAT*
β-version	The short version of the *DAT*

De eodem et diverso

SIRACUSIO EPISCOPO WILLELMO ADELARDUS SALUTEM.

Dum priscorum virorum scripta famosa – non omnia, set pleraque – perlegerim, eorumque facultatem cum modernorum scientia comparaverim, et illos facundos iudico, et hos taciturnos appello, quippe nec illi omnia noverant, nec isti omnia ignorant. Quare, sicut nec illi omnia dixerunt, <ita nec>[1] isti omnia tacere debent. Scribendum igitur aliquid vel modicum censeo, ne si invidie attrectationes[2] metuant, inscitie accusationem incurrant. Nam et ego cum idem metuens, iniuste cuidam nepotis mei accusationi rescribere vererer, in hanc demum sententiam animum compuli, ut reprehensionis metum patienter ferrem, accusationi iniuste pro posse meo responderem. Set quoniam omne disciplinare opus non artificis sui, verumtamen alienum expectat iudicium, factum est ut quicquid mecum confuse depinxi,[3] sapientis discretioni dealbandum supponam. Quoniam autem in epistola hac non unius tantum facultatis picturas intexui, verum omnium septem liberalium artium quasdam descriptiones intentionis necessitate interserui, consequens esse intellexi ut non quemlibet iudicem, immo talem qui septiformi rivo philosophie imbutus esset, eligerem.[4]

Tibi igitur, Willelme, Siracusie presul, omnium mathematicarum artium eruditissime, hanc orationem direxi, ut quicquid mea scientiola haut satis fretum in publicum prodire non audet, et examine tuo securum exeat, et nominis tui prelatione corrosionis invidiose morsus non pertimescat. Hanc autem epistolam 'de eodem et diverso' intitulavi, quoniam videlicet maximam orationis partem duabus personis – Philosophie scilicet atque Philocosmie – attribui, una quarum 'eadem', altera vero 'diversa', a Principe Philosophorum appellatur. Tuum igitur erit et superflua resecare et inordinata disponere; ego pro intentione mea nepoti respondebo.

1 ita nec *Willner add.*
2 attrectationes *Willner*, attrectiones P
3 depinxi *Willner*, depixi P
4 eligerem *scripsimus*, intelligerem P

ADELARD GREETS WILLIAM, BISHOP OF SYRACUSE.

When I examine the famous writings of the ancients – not all of them, but most – and compare their talents with the knowledge of the moderns, I judge the ancients eloquent, and call the moderns dumb. Granted, the ancients did not know everything, nor are the moderns wholly ignorant. So, just as the ancients did not say it all, the moderns should not keep silent about all. Therefore, I believe that something should be written – modest though it may be – lest the moderns, in their fear of the contagion[1] of envy, should incur the accusation of ignorance. I too, filled with such a fear, was afraid to write anything in response to a certain unjust accusation of my nephew, but finally forced myself to believe in this: that I should bear the fear of criticism patiently, that I should reply to an unjust accusation as best I could. But, since every work of instruction awaits the judgement not of its author, but of someone else, it has come about that whatever I have depicted in my own mind in a confused way, I put forward to be illuminated by the judgement of a wise man. However, because in this letter I have embroidered depictions not of one art alone, but in my sewing have figured all seven of the liberal arts, as the purpose of the work demanded, I understood the consequence to be that I should choose a judge not at random, but rather one steeped in that seven-formed stream of philosophy.

To you, therefore, William, bishop of Syracuse, most learned in all the mathematical arts, I have addressed this speech, so that whatever does not dare to appear in public because it is ill-supported by my meagre learning, may go forth confident, on account of your scrutiny, and, by sporting your name in front, may not fear the bites of gnawing envy. I have called this letter 'On the Same and the Different', since I have given the greatest part of the speech to two characters – Philosophia and Philocosmia – one of whom is called by the Prince of Philosophers[2] 'the Same', the other, 'the Different'. Your duty, then, will be both to prune away what is redundant and to rearrange what is badly ordered. I shall reply to my nephew in accordance with my aim.

ADELARDUS AD NEPOTEM DE EODEM ET DIVERSO.

Sepenumero ammirari soles, nepos, laboriosi itineris mei causam et aliquanto acrius sub nomine levitatis et inconstantie propositum[5] accusare. Quod si quis alius vel vulgaris hoc errore teneretur, incurie tradendum putarem. De te autem non minus miror quam doleo. Miror equidem, quia cum in pueritia[6] adhuc detinearis – cuius levitas quodammodo propria est – in me accuses quod in te ipso, si non dissimulas, respicies. Doleo magis, quia cum nichil te carius habeam, soleatque nobis in omni sententia communis animus esse, de hoc presertim tali proposito[7] meo dissentiamus. Quare, ne[8] illam alteram identitatem que amicorum animis inesse solet, hoc infortunio amittamus, sententias in unum conferamus. Et ego, si tibi idem videtur, causam erroris mei – ita enim vocare soles – paucis edisseram, et multiplicem labirintum ad unum honesti exitum vocabo; tu utrum recte texam animam adverte, et ea qua soles vel in sophismatum verboso agmine vel in rethorice affectuosa elocutione, modesta taciturnitate utere. Ego rem quam per biennium celavi, ut tibi morem geram, aperiam. Tu vero orationem in fine, ut diem in vespere, diiudica.

Erat preterito in anno vir quidam apud Turonim tum sapientia tum moribus gravis, adeo ut eo tam vulgares quam philosophi uterentur. Set quid plura de laude eius, cum presentis etatis auditores plerumque invidiosi sint, et te eius probitas non lateat qui una ibi mecum adesses? Hunc ego admodum colebam, studens eius prudentia doctior fieri. Cumque semel michi situs siderum, qualitates planetarum, distantias orbium, nocturnus exposuisset, 'Tu', inquit, 'utrum recte executus sim tecum expende; ego me domum recipiam.' Hic

[5] propositum *Willner*, propossint P
[6] in pueritia *Willner*, impueritia P
[7] propositio P
[8] ne *Willner*, nec P

ADELARD TO HIS NEPHEW, ON THE SAME AND THE DIFFERENT.

Time and again, dear nephew, it is your wont to wonder about the reason for my troublesome journey, and to criticize my resolve somewhat harshly as frivolous and capricious. If anyone else, or of the common crowd, were misled by this error, I should think they should be disregarded. Concerning you, however, I am just as much surprised as grieved. I am surprised because, since you are not yet free from boyhood – and frivolity is, in a way, proper to boyhood – you criticize in me what you will see in yourself, if you are honest with yourself. I grieve the more because, since I have nothing dearer to me than you, and we usually have the same views on every question, it is especially on this plan of mine that we disagree. Therefore, lest by this unfortunate pass we lose that sense of oneness which is usually found in the souls of friends, let us bring our points of view together. I, for my part, if you agree, shall describe the cause of my straying – for this is what you usually call it – in a few words, and I shall summon the many windings of a labyrinth to the single exit of what is honourable. You, take note of whether I am weaving my arguments correctly, and bring to bear that modest silence that you normally use in the face either of a wordy host of sophisms or of emotional rhetorical utterance. I shall reveal the matter which I have concealed for two years, in order to please you. But as for you, judge the speech when I have done, just as one judges the day at its close.

There was last year a certain man in Tours who was respected both for his wisdom and for his character, to such an extent that he was consulted as much by the common people as by philosophers. But why should I go on at length about his fame when one's audience these days are for the most part envious, and his merits are well-known to you who were there with me? I cultivated him a good deal, trying to become more learned from his wisdom. One night, when he had explained to me the positions of the constellations, the qualities of the planets and the distances of their orbs, he said: 'It is for you to consider in your mind whether I have described things correctly; I shall take myself home'. At this

ego, tum tractatus dignitate cum senis ammonitione occupatus, ad audita relegenda animo accingor. Et quia locus non nichil quietis turbationisve sensuali tumultu anime inferre solet, hunc michi quam quietissimum eligo, extra civitatem – scilicet ubi me nichil preter odores florum[9] et Ligeris fluminis fragores inquietaret. Itaque, cum soli relectioni[10] sententie illius operam darem, cunctis extra cessantibus, duas mulieres, unam a dextra aliam a sinistra, et aspexi et ammiratus sum. Erat autem dextra quam vulgus aspicere horreat, philosophisque numquam penitus innotescat. Unde fit ut nec illi eam querant, et hii quesitam numquam totam optineant. Stabat hec undique septem stipata virginibus, quarum facies, cum diverse essent, ita tamen intertexte erant ut nulla intuenti pateret, nisi cum omnes simul aspiceret.[11] Sinistra vero ita vulgari allectioni subiacebat ut et eam solam assequerentur. Set et hec quinque pedissequis comitata[12] erat, quarum facies cognoscere michi pronum non erat. Erant etenim quasi pudore oppresse, et oppositarum septem non ferentes aspectum. His ego perterritus, cum vicissim nunc has nunc illas subtristis visu percurrerem, sinistra illa vultum verbis accommodans hac voce exorsa est:

‹PHILOCOSMIA›[13] ‘Que te, iuvenis, in hunc errorem causa subegit, ut huic te tam inefficaci studio totum accommodes, et fallaces fugas subtilium velut umbram propriam consequi frustra labores? An ignoras quia, etiam si quam certitudinem ea que queris haberent, ex singulis tamen innumere dubitationes velut animi spine succrescant? Quin potius michi te paulisper accrede et ipse ea que explicabo diiudica, de multis quid tibi placeat eligens.

[9] odores florum *Willner*, odorum flores P
[10] relectioni *Willner*, relictioni P
[11] aspiceret *Willner*, aspicerem P
[12] commitata P
[13] Philocosmia *addidimus*, P *om.*

point I, struck as much by the dignity of the subject as by the advice of the old man, prepared to go over again in my mind what I had heard. And because the surroundings can either relax the mind or disturb it through exciting the senses, I chose this spot outside the city as the quietest possible place for me – that is, where nothing would disturb me except the scents of flowers and the rippling of the river Loire. Thus, when I was concentrating only on going over his words, blotting out all external impressions, I saw and was filled with wonder at two women, one on my right, the other on my left. The one on the right was she whom the common people shun, and whom philosophers never completely succeed in knowing. Hence it happens that the common people do not seek her out, and the philosophers never obtain the whole of her even when they have sought her. This woman was standing surrounded on all sides by seven maidens, who, although their faces were different, nevertheless were so intertwined[3] that one could not see any of them clearly, except when one looked at them all at once. But the woman on the left was so prone to attracting the common people that they pursued her alone. This one too was accompanied by five serving-girls whose faces I was not able to recognize easily. For they were lowered, as if in shame, and unable to bear the gaze of those seven maidens opposite them. When I, terrified by these maidens, ran my anxious eyes in turn now over some, now over others, the woman on the left, matching her expression to her words, began in this tone:

'Young man, what has driven you into this error, that you devote yourself entirely to a pursuit that is so useless, and, like a man chasing his shadow, you struggle in vain to catch up with fleeting fallacies of subtle arguments? Or do you not know that even if the things that you seek have some foundation, nevertheless from each one of them innumerable doubts sprout up like thorns in the mind? Rather, entrust yourself to me for a little while and judge for yourself what I shall describe, choosing what pleases you from a wide range of options. Philocosmia

‹DIVITIE›[14] Sunt michi quinque pedisseque hee, quarum uni et prime ita a primis annis nitor auri argentique varieque suppellectilis successit, ut nulli alii iam servire sciant nisi ei et cui ipsa arriserit. Sunt ei mille edificia turrita patentia, totidem subterranea, nullus eorum angulus vacuus; quocumque verteris, te diversis muneribus licet oculos pascas. Quid autem auro purius, quid argento formato habilius,[15] quid gemmarum distinctione lucidius? Set non michi, immo oculis propriis crede, qui et ideo singulis dati sunt ut de his iudicarent.

Deinde cui ista suppetit, ei nec ubertas[16] agrorum, nec copia pecorum, nec pictura pratorum, nec quicquid mortales animos ducere solet deest. Idem potens, idem dignus, idem famosus, idem voluptatibus plenissimus, idem philosophie fortissimus dicitur, cui si quid sapientie addatur, velut ficus matura findetur. Eadem absente frustra predicabere.

Postremo illa nescio que alia, quam ceci 'Philosophiam' vocant, hac absente, ostiatim mendicat,[17] et se et suos per queque vilissima distrahit, adeo ut iam communi sensu non nisi ludibrium sit. Unde fit ut eius sequaces demum resipientes, illam spernant, huic nostre adhereant, eosque deinde precipue negligant quos Philosophie intentos esse intelligant, qua ipsi diu detenti, pannosi per plateas sibilaverant, maluntque cum fortunatis de acervis suis numerare quam mendicantium philosophorum infinitas querimonias fulcire. Nec novum istud proverbium est: 'quot panes diviseris, tot philosophos habebis.' Qui, quia in se nullum solatium habent, inter se ita inferunt ut, velut pica pice, alter alteri oculos effodiat, et quamcumque partem alter sumpserit, eam alter improbet – postmodum eandem probet. Inde fit ut nec rogati, nomen suum audeant confiteri. Hiique, cum tales sint, audent tamen dicere: 'Felix qui potuit rerum cognoscere causas'? Immo, ut

[14] Divitie *addidimus*, P *om.*
[15] abilius P
[16] ubertas *Willner*, verbis P
[17] mendicat *Willner*, -cant P

Riches

I have these five serving-girls.[4] With one of them – and the first – the splendour of gold and silver and ornaments of every sort[5] has won such favour from her earliest years on, that they now know how to serve no other except her and the one on whom she smiles. She has a thousand turreted buildings which are visible, and the same number underground. No corner of them is empty; wherever you turn, you may feast your eyes on all kinds of treasures. What is purer than gold, what is more versatile than worked silver, what is brighter than the shimmering of jewels? But do not take *my* word for it: trust your own eyes, for eyes have been given to one and all precisely in order that they may judge about these things.

Add to this that the one she helps does not lack fertility of fields, abundance of cattle, the beautiful delight of meadows, and whatever is wont to capture mortal souls. That person is called powerful, respectable, famous, most abounding in pleasures, strongest in philosophy. If any wisdom is added to him, like a ripe fig he will burst. Without this maiden, your reputation will be of no use.

Finally without this maiden, that other maiden (don't ask me who!) whom blind people call 'Philosophy', begs from door to door, and drags herself and her followers through all the most vile things, to such an extent that by common consent she is now nothing but an object of ridicule. Hence it happens that her followers, finally coming to their senses, spurn her, and attach themselves to this maiden of ours. Then they make a point of avoiding those whom they understand to be keen on Philosophy, by whom they themselves were detained for a long time, and as a result had whistled in the streets wearing rags. They prefer to count money from their own piles, with the other fortunates, rather than to support the endless complaints of philosopher-beggars. An old saw runs: 'However many pieces you divide the bread into, so many philosophers will turn up.'[6] Because they find no solace in themselves, they behave towards each other in such a way that, like magpies, one pecks out the eyes of the other, and whatever side one takes, the other objects to – and soon afterwards adopts! Hence it happens that, even before they are asked, they will insist on telling you their name.[7] Yet do these men in such a situation still dare to say: 'Happy is the man who has been able to understand the causes of things

'cognoscere' concedam, infelicem tamen et miserum qui nichil quod incipit efficit, nichil quod cupit assequitur. Set nec res ipsas cognoscunt. Summi enim eorum (si 'principes' dicendi sunt) non modo in cognoscendis illis, verum etiam in inquirendi[18] modis, contrarii inventi sunt. Alius enim a sensibilibus investigandas esse censuit; alter ab insensibilibus incepit. Alius eas in sensibilibus tantum esse arguit; alter preter sensibilia etiam esse divinavit. Sic, dum uterque alterum inquietat, neuter fidem adipiscitur. Quod si hos pro summis non habes, nonne Tholomeus a Pitagora in numerorum proportione[19] dissentit, dum de diatessaron et diapason unam effici consonantiam ostendit? Nonne Epicurus aliter ac ceteri philosophi homines videre docuit, dumque presubtilis in sententia de visu fuit, in eadem ab aliis cecus esse dictus est? Quod si hii a memoria recesserunt, accipe modernos ac Latine eloquentie summos – Tullium dico et Boetium – in divisione partium sillogismi contrarios. Cui tandem eorum credendum est, qui cotidianis novitatibus aures vexant? Et assidue quidem etiam nunc cotidie Platones, Aristotiles novi nobis nascuntur, qui eque ea que nesciant ut et ea que sciant sine frontis iactura promittant. Estque in summa verbositate summa eorum fiducia. Quid enim scurrili impudentie similius quam quod ipsi se sentire simulant? Aiunt enim nullam esse certificationem sensuum, nec oculis nec auribus ceterisque credendum esse. Quo munere si priventur ipsi, aliter sentiant! Utinamque omnes ceci surdique efficiantur! Ac merito. Sequuntur enim, ut dicunt, rationem ducem, qua nichil cecius est, cum id quod nichil in actu rerum est se videre menciantur, hiique ei fidem habent. Quibus merito libet imprecari:

[18] in inquirendi *Willner*, inquirendi P
[19] proportione *scripsimus*, propositione P

(*res*)?'[8] If I agree with the 'understand', nevertheless I say he is, rather, unhappy and wretched who completes nothing that he begins, who gets nothing that he desires! They do not even understand things (*res*) themselves. For the best philosophers (call them 'princes' if you must) are found to be at odds with each other not only in the objects of their investigations but also in their methods of enquiry. One decided that things (*res*) should be investigated starting from sensibles; the other began from insensibles. One argued that things (*res*) existed only in sensibles; the other imagined that they also existed outside sensibles. Thus, since each turns the argument of the other upside down, neither wins our confidence. But if you do not consider these to be the best philosophers, does not Ptolemy disagree with Pythagoras concerning numerical ratios, when he shows that from the fourth plus the octave one consonance is made? Did not Epicurus teach men to see in a different way from the rest of the philosophers, and while he was exceedingly subtle in his opinion about sight, he was said by others to be blind in it? But if these lived too long ago to be remembered, take the moderns and those who are best in Latin eloquence – I mean Cicero and Boethius – who have opposing views about the division of the parts of the syllogism.[9] So, who can be believed of those who vex our ears with new-fangled ideas every day? All the time, even now, there arise daily in our midst new Platos and new Aristotles, who promise what they do not know as boldly as what they do know, without batting an eyelid. They place their greatest confidence in the greatest torrent of words. Can anything be more similar to scurrilous impudence than their pretence of being sensible? For they say that the senses provide no proof, and one should believe neither the eyes nor the ears, nor the other senses. If they are deprived of the use of their senses, they will change their minds! Would that they were all made blind and deaf! And deservedly. For they follow (they say) reason as a leader, than which nothing is more blind, since they tell the lie that they see that which is nothing in reality (*res*). And these people put their trust in that! Thus one can rightly call down upon them this curse:

PITAGORAS

Qui primum dignam docuit vanescere mentem,
Ut rerum falsis credat imaginibus,

MATERIAM ET FORMAM INDIVIDUAM

Dum quicquid toto iunxit natura favore,
Disiungit, ceci capta furore ducis,
Hec quoque que cernis, cum sint diversa creata,
Contexens, unam colligat in speciem –
Hic, inquam, procul elisus pellatur ab oris,[20]
Atque suos secum sub loca ceca trahat,
In tenebris tenebrosa docens tenebrosus Appollo,
Fictilibus verbis detineat socios,
Nec cuiquam credat, nec ei credatur ab ullo,
Dum verbis rerum tollit ab orbe decus.

POTENTIA

Tu vero, iuvenis, hanc potius quam verbis descripsi tibi habe. Quod si hec non placet, sunt hic alie quatuor quarum tui causa non piget naturas aperire. Et est una earum potentissima, adeo ut, quicquid ubique terrarum habitantibus[21] iacet, ipsa imperio premat, et potestatibus omnibus quicquid velit inscribat. Que inde etiam 'potentia' proprium nomen sortita est. Cui quotiens in populos irasci libet, non est qui ei resistat. Itaque nec gladius eius modum novit. Unde fit ut ex vultu eius ita omnes pendeant ut si ipsa asserat corvos albos et cignos nigros, non minus ceteri in verba eius iurent.[22] Cuius non modo facta verum etiam dicta per totum orbem triviatim predicantur, ut Ammonis et Appollinis famosas consultationes excedat.

DIGNITAS

Est et alia que quodammodo huic obnoxia est, habens omnimodam in se dignitatem. Que quicquid in toto mundo qualitercumque dignum est, id tale esse et ei tribuit et ipsa efficit. In huius iure sunt dictature, consulatus, magistratus, ceteraque his similia. Que quanta intentione hominum appe-

[20] horis P
[21] habitantibus *Willner*, habitant P
[22] iurent *Willner*, -rant P

He who first taught the worthy mind to think empty thoughts, Pythagoras
so that it believed in the false images of things (*res*),
both separating what nature had joined with full approval Matter and individual form
(enthralled as it was by the madness of a blind leader),
and intertwining and bundling into one species
what you see as distinct – since they were created
different –
this man, I say, should be crushed and banished far away from our
shores;
he should drag his followers with him down
to the regions without light:
a murky Apollo teaching murky doctrine in the
darkness.
May he detain his students with his words of clay.
May he believe no one; may no one believe him,
whilst he, by words, deprives the world of the beauty of
things (*res*).

But you, young man, would do better to take this maiden, whom I have described in words, for your own. If *she* does not please you, there are four more here, whose natures I am not unwilling to reveal for your sake. One of them is most powerful, to such an extent that, whatever part of the world is inhabited, she dominates with her rule, and marks out whatever she wishes with all powers. Hence also she has acquired as her proper name 'Power'. However many times it pleases her to express her anger against peoples, there is none who can resist her. Thus her sword knows no limit. Hence it happens that all hang on her lips in such a way that if she asserts that ravens are white and swans are black, everyone else swears on the truth of her words. Not only her deeds but also her words are praised in the streets throughout the whole world, so that she is more famous than the oracles of Ammon and Apollo. Power

There is another maiden who is in some way beholden to this one, having every kind of honour in herself. For whatever in the whole world has some sort of honour, that it has such is attributed to her, and she brings it to pass. In her power are the offices of dictator, consul, magistrate and everything else like this. You yourself see how keenly people seek these. For Honour

tantur ipse vides. Adeo etiam hii qui 'philosophie' nomen sibi preferunt, hanc tamen latenti ambitione affectant,[23] ut, dum se eam quodammodo contempnere simulant, ad eam tandem provehi gratulentur, tumque demum sibi philosophi esse videntur.

FAMA[24] Est item quarta que quot plumis tegitur totidem oculis variatur. Hec tibi quicquid seculorum preteritum est presensve aut futurum, deteget. Hec eadem est que omnes etates hominum ad famosissima facta compellit. Hec Iasonem pelagus temptare coegit, hec Herculem innumeris monstris obiecit, hec Moeneteum pro salute patrie subiecit, hec omnia que impossibilia esse videntur ad possibilem redigit facultatem. Postremo hec est illa que carientem pellit senectutem, que suos familiares et dum vivunt ubique notos, et post mortem vivos perpetuo reservat. Unde non inmerito qui eam secuti sunt dii immortales et ab antiquis vocati et posteris venerati sunt. Quamobrem si quid probe[25] factum est, magis iuvat fame tradere quam id ipsum fecisse, adeo ut etiam qui naturali probitate[26] carent in malitia alios superare nitantur, malentes vel sic quodammodo predicari quam omnino esse ignoti. Et hec quanti sit precii vides!

VOLUPTAS Est et michi quinta adhuc admodum luculenta, cui universa iocunditas subdita est. Qua absente si quid factum dictumve fuerit, nec facienti nec dicenti iocundum erit. Huius subiecte[27] sunt pulcritudo, velocitas, valitudo, alacritas et quicquid elementarem universitatem letificat. Hec ita sensibus nostris preest ut ei soli servire malint. Hec enim unguentis oblitos, floribus redimitos, odoratu pasci docuit; hec melleos Bachiosque latices gustare ammonuit; hec auro ac gemmis ceterisque rerum formis insitire oculos iussit; hec universis tinnitibus rata modulatione constantibus, quos[28] Greci

[23] affectant W, -tent P
[24] *The heading* 'Fama' *is in the text, unlike the other headings, which are in the margin.*
[25] probe *Willner*, proprie P
[26] probitate *Willner*, proprietate P
[27] subiecte *scripsimus*, subdite *Willner*, sub une P
[28] quos *Willner*, quas P

even those who openly sport the name of 'philosophy' pursue her with hidden ambition to such an extent that, while they make a pretence of despising her in some way, they are glad to have finally arrived at her. For then at last they seem to themselves to be philosophers.

Then there is a fourth maiden who is pricked out with as many eyes as she is covered with feathers. She will reveal to you whatever in the world has passed, is present or is to come. She is the one who has driven people of all ages to most famous deeds. She compelled Jason to brave the ocean, she pitted Hercules against countless monsters, she bowed down Menoeceus for the safety of his fatherland, she brings everything which seems to be impossible into the realm of possibility. Finally, she is the maiden who drives away decaying[10] old age and ensures that her followers both are known everywhere while they live and live perpetually after their death. Hence, not unworthily, those who followed her were called 'immortal gods' by the ancients, and were venerated by posterity. Therefore, if anything has been done well, people are more pleased to have become famous for this than to have done the deed itself, to such an extent that even those who lack natural goodness strive to be superior to others in evil, preferring to win some sort of renown even like this, rather than to be completely unknown. You see how valuable she is! Fame

I have, too, a fifth maiden who is also quite splendid. She has all pleasure in her control. Without her whatever is done or said will be pleasant neither to the doer nor to the speaker. Her servants are beauty, swiftness, good health, eagerness and whatever makes the elemental world happy. She rules over our senses in such a way that they prefer to serve her alone. She has taught people to feast on scent – smeared with ointments and garlanded with flowers; she has told them to taste honeyed and Bacchic draughts; she has ordered the eyes to thirst after gold and gems and the other beauties of things; she has opened the ears of animate beings to all the sounds of harmonic modulation, which the Greeks call 'symphonies'; Pleasure

ΣΥΝΦΩΝΙΑΣ[29] vocant, aures animantium adhibuit; hec demum, ne qua pars corporis voluptati non serviret, illecebros tactus toti superficiei[30] corporis subtexit. Unde haut iniuste Epicurus,[31] vir quidam et sapiens et nobis familiaris, summum bonum diffiniens voluptatem esse dixit, sine qua, ut dixi, quicquid accidit, bonum esse non videtur.

Et hee hactenus pro affectata brevitate succincte laudate sint. Tu autem quam tibi malis elige de his quinque! Nec illi in qua modo torpebas inani philosophie amplius osciteris. Est enim in verbis solis, et ea dum audiuntur tantum delectant. Sin rebus animum inducas, statim ipsa cum verbis suis simul evanescit.'

PHILOSOPHIA[32] Hic illa dextra in modestum turgorem elata, 'huncne etiam', inquit, 'inpudica, venenis tuis michi eripere conaris, dum fallacias tuas et tegis nominibus et vestis exemplis? Set apud alios. Me enim presente, numquam hodie laqueos tuos ei insinuabis. Et ut me et illum hoc infortunio liberem, prius ea quibus in me invecta es expendam. Deinde locus erit de tuis quas laudavisti meretriculis quid tenendum sit paucis docere. Et quod[33] michi primum opposuisti primum absolvam, ut, quo ordine maledicendi audaciam sumpseras, eodem ordine amittas.

Innumeras dicis spinas mentis ex questionibus meis oriri. Set nec innumere sunt, nec spine dicende sunt. Quod sic si sapis collige. Rerum Conditor optimus omnia ad sui similitudinem trahens (quantum eorum natura patitur) animam mente quam Greci NOYM vocant exornavit. Hac ipsa dum in sua puritate est, tumultu exteriore carens, plane utitur. Nec modo res ipsas, verum etiam earum causas et causarum initia assequitur, et ex presentibus futura longo[34] tractu cognoscit,

29 ΣΥΝΦΩΝΙΑΣ *Willner*, CYNΦONYAC P
30 superficiei *Willner*, -ficie P
31 iniuste Epicurus *Willner*, Epicurus iniuste P
32 'Philosophia' *is written in abbreviated form* (Pha) *in bold letters in the text.*
33 quod *Willner*, quid P
34 longo *Willner*, longa P

finally, lest any part of the body should not serve pleasure, she has covered the whole surface of the body with the enticements of touch. Hence Epicurus, a wise man and our familiar, did not unfairly define the highest good as pleasure; as I have said, whatever happens without it, does not seem to be good.[11]

Since I have striven for brevity, here let us round off the praise of these maidens. But you choose which you prefer from these five! Do not be lulled any more by that empty philosophy in which you were slumbering just now. For it exists in words alone, and those only please whilst they are heard.[12] If, however, you apply your mind to things (*res*), immediately she vanishes taking her words with her.'

At this point the woman on the right, roused to a moderate passion, said: Philosophia

'Do you, shameless one, attempt to snatch from me with your poisons this man too, by disguising your deceits with names, and clothing them with fine examples? Try with others. For whilst I am present, you will never today ensnare *this* one in your nets! So that I may free myself and him from this misfortune, first I shall weigh up those arguments you level against me. Then will come the time to show in a few words what should be thought about those little hussies of yours whom you have praised. I shall dispose of first that argument with which you first attacked me, so that you lose the boldness of your slandering in the same order as you took it up.

You say that innumerable thorns in the mind arise from my questions. But they are neither too many to number, nor should they be called thorns. Understand it in this way, if you are intelligent enough. The supremely good Creator of things (*res*), drawing all things into his likeness (as far as their nature permits), has adorned the soul with mind, which the Greeks call 'noys'. The soul uses this with clarity when she is in her pure state, lacking any disturbance from outside. She reaches not only things (*res*) themselves, but also their causes and the beginnings of their causes, and from present things understands those to come, a long time in advance. She

quidque ipsa sit, quid mens qua cognoscit, quid ratio qua inquirit, deprehendit. Eadem testeo et lutulento corporis amicta carcere, non parvam sue cognitionis amittit portionem. Set nec fex illa elementaris ad plenum potest hoc decus abolere. Querit enim quod perdidit et, memoria deficiente, utitur opinione.[35] Et per summam quam tenet, singula consequitur, reducens composita in ea ex quibus compacta sunt. Et partium naturam librans, si quam compositionem habent, enodat, et demum simplicitatem reperiens, initii speciem mira subtilitate intuetur. Rursus et eadem initia formis suis paulatim vestiens, in pluralitatem compositorum sensibilium deducit. Cum igitur et initia finita sint – neque enim aliter initia essent – et composita ex eis sensibus subiecta sint, non infinite ductu meo dubitationes oriuntur, nisi quis tuis oppressus illecebris, in ipso questionis limine[36] excecatur.

Set nec spine dicende sunt, cum adeo vis naturalis a cognoscendo non temperet ut etiam, exterioribus cognitis, ipsa secum de se disputet, et in se reciprocato quid anima sit [anima][37] diffiniat.[38]

Quod autem meos mendicare dixisti, etsi id per sequentia magis refellam, tamen hoc interim ne ignores, eos scilicet tum vera rerum speculatione non fraudari cum a possessionum curis liberi sunt. Quibus cum induuntur, nec suos neque se ipsos deinde cognoscunt, meique esse desistunt, tua contagione maculati. Tu vero callide ex scenica videri disserta temptas, dum in me causam erroris vertis cum tu potius eius causa sis, si ex quo te contingunt et se cognoscere desistunt.

Quod autem in familie mee Principes more tuo incanduisti, id non minus cassandum quam cetera. Contrarios eos in rerum inquisitione dicis. Ingeniose quidem. Unum enim altero

35 opinione *Willner*, -em P
36 limine *scripsimus*, lumine P
37 anima P, *seclusimus*
38 diffiniat *scripsimus*, diffinit P

grasps what she is, what is the mind by which she understands, what the reason by which she enquires. The same soul, when clothed in the clay and mud of the body's prison, loses not a small portion of her understanding. But not even those dregs of the elements can fully wipe out this glory. For she seeks what she has lost, and, when the memory fails, she uses opinion.[13] Through the totality of things she holds, she follows up each thing, reducing composites into what they were put together from. Then, weighing up the nature of the parts, she unravels whatever composition they have, and at length finding simplicity, she beholds the species of their beginning with wonderful subtlety. Again, also gradually clothing the same beginnings with their forms, she brings them into the plurality of composite sensibles. Therefore, both since the beginnings are finite – for otherwise they would not be beginnings – and since what is composed from them is subject to the senses, the doubts which arise through my leadership are not infinite, unless someone, overcome by your seductive words, is blinded on the very threshhold of the question.

Nor should they be called thorns, when the natural force does not cease from understanding, to such an extent that even when outside things have been understood, it disputes with itself about itself, and in itself reciprocally defines what the soul is.[14]

That you said that my disciples are beggars, I shall refute in more detail later. Nevertheless, in the meantime you should bear in mind that it is precisely when they are free from the cares of having possessions that they are not deprived of the true speculation on things (*res*). When they are encumbered by possessions, they do not know their own people and then do not know themselves. They cease to be mine, having been sullied by your touch. But you cunningly attempt to seem eloquent by using the tricks of the cabaret,[15] while you divert onto me the cause of error when you are rather its cause, if from the time my disciples touch you they cease to know even themselves.

What you have said against the Princes of my family in your usual heated way, is to be quashed no less than the rest. You say they are at odds with each other in the enquiry after things (*res*). You argue in a clever enough way. For you

vicissim collidere temptas dum non tibi confidis, ut quemquam eorum propria aggrediaris contradictione, quamquam eos a contrarietate facile absolvet si quis exiles tantorum virorum sententias non aliter atque ab eis dicte sunt intelligat. Quod quidem intelligenti breviter exponam.

PLATO Unus eorum mentis altitudine elatus pennisque quas sibi indui obnixe nisus, ab ipsis initiis res cognoscere aggressus est et quid essent antequam in corpora prodirent expressit, archetipas rerum formas dum sibi loquitur diffiniens.

ARISTOTILES Alter autem artificialiter callens, ut lectores complices facultate instrueret,[39] a sensibilibus[40] et compositis orsus est. Dumque sibi eodem[41] in itinere obviant, contrarii dicendi non sunt. Amat enim et compositio divisionem et divisio compositionem, dum utraque alteri fidem facit. Unde si quid in digitis et articulis[42] abaci numeralibus ex multiplicatione creverit, id utrum recte processerit divisione eiusdem summe probatur.

PLATO ARISTOTILES Quod autem unus ea[43] extra sensibilia, alter in sensibilibus tantum existere dixit, sic accipiendum est. Genus et species (de his enim sermo est) etiam rerum subiectarum nomina sunt. Nam si res consideres, eidem essentie et generis et speciei et individui nomina imposita sunt, set respectu diverso. Volentes etenim philosophi de rebus agere secundum hoc quod sensibus subiecte sunt, secundum quod a vocibus singularibus notantur et numeraliter diverse sunt, individua vocarunt – scilicet Socratem, Platonem, et ceteros. Eosdem autem altius intuentes – videlicet non secundum quod sensualiter diversi sunt, set in eo quod notantur ab hac voce 'homo' – speciem vocaverunt. Eosdem item in hoc tantum quod ab hac voce 'animal' notantur considerantes, genus vocaverunt. Nec tamen in consideratione speciali formas individuales tollunt, set obliviscuntur, cum a speciali nomine non ponantur; nec in generali speciales ablatas intelligunt, set

39 instrueret *Willner*, -ent P
40 sensibilibus *scripsimus*, sensilibus P
41 eodem *Willner*, edem P
42 articulis *Willner*, -culi P
43 ea P *p.c.*, erat P *a.c.*

attempt to pit one against the other in turn, while you do not trust yourself to attack any one of them with your own counter-arguments. But anyone will easily free them from being at odds with each other if he understands the unadorned utterances of such great men in the way that they were meant by them. I shall explain this briefly – at least for the intelligent listener.

One of them, borne aloft by the sublimity of his mind and by those wings[16] which he strenuously attempted to don, embarked on knowing realities (*res*) from their very beginnings and described what they were before they entered into bodies, defining the archetypal forms of realities (*res*) in an internal soliloquy. Plato

The other, being expert in the rhetorical rules,[17] so that he might instruct readers as accomplices in his eloquence, started from sensible and composite things (*res*). And, as they meet each other on the same path, they should not be called at odds. For, composition loves division and division loves composition, as long as each corroborates the other. Hence, if any product arises from multiplication on the countable digits and *articuli* of the abacus,[18] the correctness of the calculation can be proved by the division of the same product. Aristotle

That one of them said that realities exist outside sensibles, the other that they exist only in sensibles, should be understood in this way: genus and species (for this is what we are talking about), are the names of subject things (*res*) also. For if you consider things (*res*), the names genus, species and individual are imposed on the same essence, but in different respects. For, philosophers wishing to deal with things (*res*) insofar as they are subject to the senses, and insofar as they are denoted by terms in the singular and are diverse in number, called them individuals – that is, Socrates, Plato and the rest. But looking at the same individuals more profoundly – namely, not insofar as they are diverse according to the senses, but in that they are denoted by this term (*vox*) 'man', they called them a species. Again, considering the same things with respect only to the fact that they are denoted by this term 'animal', they called them a genus. Nevertheless in considering the species they do not take away the individual forms, but they forget them, since they are not posited by a name denoting a species; nor in the term denoting a genus do Plato Aristotle

inesse non attendunt, vocis generalis significatione contenti. Vox enim hec 'animal' in re illa notat substantiam cum animatione et sensibilitate; hec autem 'homo' totum illud, et insuper cum rationalitate et mortalitate; 'Socrates' vero illud idem addita insuper numerali accidentium discretione. Unde vel doctrina non initiatis patet consideratio individualis; specialis certe non modo litterarum profanos verum etiam ipsius archani conscios admodum angit. Assueti enim rebus discernendis[44] oculos advertere[45] et easdem longas vel latas altasque conspicere, necnon unam aut plures esse, undique circumscriptione locali ambitas percurrere, cum[46] speciem intueri nituntur,[47] eisdem quodammodo caliginibus implicantur. Nec ipsam simplicem notam sine numerali aut circumscriptionali discretione[48] contemplari, nec ad simplicem specialis vocis positionem ascendere queunt. Inde quidam cum de universalibus ageretur, sursum inhians, 'Quis locum earum michi ostendet?' inquid. Adeo rationem imaginatio perturbat et quasi invidia quadam subtilitati eius se opponit. Set id apud mortales; divine enim menti, que hanc ipsam materiam tam vario et subtili tegmine formarum induit, presto est et materiam sine formis et formas sine aliis, immo et omnia cum aliis sine irretitu imaginationis distincte cognoscere. Nam et antequam coniuncta essent, universa que vides in ipsa noy simplicia erant. Set quomodo et qua ratione in ea essent, id et subtilius considerandum et in alia disputatione dicendum est. Nunc autem ad propositum redeamus.

Quoniam igitur illud idem quod vides et genus et species et individuum sit, merito ea Aristotiles non nisi in sensibilibus

[44] discernendis *Willner*, -endi P
[45] advertere *Willner*, adverte P
[46] cum *Willner*, cur P
[47] nituntur *Willner*, nut- P
[48] discretione *Willner*, sine P

they understand the terms denoting species to be removed, but they do not pay attention to their being in the genus, being content with the signification of the term denoting a genus. For this term 'animal' in that thing (*res*) connotes a substance with animation and sensation; but this term 'man' connotes all that, and, in addition, rationality and mortality; 'Socrates', however, connotes that same thing with the addition of a numerical distinction by way of accidents.[19] Hence, consideration of the individual is easy even for those who have not begun a course of instruction; while consideration of the species, I admit, somewhat taxes the minds not only of those who are untouched by letters, but also of those initiated into the sanctuary itself. For, those accustomed to using their eyes for discerning things (*res*) and to noticing their length, breadth and height, and whether they are one or many, and to running their eyes over them as they are surrounded on all sides by circumscribing space, when they strive to gaze at the species, are wrapped about in some way by the same clouds of darkness <as the wholly unlettered>. Neither are they able to contemplate the simple *nota* itself without its differentiation by number or by place, nor can they ascend to a simple application of a term denoting species. Hence someone, when the subject of a discussion was universals, looking up eagerly, said: 'Who can show me where they are?' So much does imagination disturb reasoning and oppose itself to reason's subtlety as if by a kind of envy! But this is the case only for mortals; for the divine mind, which clothes this matter itself with such a varied and subtle covering of forms, it is no problem to know both matter without forms and forms without other forms: in fact, to know all things distinctly together with others but without the entanglement of the imagination. For even before they had been joined together, all things which you see were simple in *noys* itself. But how and in what way they were in *noys* should be considered more subtly and discussed in another disputation.[20] Now let us return to the matter at hand.

Since, then, that same thing that you see is genus, species and individual, Aristotle proposed correctly that these three do not exist except in sensible objects. For they are the

esse proposuit. Sunt etenim ipsa sensibilia, quamvis[49] acutius considerata. Quoniam vero ea in quantum dicuntur genera et species nemo sine imaginatione presse pureque intuetur, Plato extra sensibilia – scilicet in mente divina – et concipi et existere dixit. Sic viri illi, licet verbis contrarii videantur, re tamen idem senserunt. Nec tamen ego id ad vivum reseco, ut omnia omnium verba a falsitate absolvam. Ut enim tui erronei quandoque in veritatem raram incidunt, sic et meorum quidam erraverunt, et hoc ipso mei adversarii effecti sunt. His hactenus eradicatis, ad ea que de sensuali iudicio iaculata es accedamus.

Ut sensus extolleres, rationem cecam ducem, si recolo, vocasti. O perversa rerum conversio, cum nichil ratione certius, nichil sensibus fallacius! Primum, quia nec in maximis nec in minimis rerum sensus vigent.[50] Quis enim umquam celi spacium visu comprehendit? Quis sonum eiusdem celestemque concentum auribus clausit? Quis item athomi parvitatem oculo distinxit? Quis sonum eisdem[51] athomis collisis creatum aure notavit? Set 'pingui,' ut dicunt 'agamus Minerva.' Num si eminus coctani formam aspicias, id ideo coctanum esse necesse est? Set sunt et alia forma similia, substantialiter tamen diversa. Quod si ad tactum recurras, sunt et alia eiusdem lenitatis corpora. Odoratum advertis. Atqui potuit sibi ex coctanorum contactu odorem accidentalem attraxisse. Sic necesse est ut dentes imprimas. O argumentum perspicuum quod magis cani quam homini conveniens est! Set nec huic credendum est cum sepissime fallat. Quod enim fallax est, verum quidem incidere potest, certificare autem non potest. Unde nec ex sensibus scientia set opinio oriri valet. Hinc est quod familiaris meus Plato sensus irrationabiles vocat.

Deinde non modo verum non investigant, set etiam ab veri investigatione animum violenter extrudunt. Cum enim anima in aliqua noticia aciem fixerit, fit ut aut auribus sonus

[49] quamvis *Willner*, quavis P
[50] vigent *Willner*, vigeret P
[51] eisdem *Willner*, eiusdem P

sensibles themselves, albeit more subtly observed. But since no one without using imagination can gaze closely and purely at them inasmuch as they are called genera and species, Plato said that they are conceived and exist outside the sensibles – that is, in the divine mind. Thus, although those men in their words seem to be at odds with each other, in reality (*res*) they thought the same. But I do not wish to take this line of reasoning too far, so as to free all words of all people from falsity.[21] For while your misguided family sometimes hits upon a rare true point, so even some of mine have made mistakes, and by this have become my adversaries. Having pulled these problems up by the roots, let us deal with those darts you have hurled concerning the judgement of the senses.

To exalt the senses, you have called reason a blind leader, if I recall. Oh, perverse reversal of things, since nothing is more certain than reason, nothing more deceptive than the senses! First, because the senses have no power either in the largest or in the smallest things. For who has ever comprehended the whole area of the heavens in his sight? Who has cupped its sound and the celestial harmony in his ears? Who, again, has distinguished with his eyes the smallness of an atom? Who has noticed with his ears the sound created when these same atoms collide? But "let us discuss the matter", as they say, "with a well-fed Minerva."[22] Say you see the *shape* of a fig from a distance, surely it is not necessarily a fig because it looks like one? For other things are similar in shape, but differ in regard to their substance. What if you have recourse to touch? There are other bodies of the same smoothness. You turn to smell. Well, it might have picked up the smell of figs accidentally from contact with them. Thus it is necessary to take a bite. Oh, what a wonderful argument – that suits a dog better than a man! But not even taste should be trusted in, since it very often deceives. What is deceptive can indeed hit the truth, but it cannot give proof. Hence from the senses can arise not knowledge, but only opinion. This is why my friend Plato calls the senses irrational.

Secondly, the senses do not only *not* seek out the truth: they even forcibly drive the mind away from the investigation of the truth. For when the soul has fixed its attention on some thought, it happens that either a sound strikes the ears or

instrepat aut lux oculos feriat aut aliquo tali motu inquietetur.[52] Itaque tumultu sensuali usque ad sedem anime occursante, ipsa ab investigatione sua fatiscit. Hinc est quod quotiens in altum iudicium vocamur, locum nobis solitarium eligimus ubi minus sensibus impediamur. Hinc illud aliud quod in sompnis anima quia quodammodo tunc liberior est a vexatione sensuum, aciem stringit, et de futuris etiam aut verum aut verisimile quandoque deprehendit, et sub aurora minus fallitur, utpote iam digestis cibis expedicior. Adeo hii ubi dominantur, verum suffocant. Unde meus Plato:

> Est hec eorum plena erroris et falsitatis opinio orta ex depravatione sensuum, nec habet ullum certum ducem talis peragratio.

Que cum ita sint, cautio Artificis mederi eis volens, rationem et ducem et dominam eorum in cerebro locatam superposuit. Hec igitur sua diffinitione quicquid ab illis peccatur dirigit, et quid unaqueque res sit, aut substantialiter diffiniendo aut accidentaliter describendo, demonstrat. In contrarium igitur sententie tue res relapsa est – ut videlicet sicut sunt sensus hebetes et ratio dominatrix, ita et deprehensum sit.

Postremo sensus quomodo sentiant vel quid ipsi sint sentire nequeunt. Idque non vulgo quod dubitare nescit, set solis philosophis ratione duce pronum est. Tandemne igitur pudet te adeo impudenter sensus extulisse, cum nec etiam quid ipsi sint sentire queant? Quare ipsi potius imprecatione et odio digni sunt.

Quisquis dissimulans oculi lumen melioris,
  His que[53] non sentit nescit habere FIDEM,[54]
Qua precellebat rationis luce relictus,
  Det sua fortune colla premenda iugo.
Possideat dum possessis numquam potiatur,
  Non alii largus, non sibi proficuus.
Ignoret pariter causas et semina rerum,
  Seque simul blandi captus amore mali.

[52] inquietetur *Willner*, inquietur P
[53] que *Willner*, qui P
[54] FIDEM *capitals in* P

light hits the eyes or the soul is disturbed by some such movement. Thus when a sensory excitement rushes up as far as the seat of the soul, she herself is not strong enough to continue her investigation. Hence it is that, whenever we are called to make an important decision, we choose a solitary spot for ourselves, where we may be less hindered by the senses. This too is why in sleep the soul, because it is then in some way freer from the harassment of the senses, sharpens its wit and sometimes grasps the truth or something like it even concerning future events. It makes fewer mistakes just before dawn, since it is less encumbered once the food has been digested. This proves how much the senses suffocate the truth when they dominate. Hence my Plato says:

> This opinion of theirs, full of error and falsity, has arisen from the wantonness of the senses, and such a straying does not provide any reliable leadership.[23]

This being so, the Artist, in his providence, wishing to provide a remedy for the senses, placed in a superior position in the brain reason as guide and mistress over them. She, then, by her ability to define, corrects what is wrongly interpreted by them and demonstrates what each reality (*res*) is either by defining the substance, or by describing the accidents. So the matter has turned out to be the opposite to what you stated: namely, that just as the senses are blunt and reason is the mistress, so (through reason) this too has been discovered.

Finally, the senses cannot sense how they sense or what they themselves are. But this is obvious not to the common people, who do not know how to doubt, but only to philosophers who have reason as their guide. Are you, then, I ask, not ashamed of having so impudently extolled the senses, when they are not able to sense even what they are? This makes them rather worthy of hatred and a curse.

He who, ignoring the light of a better eye,
  Knows not how to have trust in what he does not perceive,
Abandoned by the light of reason, by which he used to excel,
  Should put his neck under the heavy yoke of fortune.
May he possess, but never be in control of what he possesses,
  Neither benefiting another nor profiting himself.
May he be ignorant equally of the causes and of the seeds of things,
  And of himself too, enthralled by his love of smooth evil.

Visibus ignoret nostris cur sidera quedam
Invideant, visus cetera non fugiant,
Cur tellus medium teneat, dum nescia tantis
Rerum ponderibus cedere, pressa iacet,
Ver, autumnus, hiemps, cur pingat, compleat, artet,[55]
Prata, domos, latices, gramine, farre, gelu.
Lumine privatus pro veris falsa requirat,
Dum rerum causas disputat esse – nichil.'

Hic, illa paululum exhalante, ego qui dudum non parvo affectu locum dicendi expectaveram, in hec verba prorupi:

'Me potius, o vere recognitionis dux, cum hac ostentatrice confligere permittas oro. Nam et michi eam devincere non parvi muneris instar habebitur, et tu quantum supra ipsam possis eadem victoria haut obscure ostendetur. Neque enim michi quin superior discedam diffido, cum et ea que nuper in familiaris tui consolatione haut imprudenter disseruisti mente teneam, et ex illis alia non minus probabilia ratione duce argumenta contexere non torpescam. Ideoque[56] hic non iam eam cum suis procul hinc fugare intendam, verum etiam illas ipsas quibus astantibus comitata superbit quarumque illecebris desiderabilibus incautos allicit, tui iuris potius esse debere arguam. Quod sic colligo.

Corporea mundi sensilis mole divina Fabricatoris providentia diversis qualitatibus informata, aliud quoddam quod eidem motum incrementumque ceteraque id genus accommodaret extrinsecus petendum restabat. Licet etenim ipsa inter se corpora alia aliis leviora gravioraque essent, non tamen etiam incrementum necessarium motumque voluntarium ex se habere poterant, quippe quorum principia inertia immobiliaque extiterant. Sive igitur semel simulque Voluntas omnipotens id cum ceteris digessit – quem modum ratione[57] persequi difficile est – sive aliquam ipsis rebus faciendis moram immutabilis ipse concessit, naturam universis prepotentem muneribus corporibus egentibus adesse voluit,

55 artet *Willner*, arcet P
56 Ideoque *Willner*, idque P
57 ratione *scripsimus*, oratione P

May he be ignorant of why certain constellations begrudge our sight,
While the others do not flee our vision,
Why the earth holds the middle position, and refuses to yield,
Though pressed down by the mighty weight of things,
Why spring, autumn, winter, colour, fill, grip
Meadows, homes, brooks, with grass, with corn, with ice.
Deprived of light, may he seek false things for true,
While he argues that the causes of things are – nothing.'[24]

At this point, whilst she was taking a short breath, I, who for a long time had been waiting with some impatience for an opportunity to speak, broke forth into these words:

'I beg you to allow *me* rather, you guide of true understanding, to battle with this boastful woman. For me, it will be considered not a small boon to defeat her, and, as for you, the same victory will show not obscurely how much power you have over her. For I have every confidence that I shall come out of this the victor, since I hold in my mind the not unwise arguments you have given just now as a consolation for your friend,[25] and I shall not be sluggish to weave from them other no less convincing arguments, with reason as my guide. So my present aim is no longer to drive far away from here this woman with her serving-girls, but rather to convince even these maidens themselves, whom she is proud to be accompanied by and by whose desirable enticements she ensnares the incautious, that they ought rather to be under your control. I argue in this way.

When the corporeal bulk of the sensible world had been informed with its diverse qualities by the divine providence of the Artificer, there remained the necessity to seek some other thing from outside to give the world movement and growth and everything else of this kind. For, although the bodies themselves were among themselves lighter or heavier than one another, they were not able to derive from themselves the growth and voluntary movement necessary to them, since their principles had been inert and motionless. Therefore – whether at one and the same time the all-powerful Will arranged this with everything else (a process difficult to explain by reason) or whether it allowed there to be some delay in making things themselves, while being unchangeable itself – it wished that a nature possessing all facilities to a preeminent degree should be available to the bodies that

eandemque animam fore, cuius potentie quia supra naturam corporis sunt, ipsam[58] quoque incorpoream esse. Unde Philosophus in Thimeo:

> At vero Deus tam antiquitate quam virtutibus preire animam nature corporis iussit, dominamque eam et principali iure esse voluit circa id quod tuetur.

Si igitur, ut Philosophus promittit, Deus eam dominari corpori voluit, non est dubium quin id ipsum ea facere potuerit. Alioquin Deus id noluisset. Natura autem corporis essentialiter mutabilis, maioris et minoris et eorum medii susceptrix erat. Quid? Hanc[59] igitur animam ira et cupiditate munitam esse, haut dubium est, ut extrema superfluitatis et diminutionis irascens corrigere mediumque tenere cupiditate affectaret, quippe cum eius ipsius essentia sine particulari discretione tota sibi equalis nec magnum nec parvum in se cognosceret.[60] Set et ipse eius potentie – ira dico et concupiscentia – a sua moderatione quandoque aut declinature aut casure erant, utpote a corporeis passionibus irritande. Ratio igitur huic ad has revocandas et necessaria erat, et ab ea non aberat. Quod et Philosophus non ignorat dicens:

> Primo quidem sensum ex violentis passionibus excitari, post quem[61] mixtam ex voluntate tristiciaque cupidinem nasci, tum vero metum atque iracundiam ceterasque pedissequas earum perturbationes diverso affectu pro natura sui promoventes. Quas quidem si frenarent ac subiugarent iustam his lenemque[62] vitam fore.

Anima igitur ex omni parte perfecta corporis membris, ut universum elementarium regeret, insita est. Set hoc ipsum ei nequaquam, ut dixi, pronum ac facile erat. Primo enim ingressu non modicam sue divinitatis patitur iacturam, adeo ut deinde hanc ipsam fecem elementariam quam ut domina tueri debebat, sibi ipsi Auctorique suo excecata preponat, oblita sui tam principii quam finis. Tum illas opes quas paulo

58 ipsam *Willner*, ipsa P
59 Hanc *Willner*, Hec P
60 cognosceret *Willner*, cognoscere P
61 post quem *scripsimus*, postquam P
62 lenemque *Willner*, levemque P

needed it, that this nature should be the soul, and that, since her powers were above corporeal nature, she also should be incorporeal. Hence the Philosopher in the *Timaeus* says:

> God ordered the soul to precede corporeal nature, both in age and in powers, and he wished her to be mistress and to have the principal jurisdiction over that which she protected.[26]

If, then, as the Philosopher asserts, God wished her to dominate body, there is no doubt that she was able to do this very thing. Otherwise God would not have wished it. Corporeal nature was, in essence, changeable: susceptible of 'too great', 'too small' and the middle path. Well then, undoubtedly this soul is endowed with anger and desire, so that, by showing her anger she can correct the extremes of 'too great' and 'too small' and, by her desire, hold to the middle path, since her own essence, without bias in any direction, but totally equal to herself, knows in herself neither the great nor the small. But her very powers – I mean, anger and desire – were sometimes prone to droop or to fall from their moderating activity, since they were to be harassed by the corporeal passions. Reason, therefore, was necessary to her for calling back these powers, and was not absent from her. This, too, the Philosopher is aware of when he says:

> First the sense is moved by violent passions, after which a desire arises, mixed from the will and pain, then fear and anger and the other disturbances, their servant-girls, accompanying them advance with feelings which differ according to their individual natures. But if they reined in and kept under control these feelings, they would live a just and calm life.[27]

The soul, therefore, perfect in all its parts, was grafted[28] into the members of the body to rule the elementary universe. But this very duty, as I have said, was not at all easy or natural for her. For, on her first incorporation she suffers a considerable loss of her divinity, to such an extent that thereafter these very elementary dregs which she was supposed to protect as mistress, she, in her blindness, prefers to herself and her Creator, forgetful of both her beginning and her end. Then, those riches which you shamelessly extolled

ante impudice extulisti, inebriata non moderate expostulat, eo usque ut hec ipsa mala preciosa – quibus ipsa precium ascripsit, neque enim ullum ex se habent nisi dum habere putantur – hec ipsa, inquam, dum accumulet, nulla sibi rationis, nulla veritatis falsitatisve discretio. Postremo – quod in omnibus erroribus pessimum est – hac ipsa scilicet miseria feliciter prospereque secum actum esse persuadere non dubitat.

Et de his hactenus. De aliis vero quas subsequenter non minus persuasorie exposuisti – potentiam dico, dignitatem, famam, voluptatem – non minus reprobandam improbitatem quam patefaciendam[63] habeo.[64] Enervant enim lucem illam rationalis excellentie atque in pecudem ex homine redigunt, inferentes illam sibi ignorantie noctem, ut quid ipsa sit anima intueri non possit. Unde, si quando eorum erronei resipiant,[65] singula istorum plena acutissimis passionibus non diffiteantur.

His igitur anima in corporis carcere vinculis oppressa, unum inter universa remedium est quo[66] eadem se sibi reddit domumque reducit – doctrine videlicet huius philosophie, artesque quas vocant liberales. Nam et has ipsa eadem, quondam tota apud se a devio errore libera, prudentia sue subtilitatis elicuit, litteralique memorie, non satis vocibus credens, mandari iussit, ut per eas lumen suum, si quando hebetaretur, refulgeret, metuens id quod deinde (non fallax prenuntia) sibi accidere sensit. Has igitur quotienscumque a vero decidit, si respiciat, a casu resurgit. Habet enim in istis quid ipsa Auctori, quid sibi cetera debeant, rerum scilicet[67] communitates et differentias absolute distinctas. Quibus attente relectis[68] ipsa se recognoscens non tantum a sarcina rerum premente exoneratur,[69] verum etiam omnia corporea, quantum natura patitur, perosa despectat, omni beatitudine referta dum se Auctoremque suum contemplando cognoscit. Habemus igitur contrariam errori tuo verissimam diffinitionem, tum scilicet me ceterosque beatos esse cum supradictis

[63] patefaciendam *Willner*, patefaciam P
[64] habeo *Willner*, ab eo P
[65] resipiant *Willner*, respiant P
[66] quo *Willner*, quod P
[67] rerum scilicet *scripsimus*, rerumque P
[68] relectis *Willner*, relictis P
[69] exhonoratur P

just a little while ago, she demands in an immoderate way in her drunkenness, to such an extent that these 'valuable' ills – it is she who has ascribed 'value' to them, for they do not have any value in themselves, but only in that they are thought to have[29] – whilst intent on heaping up these 'valuable' ills, I say, she can make no rational distinction between truth and falsehood. Finally – and this is the worst of all these errors – in this very wretchedness she does not hesitate to argue that things have turned out happily and prosperously for her.

But enough of this! As to the other maidens whom you went on to describe no less persuasively – I mean power, honour, fame and pleasure – I must reprove no less than expose their perversity. For they weaken that light of rational excellence and reduce man to the level of a beast, bringing to him that night of ignorance, so that the soul is not able to view what she is. Hence, if those they have led astray ever come to their senses, they will admit that each of them is full of the sharpest passions.

When the soul, then, is oppressed by these bonds in the prison of the body, among all possible remedies there is one by which she can restore herself to herself and bring herself back home: that is, the teachings of this philosophy, and what they call 'the liberal arts'. For she herself once, when completely at home with herself and free from misleading error, drew forth these arts by the wisdom of her subtlety, and ordered them to be committed to the memory of the written text, not sufficiently trusting in oral transmission, so that through them her own light, if it should ever grow dim, should blaze up again, fearing that which she felt would happen to her afterwards (being no false prophet). Whenever, therefore, she falls from the truth, if she considers these, she rises from her fall. For she has in these what she herself owes to the Author, and what everything else owes to her: namely, the common elements in things and their differences made absolutely distinct. When these have been re-read attentively she, recognizing herself, not only is released from the burden of things pressing down on her, but also despises all corporeal things, hating them as much as nature allows, and she is replete with all blessedness while she recognizes herself and her Author by contemplation. Therefore, we have a very true

adheremus artibus, infelices vero ac perditos si tuis quas iactas subiaceamus, easque ipsas rationi subiacere debere non ultra dubitamus. Quare huic suisque eas regendas relinquas.'

His ego cum voce currente animique affectu instarem, illa quodam[70] pudore stupida, veste vultui preposita, simul cum suis paulatim retrograda cessit, meque adhuc dicere parantem in medio sermone reliquit. Tum illa dextra cum suis haut immoderate gavisa, me recipientem easdem pleno favoris oculo percurrit, animumque tali voce secuta est:

'Haut ignobili, o iuvenis, me ditasti triumpho, idque ipsum de te iamdudum speraveram[71] cum studia mea tanto affectu adhuc puer arriperes. Nunc ergo quia et me et te tanto monstro liberasti, indonatum abire haut conveniens est. Michi nempe quid super te de ceteris confidam, ex tali principio non dubium est.

Sunt igitur michi septem virgines hee, quarum naturam moresque singillatim expediam, ut quam omnium tibi malis proprio arbitrio eligas. Omnes enim simul amplecti maius quam quod tuo ingenio competeret esset.

GRAMMATICA Hec itaque prima[72] quam vides dextra ferulam gestantem, leva codicem lituris innumeris distinctum, una liberales artes ingredientes cunabulis suis nutrit, primoque lacte imbuit, sine cuius nutrimento frustra ad sapientiam tendas, adeo ut eius expertes, quacumque licet prudentia prediti, balbutientes tamen non immerito vocentur. Qui vero irretiti fuerint ea, quantum homines pecudibus munere rationis prestant,[73] tantum eos hominibus ceteris rationabili loquendi arte prestare non dubites. Hec nempe, cum primum mortales per

[70] quodam *Willner*, quondam P
[71] speraveram *Willner*, superaveram P
[72] prima *Willner*, -am P
[73] prestant *Willner*, -stat P

definition which is opposite to your erroneous view, namely that both I myself and everyone else is blessed when we adhere to the arts I have mentioned, but we are miserable and lost if we submit to the maidens whom you boast about. We do not doubt any more that these maidens themselves should submit to reason. Therefore, you should leave them to be ruled by this woman and her followers.'

When I pressed home these points on which I spoke fluently and with passion, Philocosmia, dumbstruck with a kind of shame, covered her face with her gown, withdrew backwards a little with her maidens, and left me in the middle of my speech still ready to say more. Then that woman on the right with her maidens, rejoicing not intemperately, glanced at me with an expression full of favour as I welcomed them, and followed the intention of her mind with these words:

'You have enriched me, young man, with a not ignoble triumph. I had long hoped for this result from you, ever since the time when you, still a boy, seized on my studies with such passion. So now, because you have freed both me and yourself from such a monster, it is not appropriate that you should go away without a gift. Judging from such a beginning, there is no doubt in my mind about what I can expect from you in everything else.

I have these seven maidens whose nature and character I shall explain individually, so that you may choose by your own judgement which out of all of them you prefer to make your own. For to embrace them all at the same time would be more than your wit is capable of.

As for this first one, whom you see wielding a rod in her right hand, a book marked with innumerable corrections in her left, she nourishes in their cradles those entering the liberal arts, and instils the first milk into their mouths. Without her nourishment you may stretch out your hands to wisdom in vain, to such an extent that those without experience of her, however much they may be endowed with good sense, none the less are quite rightly called mere babblers. But as for those who have been caught in her net, you should have no doubt that they excel other men by the rational art of speaking as much as men excel beasts by their gift of reason.[30] For this maiden, when mortals first wandered about here and there in the countryside like beasts, without

Grammar

agros passim et ferali more sine mutua affabilitate muta ratione vagarentur,[74] nec posset alter alteri quid vel de se vel de rebus subiectis intelligeret insinuare – hec, inquam, primum nomina rebus singulis imponens, mortales mutue allocutionis honestate beavit.

Quod quam difficilis opere sit, animo si ades, intelliges, quippe cum et res infinite sint, eedemque[75] aliter per se, aliter in coniunctione, aliter secundum quod universaliter intelliguntur, aliter secundum quod sensibus subiecte individuali proprietati subiacent, appellande[76] sint. Verbi gratia: id quidem quod in subiecto corpore accidens est, secundum quod per se quasi quedam substantia sine corporis respectu concipitur, 'albedo' vocatur; secundum hoc vero quod in subiecto est 'albi' nomine designatur. Item, animal rationale quod cernis, secundum quod in universalitate diffinitionis substantialis intelligitur, 'homo' iudicatur; in eo vero quod singularitate descriptionis discoloratur, 'Socrates' appellatur. Rebus itaque nominibus vestitis, ipsis vocibus voces alias imposuit, ut et earum inter se differentia non lateret et infinitas illa multiplex sub numerum redacta scientie comprehensu non careret. Universitatem igitur illam octo partium orationis[77] numero mira sagacitate subiecit; adeo a perfectione impositionis non temperans, ut sicut vox rei, ita et vox vocis, significativa sit. Verbi gratia: in hac oratione – 'homo ambulat' – vox subiecta rei que ambulat indicativa est. Cum vero propono 'homo est nomen', non res set vox a subiecta voce significatur. Hanc enim vocem que est 'homo' per se ipsam nominans, proprietatem nominis habere significo. Unde etiam ille labirintus oritur ut in subiecto idem sit et significans et significatum – quod quidam postea 'materiale impositum' vocaverunt.

Hec eadem quendam familiarium suorum non improbe preceptis suis imbuens et a littera rudimentum incipiens, litterasque in sillabis, sillabas in dictionibus, dictiones in

[74] vagarentur *Willner*, -etur P
[75] eedemque *Willner*, eodemque P
[76] appellande *Willner*, -ende P
[77] orationis *Willner*, rationis P

mutual courtesy and with their reason silent, and no one could convey to his fellow what he understood either about himself or about the objects around him – this maiden, I say, by putting names to single objects (*res*), first blessed mortals with the distinction of mutual conversation.

You will appreciate what a difficult task that is, if you put your mind to it. For not only are things (*res*) infinite in number, but also they must be named, now in themselves, now in conjunction, now according to how they are understood as universals, now according to how they are subject to an individual property, being subject to the senses. For example, that which is an accident in a subject body, in being conceived in itself without respect to body, as if a certain substance, is called 'whiteness'. But in respect to its being in a subject, it is denoted by the name 'white'. Similarly, the rational animal you are looking at, understood in respect to the universality of the substantial definition, is judged to be 'man', but in that it is highlighted by the singularity of a description, is called 'Socrates'. Having clothed things (*res*) with names in this way, she gave different terms (*voces*)[31] to the terms themselves, so that even their difference amongst themselves should not be hidden and that kaleidoscopic infinity, having been brought under a number, should not lack the ability to be comprehended by a science. Thus with wonderful sagacity she subjected that total multiplicity to the number of eight parts of speech; to such an extent did she not compromise the perfection of the application of these terms that, just as a term can signify a thing (*res*), so a term can signify another term as well. For example, in this phrase – 'a man walks' – the term as subject is indicative of the thing (*res*) that walks. But when I put forward the phrase '"man" is a noun', it is not a thing (*res*) but a term which is signified by the subject-term. Naming this term 'man' through itself, I indicate that it has the property of a noun. Hence the enigma arises that in the subject both the signifier and the signified are the same thing – which some people afterwards called 'the material imposit'.[32]

This same maiden, not unworthily instilling her precepts into one of her family, by beginning the teaching from the letter, then instructing him to weave letters into syllables,

constructionibus perfectis contexere docens, famam sibi late perhennem peperit. Set hoc apud Latinos. Si quis vero esset qui huius artem totam amplecti posset, non illum quicquid a Grecis barbarisve universisque mortalibus communi scripto celebratur, lateret. Hec igitur quanti sit vides possemque adhuc multa de laude eius non minus digna proferre, nisi quia et tibi non penitus ignota est, et michi de ceteris non minus breviter percurrendum est.

RETHORICA Hanc igitur quam cernis secundam hilari vultus iocunditate elateque frontis confidentia quasi aliquid dicere paratam, in cuius veste quicquid sub civilem disceptationem cadere potest intextum est, nec michi nec alii, credo, sufficienter laudare pronum est. Hominibus enim, ut supra dixi, mutua animorum expositione initiatis, hec consequenter operam addens, suaviloquo[78] ornatu verborum et sententiarum universas peritorum orationes non segniori subtilitate depinxit[79] quam invente sunt. Unde factum est ut illi qui prius indiscrete et sine legali iure vivebant, in civitatem communemque iusticiam tam potenti ammonitione tracti sint. Quare quicquid universe honestatis ex civilibus consensibus ortum est, huic ascribendum esse diiudico.

Deinde cui dubium est qua vi modernos tirannos ab irrationabili impetu adhibita refrenet, cum primo mortales omni feritate rigidos in humanitatem cetumque compulerit? Hanc eius potentiam Socrates non ignorans, universam rei publice rationem huic regendam distribuit, trino rethorice generi tres contentionis modos supponens. Aut enim in Capitolio agebatur causa et ‘deliberativa’, aut in Martio Campo et ‘demonstrativa’, aut in foro et ‘iudicialis’ vocabatur. Deinde idem philosophus, causam prosequens, partes constitutionis in quinarium digessit, docens argumenta invenire, inventa disponere, disposita eloqui, in elocutione memoria et pronuntiatione uti. His equidem si quem ad plenum hec instituerit, quamcumque rationabilem causam ingredietur,

[78] suaviloco P
[79] depinxit *Willner*, depixit P

syllables into words, and words into perfect constructions, brought forth widespread and perpetual fame for herself. But this applies only to Latin-speaking people. If there were anyone who was able to embrace her art in its entirety, whatever abounds in the writings shared by Greeks and barbarians or mortals wherever they may be, would not be unknown to him. You see how important she is. I could put forward many more not less worthy points in her praise, were it not that she is not entirely unknown to you either, and I must run through the attributes of the other maidens no less briefly.

As for this one, whom you discern in second place, with a joyful attractiveness in her face and a confident demeanour, as if ready to say something, on whose clothing whatever can fall under civil debate[33] is woven, it is not easy to praise her sufficiently, either for me or for anyone else, I believe. For when people had been initiated into telling each other what was in their minds, as I said above, this maiden, adding her contribution in turn, adorned all the speeches of those who had acquired the rhetorical skill with a sweet-sounding ornament of words and phrases,[34] with no less quick-wittedness than that with which speaking was first invented. Hence it happened that those who first lived without law and order, were brought into citizenship and communal justice by such a powerful prompting.[35] So, whatever honourable thing has arisen from the agreements of citizens, I judge should be ascribed to this maiden. Rhetoric

Secondly, who can doubt the force she uses to keep present-day tyrants from irrational impulses, when it was she who first pressed mortals, when obdurate in complete wildness, into humanity and fellowship? Socrates,[36] not unaware of this power of hers, assigned the whole organization of the commonwealth to her rulership, subjecting three kinds of dispute to the three-fold genus of rhetoric. For either a case was judged on the Capitol, and was called 'deliberative'; or in the Campus Martius, and was called 'demonstrative'; or in the Forum, and was called 'judicial'. Then the same philosopher,[37] developing the case, arranged the parts of the issue into five, teaching how to find arguments, to arrange them once found, to put into speech what was arranged, and to use memory and correct pronunciation in delivering the speech. If she instructs anyone fully in these, whatever reasonable case

superiorem discedere facit, adeo persuasioni indulgens ut cum iudicum animos ad intentionis favorem applicaverit, si in oppositum se convertat, non minus depulsionis partem probabiliter persuadeat. Unde et eam pre ceteris artibus Greci 'rethoricam' – id est eloquentiam – vocaverunt. Hac nempe quicumque munitus fuerit, et amicis iocundus et inimicis metuendus erit, utpote que non solum in philosopho spectabilis est, verum etiam inter litterarum prophanos is sepissime magis consulitur qui maiori quadam verbositate pre ceteris facundus videtur. Inde fit ut in quampluribus curiis rustici loquaces tacitis philosophis preferantur.

Quid enim in philosopho possit dubitatione dignum non est – tibi presertim cum non ignores quanta Tullium fama titulavit, honore pretulit, dignitate sublimavit, dum tamquam gladium linguam eius exacuit universisque concivibus timendum fecit. Quid autem apud imperatores potuerit, non minus ambigis, cum in orationibus suis legeris quid apud Cesarem effecerit se – ipsum scilicet qui ex adversa parte senserat Cesaremque hostem rei publice iudicaverat, sola conciliante eloquentia in amicitiam recepit. Deinde Marcum Marcellum orator dignitati restituit; Quintum Ligarium a capitali sententia persuasor resolvit; Deiotarum odio Cesariano liberavit. Quorum quidam laqueum[80] prodicionis imperatori struxerant, quidam autem contra arma eius senserant. Set non habuit Cesar quid accusaret ubi Tullius defenderet, quid vetaret ubi tanta facundia veniam oraret. Oblitus itaque sui Cesar, quem mille acies non perturbaverant, coram rethore mentis inops[81] factus est. Quantum igitur ipse victor victis[82] prelatus, tantum sibi victo rethor preferendus. Unde et ipse sibi quandoque applaudens ait: 'Cedant arma toge! concede, laurea, lingue!' Nam et apud inimicos hec lingua adeo valuit ut precium pacis tocius imperii fuerit, nec Cesari Antonius

[80] laqueum *scripsimus*, iugulum P
[81] inobs P
[82] victis *Willner*, victus P

he will embark upon, she makes him come out the victor, putting so much store on persuasion that when she has pushed the minds of the judges towards supporting the plea, if she then changes to the opposite side, she can just as persuasively advocate the rejection of the charge. Hence the Greeks called *her*, rather than any of the other maidens, 'rhetorica', that is 'eloquence'.[38] Obviously, whoever is armed with this art should be both loved by his friends and feared by his enemies, in that not only is rhetoric respected in a philosopher, but also, among those ignorant of letters, time and again the advice of one who seems to be more eloquent than the others is the more sought after because of his greater command of language. Hence it happens that in most courts eloquent churls are preferred to silent philosophers.

What power she has in a philosopher is not to be doubted – especially not by you, since you cannot be ignorant of how great was the fame with which she blazoned Cicero, with what honour she elevated him, and with what dignity she raised him up, sharpening his tongue like a sword and making him feared by all his fellow-citizens. And what power he had in the face of emperors you can as little doubt, when you read in his speeches what he made of himself in Caesar's presence – that very man who had declared his allegiance to the opposition party and judged Caesar to be the enemy of the Republic – by the conciliation of eloquence alone, Caesar received him into his friendship. Then, as an orator, he restored Marcus Marcellus to his position; as an advocate, he released Quintus Ligarius from a death sentence; he freed Deiotarus from the hatred of Caesar.[39] Of these, some had laid a net of deceit for the emperor, others had voted against supplying him with an army. But Caesar did not have anything to accuse them of when Cicero was defending them, anything to forbid when such eloquence begged pardon. Thus Caesar, whom a thousand armies had not upset, in the presence of the orator forgot himself and became mentally at a loss. As much as he, when victor, surpassed those he had defeated, so much was the orator to surpass him, now defeated. Hence the same Cicero sometimes boasted with these words: "Let arms yield to the toga! Yield, laurel branch, to the tongue!"[40] For even among enemies this tongue acquired such a power that it became the price of peace for the whole

federari dubitat, dum sibi linguam oratoris concedat. Magis enim facundiam eius quam arma Romana timebat, neque in pace Cesariana tutus esset, dum in eum Tullius scriberet.

Hec igitur quanti precii sit que sic suos extollit dubium non est. Hoc ipsum non vane intelligens Aristotiles eius preceptis plurimum incubuit, docens omnem causam aut honesti aut ammirabilis aut humilis aut ancipitis aut obscuri generis esse, singulorum membrorum communitates et differentias ostendens, qualiterque in singulis – neque enim similiter in omnibus – exoriendum sit. In qua pagina perlucidus ille tractatus[83] subscribitur qua ratione scilicet in turpi ammirabilis generis specie auditores in benivolentiam, docilitatem, attentionem, adversarii vero in odium, invidiam, contemptum, ducendi sint, quod in omni huiusmodi disceptatione non parvum effectum optinet. Set et etiam subdivisionem earundem specierum idem philosophus non tacet, dicens omnem cause constitutionem vel coniecturalem vel diffinitivam vel generalem vel translativam esse, singula quorum si quis sine invidia relegat, grata affectuositate plena inveniet. His omnibus ceterisque eius facultatis preceptis diligenter diffiniendis vir sapiens operam dedit, intelligens quanta in rerum nobilitate et utilitas et honestas artifici per ea acquirantur. Hactenus autem de ista dictum sit; deinceps ad reliquas transeamus.

DIALECTICA Hec autem tercia quam vides dextra serpentem, leva vero codicem gestantem, in quo decem quasi distinctiones depicte sunt, non modice post supradictas mortalibus necessaria est. Hec enim, quicquid ambiguitatis dubitationisve ex supradictarum preceptis nasci solet, discretione sua determinat, multiplicemque ambagem in certitudinem redigit. Hec certis divisionibus omnem deceptionem[84] locutionis excludens, nodos sophisticos aperit verumque excudit, docens quicquid in verbis decipere potest, aut equivocum esse aut univocum aut diversam partem aut diversum[85] tempus aut respectum diversum aut diversum modum; quod Boetius in Cathegoricis pretereundum esse non censuit. Hec eadem subdita sibi habet inventionem et iudicium, quibus universas lites in negotio suo

[83] tractatus *Willner*, tractus P
[84] deceptionem *scripsimus*, deceptationem P
[85] diversum *Willner*, -sam P

empire; nor did Antony hesitate to make an alliance with Caesar, if only he would give him the tongue of the orator. He feared his eloquence more than the arms of Rome, nor could he be secure in Caesar's peace while Cicero wrote against him.

There is no doubt, then, how valuable this maiden is, who exalts her own people in this way. Aristotle, fully understanding this, paid much attention to her precepts, teaching that every case is either of an honourable, a discreditable, a petty, a dubious or an obscure genus,[41] showing the shared and different elements of each, and how one should begin a speech according to each – for one does not do it in the same way in every case. In this writing that very clear text is added concerning how, in the scandalous[42] species of the 'discreditable' genus, the auditors should be led into being well-disposed, receptive and attentive, but the opponents should be led into hatred, envy and contempt – which produces quite a significant effect in every discussion of this kind. But the same philosopher is not silent either about the subdivision of the same species, saying that the issue of every case is either 'conjectural', 'defining', 'general' or 'transferring'.[43] Each of these, if one goes over them without a grudging feeling, one will find full of pleasing emotional effects. The wise man took care in defining all these and the other precepts of this art diligently, understanding how much usefulness and honour (considering the nobility of the subject) are acquired for the practitioner through these precepts. Here I come to an end of what I want to say about this maiden; now let us pass on to the remaining ones.

This third maiden, whom you see bearing in her right hand a serpent,[44] in her left a codex in which ten distinctions, as it were, are depicted, is more than somewhat necessary to mortals after the maidens mentioned before. For this one determines by her discrimination whatever ambiguity or doubt is wont to arise from the precepts of the first two arts, and she brings a clear solution out of a complicated crux. She, keeping out every verbal deception by using strict divisions, unties sophistical knots and hammers out the truth, teaching that whatever can deceive in words is equivocal or univocal, or of a different part, time, respect or mode; Boethius in his *Categories* decided this should not be passed by.[45] The same maiden has under her control invention and judgement, by which she resolves all disputes arising in her

Dialectic

ortas dissolvit. Questione enim proposita,[86] que pars eius tenenda sit, que reprobanda, prima contemplatione diiudicat. Quo autem iudicium illud confirmet, locos cum argumentis suis invenire non fatiscit, licet quibusdam videatur inventionem iudicium precedere. Hec ipsa universitatem omnium que[87] videntur in decem naturas subtili indagatione partita est, decem diversitates rerum tantum omnino diversarum natura esse arguens, quas etiam 'decem predicamenta' posteri vocaverunt, singulisque predicamentis et genera et species et individua ascribit. Ipsorumque generum alia generalissima, alia subalterna esse disserit; specierum item alias specialissimas, alias subalternas, nobilissimum illud iter alternans ut nunc ab individuis per media ad generalissimum, proprietates spoliando – immo obliviscendo – ascendat, nunc a generalissimo eodem gradu, singula formis suis retexens, ad individua usque descendat. Hec quippe universales rerum naturas mira luminis subtilitate perstringens, eo modo quo in Artificis mente ante tempus concepte sunt intueri nititur; accidentia sine respectu substantiarum considerat; ipsorum etiam accidentium differentias separat, hac quippe subtilitate confisa, ut et alios infestantibus scrupulis suis velut serpens ultro illaqueet, et aliorum iacula determinationis defensione tuta[88] contempnat.

Inde fit ut, huius affectum si quis adeptus fuerit, que propositio vera, que falsa, earundem etiam que veritatem vel falsitatem determinatam habeat, que vero incertam dubitabilemque, non ignoret; argumentorum item, quod probabile, quod necessarium, quod sophisticum; locorum, qui inherentes, qui extrinseci, qui medii; divisionum, que secundum se, que secundum accidens; diffinitionum, que substantiales, que accidentales; sillogismorum etiam varietates, figuras atque modos. Huius argutiis Aristotiles imbutus, quotiens iocari voluit, non minus coram aliis sophistice callens falsitatem astruebat, quam ceteris coram eo veritatem ipsam defendere

[86] proposita *Willner*, proposito P
[87] que *Willner*, qui P
[88] tuta *Willner*, tua P

business. For when a question has been posed, she judges on first perusal which part of it should be retained and which should be rejected. In order to confirm that judgement she does not tire in finding topics with their arguments, although some people prefer to make invention come before judgement. By a subtle investigation she herself has divided the totality of everything that is seen into ten natures, claiming that there are only ten ways in which completely different things (*res*) differ in nature, which people afterwards also called the 'ten categories'; and she ascribes genera, species and individuals to each category.[46] Further, she argues that, of the genera, some are 'most general', others 'subaltern'; similarly, of the species, some are 'most special', others 'subaltern', taking that most noble alternating path so that now she might ascend from individuals through intermediates to the most general, by stripping off – indeed forgetting – the properties; now she might descend by the same steps, weaving back forms into each one, from the most general genus to individuals. Indeed, she, panning the surface of the universal natures of things (*res*) with the wonderful subtlety of her light, strives to behold them in the way by which they were conceived in the mind of the Artist before time. She considers accidents without their relation to substances; she also separates the differences of the accidents themselves, confident as she is in this subtlety, so that, by using sharp and irritating arguments, like a serpent she ensnares others unawares, and, secure in her defence of 'determination', she despises the darts hurled by others.

Hence it happens that, if anyone has acquired her state of mind, he would not be ignorant of which proposition is true, which false; of the same propositions, too, which has a determinate truth or falsehood, and which has an uncertain and dubious one; also, of arguments, which are probable, which necessary, which sophistical; of topics, which are intrinsic, which extrinsic, which intermediate; of divisions, which are according to themselves, which according to an accident; of definitions, which are substantial, which accidental; and the varieties, figures and modes of syllogisms. Aristotle, imbued with her subtlety, whenever he wished to make a joke, could just as much draw up a false statement in front of others, knowing very well the tricks of sophistry, as others were able to defend the truth itself in his presence.

facultas erat. Ut enim cetere artes auxiliis huius nixe in sententia immote constant,[89] ita sine ea titubantes stabilitatem habere ignorant. Unde fit ut apud modernos etiam in disserendi facultate hii precipue spectentur qui pre ceteris ea notabiles sunt. Et quia hactenus de his tribus quarum precepta circa voces versantur demonstratum est, de reliquis quatuor que circa res ipsas disputant opere precium est deinceps non tacere.

ARITHMETICA Harum ergo prima que quadam excellentia reliquas tres superat eique ille acclines[90] sunt, cuius in veste innumeros caracteres numerorum significativos intextos vides, quicquid sub numeralem discretionem venit et moderatur et digerit. Unde fit ut universa visibilia numero subiecta sint et huic etiam eadem subici necesse sit. Quicquid enim est, aut unum aut plura[91] est. Ipsam illam etiam immensam universitatem[92] termini limite diffinit sibi certo atque nature. Qua pretermissa, singularum rerum pluralitas incerta confusaque est, quod plane intelligere potest, si quis rerum suarum numerum oblivisci perfecte queat easque confuse et sine discretione concipiat. Inde fit ut non solum his quas astantes vides necessariam esse dicam, verum etiam cunctis essentiis preferendam esse non ambigam, cum eis confusionem auferat, discretionem tribuat. Numerus quippe, cuius ipsa naturam et cognoscit et explicat, adeo ipsis rebus incumbit ut quecumque in ordinem a prima quadam confusione digesta sunt, ex eius similitudine exemploque decorem accepisse videantur. Quod et Zenocrates[93] intelligens, ipsam animam numerum esse diffinivit, dicens: 'Anima est numerus se movens.' Quid igitur tantus philosophus tanta brevitate intellexerit, tibi ipsi forsitan dubitabile est. Numquid numerum ab anima non distare intellexit? Numquid accidens cum suo subiecto idem esse essentialiter iudicavit? Numquid, cum numerus compositus sit, ipsam quoque animam compositam esse aberravit? Absit! Quia vero in rebus nichil tam digne essentie invenit ut numerum, querens anime comparationem, numerum sibi

89 constant *Willner*, -stat P
90 acclines *scripsimus*, addicte *Willner*, addines P
91 plura *Willner*, plures P
92 universitatem *Willner*, -tate P
93 Zenocrates *sc.* Xenocrates

Indeed, as the other arts, when they depend on the support of this one, remain unshaken in their judgement, so without it they totter and have no experience of constancy. Hence it happens that, even amongst the moderns, in disputational skill those are especially respected who are more notable than the others in this art. And because up to now a demonstration has been given concerning these three arts whose precepts deal with words (*voces*), it is worthwhile to go on hereafter to speak about the remaining four, which dispute about things (*res*) themselves.

Arithmetic

The first of these is superior by a certain excellence to the other three, who depend on her. On her cloak you see woven innumerable characters signifying numbers. She both directs and orders whatever comes under distinction by numbers. Hence it happens that, as all visible things are subject to number, they must also be subject to her. For, whatever is, is either one or many. She delimits the unmeasured totality itself by the boundary of a term which is certain to herself and to nature. If she is overlooked, the plurality of individual things (*res*) is uncertain and confused, as anyone can easily understand who can completely forget the number of his own possessions and thinks of them in a confused way and without distinction. Hence it happens that not only do I say that she is necessary to these maidens whom you see standing beside her, but also I have no doubt that she should be preferred to all essences, since she takes confusion away from them, and gives them distinction. For number, whose nature she herself both knows and unfolds, is so latent in realities (*res*) themselves that whatever was brought into order from a kind of primeval confusion, is seen to have received its comeliness from her likeness and example. This Xenocrates too understood when he defined the soul herself as number, saying: 'The soul is number moving itself.'[47] What such a great philosopher understood with such brevity is perhaps not clear to you. Surely he did not think that number was the same as soul? Surely he did not judge that the accident was essentially the same as its subject? Surely, when number is composite, he did not err by saying that the soul herself was also composite? Far from it! Rather, because he found in realities (*res*) nothing of so worthy an essence as number, when seeking something to compare the soul with, he likened

assimilavit – immo similitudini incumbens, numerum eam vocavit. Ne vero aliquis verborum venator eam id ipsum quod dixerat essentialiter esse intelligeret, consequenter subdidit 'se movens', ut, cum nullus numerus se ipsum moveat, anima autem se ipsam moveat, eam numerum esse non intelligamus, set tante dignitatis pre ceteris quanta pre ea numerus esse dicitur. Set ad propositum revertamur.

Huius ipsius numeri, dico, hec quam describo magistra ostentatrixque est. Ut autem familiarius de ea agamus, illa ipsa est quam Greci, omnium arcium executores, 'arithmeticam' quasi 'numeri virtutem' vocaverunt, intelligentes et rerum virtutem precipuam in numeris esse, et ipsam dignitati tante ut dominam preesse. In primis etenim numerum diffinit et eum collectionem unitatum vel quantitatis acervum ex unitatibus profusum esse determinat. Post diffinitionem vero, que prima omnium tractatuum precedere debet, divisionem subiungit, in parem atque in imparem disiungens, ipsasque has species diffinitione subtili consequenter discernit, secundum duorum generum contrarias qualitates – multitudinis dico et magnitudinis. Deinde ipsarum specierum subdivisionem ponit, docens parem numerum alium pariter parem, alium pariter imparem, alium impariter parem; primam subdivisionis speciem vocans eam cuius partes subdividentes usque ad unitatis occursum in paria dividuntur; pariter vero imparem intelligens qui primo loco quidem in equa dividitur, dividentia vero mox indivisibilia reperiuntur; impariter vero parem ipsum etiam in equa dividi eiusque dividentia non minus paria esse, non tamen hanc eius divisionem ad unitatem usque pertingere, priusque deficere; itaque nec solum primum eius terminum in paria dividi, nec solum extremum a divisione secludi.[94] Que quantam iocunditatem intelligentibus addant in ea figura descriptionis aperte oculis subicitur quam Nichomacus eiusque sequaces ammirabilem[95] vocaverunt. Aliam quoque numeri speciem non minus diligenter in tres partes mensuratur, dicens:

[94] secludi *Willner*, seduli P
[95] ammirabilem *scripsimus*, ad mirabilem P

number to her – indeed, dwelling on the likeness, he called her 'number'. But lest any quibbler might understand that she is in her essence exactly what he said she was, he added to the definition 'moving itself', so that, since no number moves itself, but the soul does move herself, we understand her not to be number, but of such worth in respect to everything else as number is said to be in respect to her. But let us return to our theme.

Of this very number, I say, this maiden whom I describe is the teacher and demonstrator. To deal more intimately with her: she is that maiden whom the Greeks, the practitioners of all the arts, called 'arithmetic', as if to say 'virtue of number', understanding both that the principal virtue of realities (*res*) is in numbers, and that she herself presides over such worthiness as a mistress. For, first,[48] she defines number and establishes that it is a collection of units or a heap of quantity poured out from units. After the definition, which must precede the first topics of any treatise, she adds a division, separating number into even and odd, and these species she goes on to distinguish by a subtle definition according to the contrary qualities of the two genera – I mean, multitude and magnitude. Then she makes a subdivision of these very species, teaching that one even number is 'evenly even', another 'evenly odd', another 'oddly even'. She calls that first species of subdivision the one whose parts, when subdivided, are divided into even numbers until they reach the unit. But she understands by 'evenly odd' the number which is divided in the first step into equal numbers, but soon the numbers as they divide are found to be indivisible. In the case of 'oddly even', on the other hand, they can be divided into equal parts, and the numbers as they divide are still even, but this division cannot reach as far as the unit, and fails before it gets there: thus its first term is not the only one that can be divided into equal parts, but neither is its last term the only one incapable of being so divided. What pleasure these provide for intelligent people is open to view in that diagram which Nicomachus and his followers called 'admirable'.[49] She divides the other species of number, too, into three parts, just as carefully, saying:

Impar numerus alius primus et incompositus,[96] alius secundus et compositus, alius ex utrisque quodammodo constans atque compactus. Est enim per se quidem secundus et compositus, ad alios vero comparatus primus et incompositus invenitur.

Earumque ipsarum partium descriptiones haut imparatas habet, originemque singulorum miro genere scaturiginis a ternario procedere submonstrat, quam et Eratosthenes[97] et eius sequaces 'cribrum' appellaverunt. Item, etiam eiusdem paris numeri aliam non minus dignam particionem supponit, docens alios diminutos, alios superfluos, alios inter utrosque perfectos esse. Que sententia quante gratie fecunda sit nulli nisi cui ipsa demonstrat patere potest. Ceterique itidem eiusdem flores quos et michi enumerare tempestivum non est, nec tibi id ipsum nisi in lectione adesses facile concipere. Hoc autem unum scio, eius precepta artificem totum sibi vacantem exterioraque omnia obliviscentem exposcere – quod et eius subtilitatis pre ceteris non minimum est indicium. Nos autem ad ceteras describendas transeamus.

MUSICA Hec igitur que quadam leticia ceteras excellere videtur, timphanum dulcissonum dextra gestans, leva vero libellum universorum tinnituum precepta naturasque continentem – hec, inquam, universis consonantiis preesse cognoscitur, nec solum in vocibus humanis[98] ratam modulationem natureque acceptabilem facit, verum etiam in cordis sive eribus atque malleis vim suam et habet et ostendit. Unde fit ut omnes animas hec in se non parva potentia quadam convertat. Pueriles enim mentes noviterque in corpora occupatas tanta vi astringit ut primis vocibus quas in corpore audierunt et delectentur si concordes sint, et offendantur si convenientes non sint. Neque id usus nec[99] natura efficit, nisi si olim celitus aliquid tale mente hauserint. Id quondam nutrices eorum haut prave intelligentes, ex quo eos quacumque causa in fletum commoveri sentiunt, statim eis musica alicuius soni

[96] incompositus *Willner*, -tis P
[97] Erastotenes P
[98] humanis *Willner*, humani P
[99] nec *scripsimus*, immo P

> Of odd numbers one is prime and not composite, another is secondary and composite, another in a certain sense consists of, and is put together from, both. For it is in itself secondary and composed, but when compared to others it is found to be primary and not composed.[50]

She has not-ill-thought-out descriptions of these parts too, and adds a demonstration of how they gush forth from their source in the number three like a wonderful stream of bubbling water, which both Eratosthenes and his followers called the 'sieve'.[51] Similarly, she also adds another no less worthy division of the same even number, teaching that some are deficient, others superabundant and others, coming between the two, are perfect. How much pleasure this idea abounds in is clear only to the man to whom she herself gives a demonstration. There are, too, her other blossoms, which neither is it the right time for me to relate, nor would it be easy for you to conceive unless you were present at the lecture. This one thing I know, that her precepts demand a practitioner who makes himself entirely free for her and forgets all outside things – which is very good evidence of her subtlety compared with that of the other maidens. But let us pass on to describing these.

This maiden seems to excel the rest in a kind of joyfulness. Music
She holds a sweet-sounding cymbal in her right hand, but in her left is a little book containing the laws and natures of every sound. This maiden, I say, is recognised as being in control of every consonance, and not only produces melodies in the human voice which are controlled and conform to nature, but also is powerful and manifests her force in strings, brass instruments and percussion. Hence it is that she draws all souls to herself with no mean force. For she captivates the minds of children soon after they are wrapped in bodies with such force that they are both delighted by the first sounds which they hear in their embodied state if those sounds are harmonious, and are upset if the sounds are dissonant. This cannot be brought about by experience, nor can nature cause this, unless they have once drunk deeply in their minds something similar in heaven. Their nurses, having once understand this not improperly, the moment they sense that children are about to burst into tears for some reason or other, rush to them with the musical modulation of some

modulatione occurrunt,[100] quippe id sibi unicum et familiare solatium anima reservavit, quod quidem dummodo sibi offeratur, cetera universa huic ut attendat obliviscitur. Et ne exempli postulatio longe quesita diutius nos fatiget, tu ipse, si recolligis, cum preterito anno in eadem musica Gallicis studiis totus sudares adessetque in serotino tempore magister artis una cum discipulis, cum eorum Regineque rogatu citharam tangeres, puerulus quidam non certe locutionis sono[101] irretitus, ex cithare sonitu tanta hilaritate affectus est ut et manus digitosque suos simili[102] nisu movere aggrederetur,[103] omnibusque astantibus risum moveret. Set non omnes eque vim causamque motus illius intelligebant.

Adolescentiam vero iuventutemque ita huic alligatas naturaliter intellige, ut non solum in toga sibi se totos accommodent, verum etiam in bellicis negotiis et incipiendis et remittendis ipsa stimulos ac frena prestet. Nam et in aggrediendo prelia litua acutiora animos stimulant, et in remittendo tuba gravior eosdem resides facit. Adeo hec vis anime imperativa est! Nam, ut fabulosa pretermittam – que musicis instrumentis muros crevisse, silvas artificem secutas esse asserunt[104] – id saltem dubium non est, si quis delectationis sue reminisci volet, quin et animum ex turbiditate in quietem, ex quiete vero in pietatem, hec ipsa constituat. In senectute vero tantam hoc decus efficatiam obtinet ut non solum vocum concordiam hec etas exposcat, verum et mores et facta universa in ethicam consonantiam redigere et gaudeat et nitatur. Unde et, hac intentione firmiter constituta, a quampluribus sapientibus hec etas 'gravis' appellatur. Cum itaque omnem etatem non modo ab avaritia in largitatem, verum etiam ab egritudine in sanitatem provocet, non minimum philosophis argumentum erat animam a superioribus stellis in corpora descendisse eamque quam ibidem simphoniam assueverat in corpore vel recenter receptam recognoscere. Set hec apud rationales. In mutis vero animalibus vis

100 occurrunt *Willner*, occurrit P
101 sono *Willner*, dono P
102 simili *Willner*, similis P
103 aggrederetur *Willner*, egr- P
104 asserunt *Willner*, asserit P

melody. For the soul has reserved for herself that unique and intimate consolation, and no sooner is this available than she forgets everything else in order to enjoy this. And, lest the long-sought request for an example wear us out any longer, you may recall the occasion last year when you yourself in your studies in France were putting all your efforts into the same science of music, and a master of the science was present one evening with his pupils. These asked you – and the Queen herself asked you – to play the cittern. And then a little boy, who was still stumbling over his words, was so captivated by the sound and moved to such joy by the notes of the cittern that he started to move his hands and fingers in time to the music and made all the bystanders laugh. But not all of them understood the force (of music) and the cause of the child's reaction equally.

Look how adolescents and young men are by nature so captivated by it that in times of peace they dedicate themselves to it entirely, and in times of war they are goaded into battle or restrained from the attack by it! For in the attack, the higher-pitched sounds of the trumpets goad their spirits, and in the retreat the deeper tuba quietens those same spirits. Such is the command over souls that this force has! I need not mention examples from mythology – of walls springing up at the sound of musical instruments, or of woods following the musician as he played. This, at least, is in doubt to no one who cares to remember his own enjoyment of music: that this force both makes the anxious mind relax, and moves the relaxed mind to a sense of reverence. In old age this wonderful thing has such effectiveness that people at this stage in life not only demand to hear harmonious singing, but also are glad to strive to bring their own characters and all their actions into an ethical harmony. Hence, because it is firmly grounded in this aim, this stage in life is called by most wise men the *grave* age. Since, therefore, the force of music calls people of all ages not only from greed to generosity, but also from sickness to health, this was good evidence for the philosophers that the soul descended into bodies from the stars above and that she recognised that symphony which she had got accustomed to in the heavens when she had been in the body even for a very short period of time. This you can say about rational beings. But even in mute animals the same

eadem non parvum effectum habet. Apud Anglos enim pisces ipsi dulci cimbali sonitu super aquam natantis in retia ire coguntur. Apud Parthos vero cantus ipsorum hominum dammas sine omni retis laqueo in sompnos solvi cogunt. Volucres cantibus in laqueos duci haut dubitabile est. Teque ipsum etiam, dum de eadem disputares, id asserere audivi, nisum scilicet silvestrem – quem non alia ratione domare poteras – musico instrumento domesticum fecisse. Hinc et idem supradictus philosophus animam mundanam ex talibus convenientiis compactam esse dixit, non quod eo modo simpliciterque intellexerit, set id involvens, quia, cum[105] plures potentias habeat ea,[106] eius inter dignissimas est, quod sibi concors est corporibusque idem inferre gaudet, et secundum hoc quod partes corporis armoniatas videt, suas in eis potentias exercet. Unde fit ut hoc quoque animare non cesset, illud vero inanimum diutius relinquat.

Hoc idem Pitagoras non preteriens cum eius distinctiones non parva cura inquisivisset – ne id ipsum posteris lateret – officio stili usus est, docens primo musicam aliam mundanam, aliam humanam, aliam instrumentalem – nam et elementorum qualitates ponderaque temporumque distantie ab hac concordia remote non sunt; deinde disputans que quidem de innumera vocum multitudine consonantiarum dignitati apte sint, redigens eas in paucum diffinitumque numerum (iuxta quod omne preciosum rarum est); easque ipsas non sensuali iudicio – quo nulla subtilitas cernitur – set rationabili arbitrio dividi, quippe cum ipse eandem scientiam non nisi ratione duce a malleorum sonitu comprehenderit, et exinde se ad cetera instrumentorum genera mathematice transtulerit. In omnibus etenim una concordie ratio est. Quod enim proportiones ponderum faciunt in malleis vel cimbalis, idem eedem in fidibus ratione longitudinis brevitatisque, intensionis remissionisque faciunt, idem etiam in fistule foraminibus ratione profunditatis sive amplitudinis, idem in omni elementario

105 cum *Willner*, tum P
106 ea *Willner*, eam P

force has a not inconsiderable effect. Among the English the very fish are driven into nets by the sweet sound of a bell floating on the surface of the water. Among the Parthians the songs of men themselves force the deer to be dissolved into sleep, so there is no need for nets. One cannot doubt that birds are lead into snares by songs. I have heard even you, while you were disputing about music's force, make that claim, saying that you made a wild hawk – which you could not tame in any other way – docile by playing a musical instrument. Hence that same philosopher said that the World Soul was put together from such consonances, not because he understood it in that way in a simple sense, but rather, he was using a metaphor to imply that, whilst the soul has many powers, that power is among her most worthy, that she is in harmony with herself and is glad to pour that same harmony into bodies; and, according to the extent to which she sees the parts of that body are in harmony with themselves, she exercises her powers in these parts. Hence it also happens that she animates this body incessantly, while she leaves that one without soul for a longer period of time.

Pythagoras did not overlook this same fact when he enquired diligently into music's divisions and – lest his research should be lost to posterity – wrote down his findings.[52] He taught first that one kind of music was cosmic, another human, a third instrumental – for neither the qualities of the elements, nor weights, nor the divisions of the seasons are immune from this concord. Then he discussed which of the innumerable number of intervals were worthy of the name of 'consonances', and reduced them to a small and definite number (according to the principle 'whatever is precious is rare'[53]). He said that these consonances were to be divided not by the judgement of the senses – by which no subtle distinctions can be made – but by the arbitration of reason. For he himself had come to understand the science of music as a result of applying reason to the different sounds produced by hammers, and from there had turned to investigating the other kinds of instruments in a mathematical way. For in all things there is the same law of harmony: particular ratios of weights in hammers or cymbals produce the same intervals as the equivalent ratios of length and shortness, or of tautness and slackness in strings, or of depth and width in the case of

quod quacumque consona discordia convenit. Que omnia vir sapiens diligenter prosequitur ipsarumque consonantiarum differentias deinde supponit. In qua pagina illa haut ignobilis sententia subnotatur, in qua quomodo tonus ex duobus semitoniis et commate constet, ipsaque inter se semitonia et a commate quid distent (nec enim equalia sunt) ostendit. Que si quis facile intellexerit, cerebro gratias habeat! De monocordi quoque divisione non minus mellita subiecit, ceteraque id genus. Que[107] si enumerare pergam, levius me facultas quam rerum numerus deseret. Nunc de reliquis ordine prosequamur, illud tamen non pretermittentes quod, cum utreque supradicte numerorum rationem in tractatibus suis recipiant, in hoc tamen differentia est, quod illa numeros, hec proportiones principaliter investigat, licet corporibus ea coaccidere constet.

GEOMETRIA Hec autem due que restant licet impares quidem sint quantitate, tamen minor maiori suffragium parat. Set et in hoc differunt, quod, cum utreque magnitudinem corporalem officialiter profiteantur, hec quidem minor magnitudinis immobilis scientiam exercet, maior vero mobilis celestisque molis habitudinem scrutatur. Unde et minorem, de qua nunc sermo est, Greci 'geometriam' a 'terre mensura' vocaverunt. Hec autem quanta efficatia mortalibus necessaria sit[108] facile colligere potest quicumque rationem terrarum mente concipiat. Inter primos enim rationabiles hominum conventus, cum id iusticia politica effecerit ut in occupationibus terrarum non tirannica invasione, verumtamen certa mensura uterentur, id demum astrictum est, ut quisque sue occupationis terminum poneret. Sic itaque factum est ut et litigia ex priori indiscretione orta bellaque eisdem de causis quondam sumpta cessarent. Id intelligentes, cetere nationes et imitando[109] artificium commoda secuti, et inventoris ingenium ammirati sunt. Contigit autem ut aliquando eedem terminorum positiones rapidis quibusdam motibus delerentur: Libie quippe

107 Que *Willner*, qua P
108 sit *Willner*, fit P
109 immittando P

the bores of pipes – in short, one finds the same situation in every medium in which there is agreement by a kind of consonant discord. The wise man has a keen nose for all these matters and then adds the differences between the consonances themselves. In this book that not ignoble statement is noted down in which he shows how a tone consists of two semitones and a comma, and what the difference is between the two semitones themselves and the comma (for they are not equal). If anyone grasps this easily, he should be thankful to his brain! He added no less honeyed words concerning the division of the monochord and the other matters of this kind. If I were to go through them all, my powers will come to an end more easily than the number of topics. So now let us tackle the remaining arts in order. But I must not forget to mention that, although both arithmetic and music use the reasoning of numbers in their exposition, there is a difference in that the first investigates numbers principally, the second, ratios, although numbers and ratios cannot be separated in their embodied state.

Geometry

The two remaining maidens are unequal in size; nevertheless the smaller provides support for the larger. But they also differ in this, that, while each professes, according to her office, corporeal magnitude as her subject, this smaller one practises the science of immobile magnitude, whereas the larger one examines the state of the mobile, celestial structure. Hence it was the smaller one, about which our present discussion is, that the Greeks called 'geometry' from 'measurement of the earth'. By how much efficiency she makes herself necessary for mortals can easily be deduced by anyone who reflects in his mind upon the way land is apportioned. For, amongst the first meetings of men that were based on reason, when political justice made sure that, in occupying land, men used not a tyrannical form of taking over, but a fixed measure, it was finally determined that each person should put a boundary to the land that he occupied. Thus it was achieved that both legal disputes arising from an earlier lack of distinction and wars which had once been started for the same reasons, ceased. The other nations,[54] noticing this, both gained advantages by imitating the invention and marvelled at the genius of the inventor. It happened that at times these positions of the boundaries were destroyed by some violent

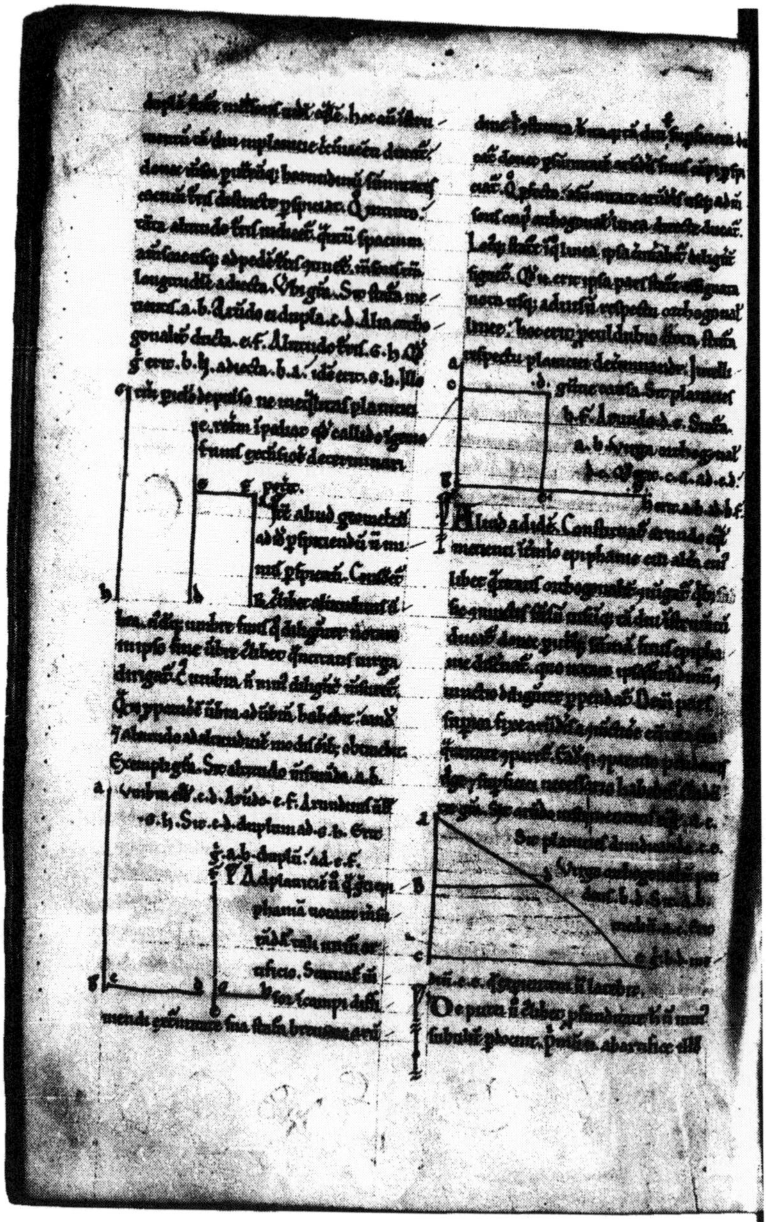

The geometrical figures in De eodem et diverso. *MS Paris, Bibliothèque nationale, 2389, fols 89v–90r. Reproduced with permission.*

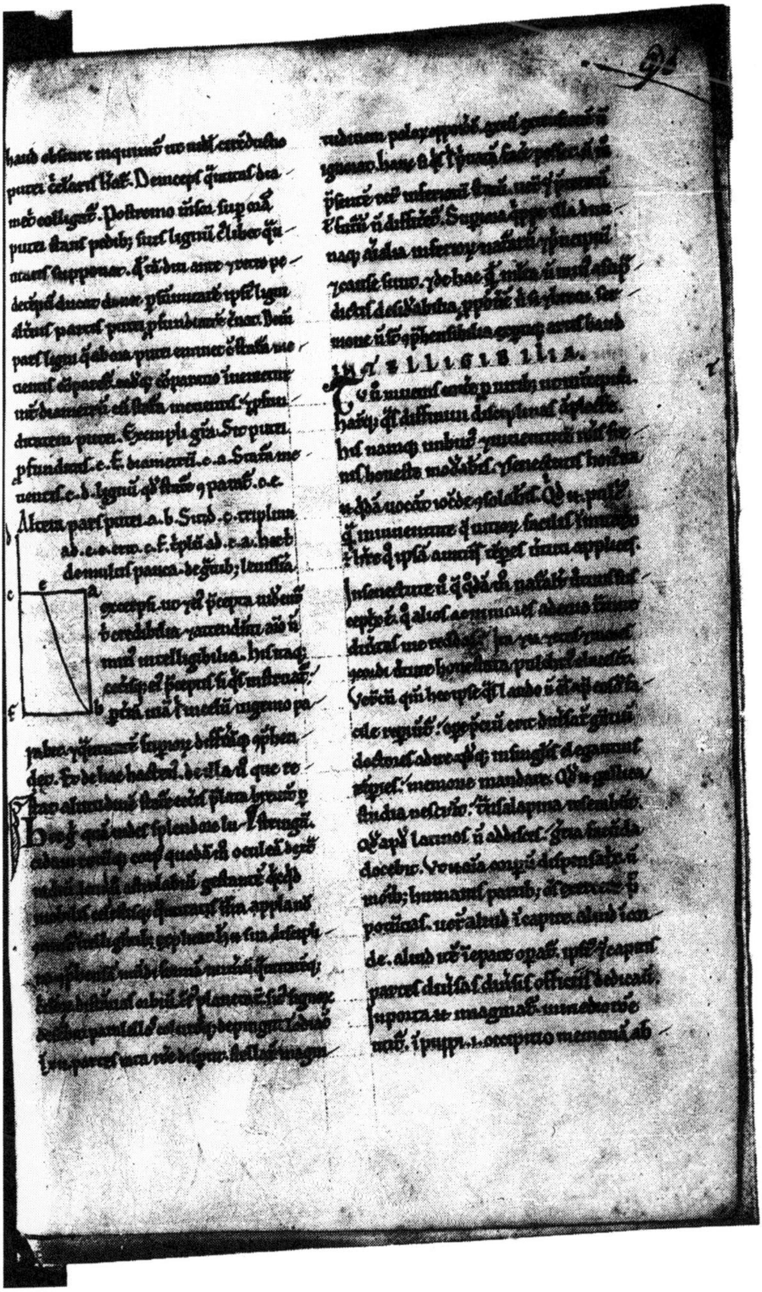

harenose accumulationes, Egipti vero undantis Nili superationes supradicta signorum vestigia superabant. Unde factum est ut sapiens quidam vir, subtilitate mentis elatus, in secretum se transtulerit id expendens: qua ratione regulam omnibus seculis perhennem de terre mensura habere posset; immo illud quod anima in sue divinitatis thesauro a principio possederat elicere satagebat.

Hic itaque, tanti voti compos, hac auxiliante, a puncto incipiens lineatimque ascendens superficie mediante in soliditatem usque progressus est. Deinde quot punctis singule linee terminentur, quot item lineis[110] superficies, quot superficiebus soliditas perficiatur, hac duce consecutus est. Subsequenter ergo orbem in partes, partes in provincias, provincias in regiones, regiones in loca, loca in territoria, territoria in agros, agros in centurias, centurias in iugera divisit, quid centrum, quid diametrum, quid rotundum, quid angulare, quid basis, quid cathetus, quid ypotenusa, quid perpendiculare, diligenter contemplans. Postremo tante speculationi diligentius incumbens, remotorum etiam corporum usu geometrico quantitatem apprehendit. Itaque factum est ut certas proponeret regulas quibus altitudo corporum, et longitudo vel latitudo planicierum, profunditasque puteorum sine artificis accessione comprehendi possent. Que quia tibi forsan incredibilia videntur, de multis pauca proponere non deficiam, ut et id fieri non impossibile iudices et ab eadem facultate intelligentie pauca prelibatus non deterrearis.

Proposita sit tibi turris cuiuslibet altitudinis mensuranda. Accipies itaque arundinem stature metientis duplam, cuius medio aliam orthogonaliter coniungere calle predicte subduplam, stature mensoris videlicet equalem. Hoc autem instrumentum tamdiu in planicie circumiacenti ducatur, donec mensor per utrarumque arundinum summitates cacumen turris distincte[111] perspiciat. Quo intuito, tanta altitudo turris iudicetur quantum spacium a mensore usque ad pedem turris

110 item lineis P *a.c.*, interlineis P *p.c.*
111 distincte *Willner*, districte P

movements: for example, the heaping up of sand in Libya, or the overflowing of the Nile in Egypt, wiped out the traces of the abovementioned boundary-marks. Hence it happened that a certain wise man,[55] superior in the subtlety of his mind, took himself into a secret place, considering in what way he might be able to find a rule for the measurement of land which would last for all time; rather, he was concerned to elicit what his soul had possessed from the beginning in the treasury of her own divinity.

This man, achieving his great aim, with the help of this maiden, beginning with the point, and ascending through the line, arrived by way of the surface at a solid body. Then, under her guidance, he went on to find out by how many points single lines are bounded, by how many lines, a surface, by how many surfaces a solid is perfected. Subsequently he divided the world into parts, parts into provinces, provinces into regions, regions into places, places into territories, territories into fields, fields into centuries, and centuries into acres, considering carefully what is a centre, what is a diameter, what is round, what is angular, what is the base, what is the side, what is the hypotenuse, what is the perpendicular. Finally, applying himself more carefully to such an important speculation, he grasped also the size of distant objects by the use of geometry. Thus it happened that he proposed fixed rules by which the height of objects, the length and breadth of flat areas, and the depth of wells could be obtained without the practitioner having to approach them. Because this perhaps seems unbelievable to you, I shall not fail to put forward a few of the many available examples,[56] so that you may judge that it can be done and will not be deterred from the same kind of intellectual pursuit having tasted a little of it.

Suppose you have to measure a tower of a certain height. You take a cane which is twice the height of the measurer, to the middle of which be careful to join at right angles another cane half the length of the first, that is, equal to the height of the measurer. This instrument should be moved about in the area surrounding the tower until the measurer can see distinctly the top of the tower in line with the tops of both canes. When this is seen, the height of the tower should be judged to be as much as the distance from the measurer to the foot of the tower, plus the height of the measurer. For

continetur, mensoris tantum longitudine adiecta. Verbi gratia: sit statura metientis AB, arundo ei dupla CD, alia orthogonaliter ducta EF, altitudo turris GH. Quod igitur erit BH adiecta BA, idem erit GH, illo tantum periculo depulso ne inequalitas planiciei rationem impediat, quod callido ingenio funis extensione determinari poterit.

Item aliud geometricum ad idem perspiciendum non

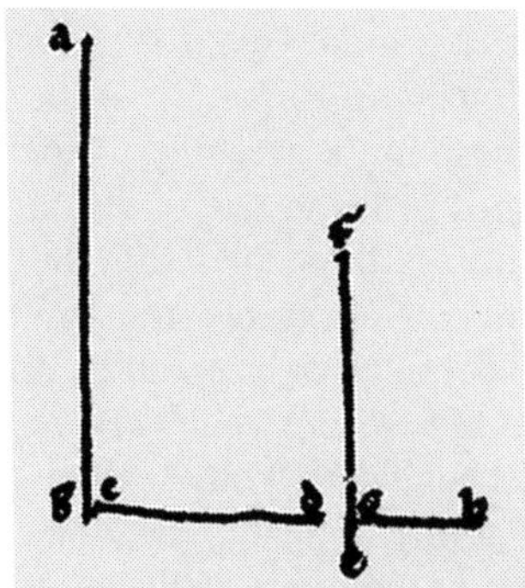

minus perspicuum. Consideretur cuiuslibet altitudinis umbra, eiusdemque umbre finis. Quo diligenter notato, in ipso fine umbre cuiuslibet quantitatis virga dirigatur, cuius umbra non minus diligenter mensuretur. Quam enim proportionem umbra ad umbram habebit, eandem etiam altitudo ad altitudinem modis omnibus obtinebit. Exempli gratia: sit altitudo mensuranda AB, umbra eiusdem CD, arundo EF, arundinis umbra GH. Sit CD duplum ad GH. Erit igitur AB duplum ad EF.

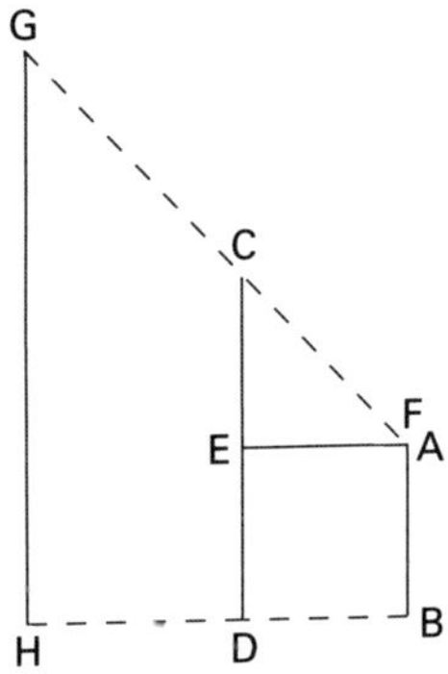

example, let the height of the measurer be AB, the cane twice this, CD, another cane at right angles, EF, the height of the tower GH. The length BH with the addition of BA will be equal to GH, provided that the unevenness of the surface of the ground is not an obstacle to the calculation, which a lively intelligence can determine by stretching a rope over the area.

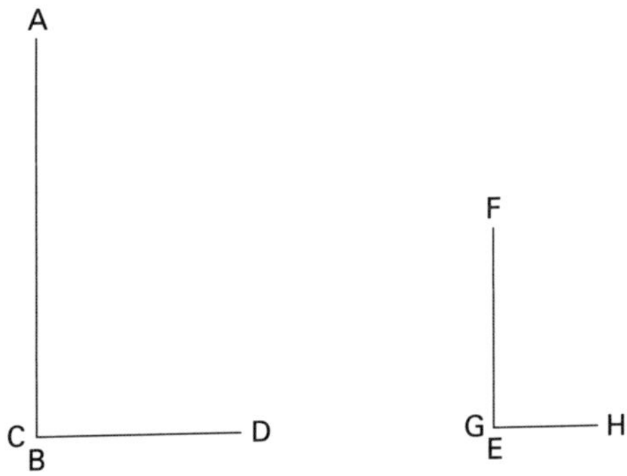

Another geometric means for solving the same problem is no less obvious. Let the shadow cast by any height be considered, and the end of that shadow. When this has been carefully noted, at the very end of the shadow let a rod of any size you like be erected. Measure its shadow no less carefully. For, whatever ratio the shadows have to each other, the same ratio the two heights will have, in every respect. For example, let the height to be measured be AB, and the shadow of the same object be CD; the cane EF, and the shadow of the cane GH. Let CD be twice GH. AB will therefore be twice EF.

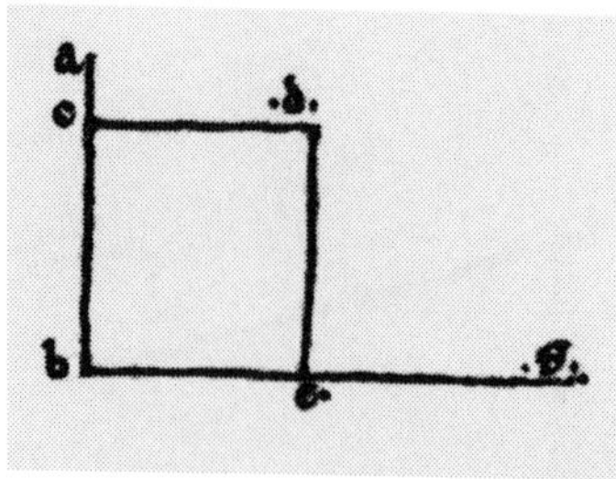

Ad planiciem vero, quam Greci 'epiphaniam' vocant, mensurandam tali utimur artificio. Statuatur mensor in campi diffiniendi extremitate, sua statura breviore arundine sibi constituta. Hec itaque tamdiu per superficiem ducatur donec per summitatem arundinis finis campi perspiciatur. Quo perfecto, a summitate arundinis usque ad mensoris corpus orthogonalis linea directe ducatur, locusque stature in quo linea ipsa terminabitur diligenter signetur. Quod enim erit ipsa pars stature a signata[112] nota usque ad visum respectu orthogonalis linee, hoc erit procul dubio tota statura respectu planiciei determinande. Intelligentie causa: sit planicies BF, arundo DE, statura AB, virga orthogonalis DC. Quod erit CA ad CD, hoc erit AB ad BF.

Aliud ad idem. Constituatur arundo equalis metienti in termino epiphanie, cui altera cuiuslibet quantitatis orthogonaliter coniungatur. Quibus sic coniunctis, susum[113] iusumque tamdiu instrumentum ducatur donec per utriusque summam finis epiphanie discernatur. Quo notato, ipsarum arundinum coniunctio diligenter perpendatur. Deinde pars superior fixe arundinis a coniunctione cum tota sui quantitate comparetur, eademque comparatio pendentis virge et super-

112 a signata *Willner*, assignata P

113 sūsum P (*compare* Macrobius, *Comm. Somn. Scip.*, 2.5.25: 'susum iusumque')

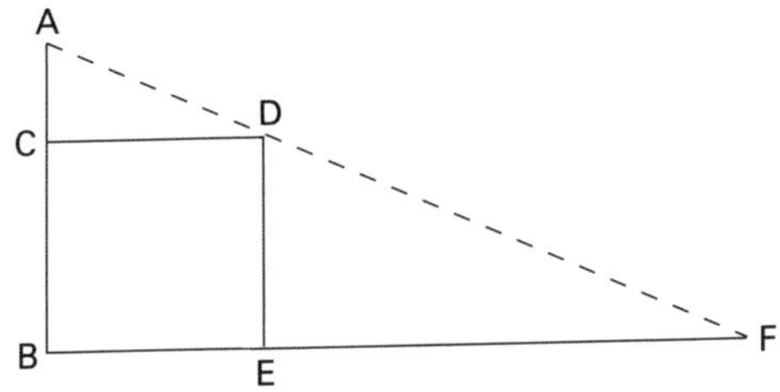

For measuring a flat surface, which the Greeks call an 'epiphania', we use this method. The measurer should stand on the very edge of the area to be measured and should have provided for himself a cane shorter than himself. This should be moved over the area until the edge of the area can be seen in line with the top of the cane. When this is done, a string at right angles to the cane should be pulled straight from the top of the cane up to the body of the measurer, and the place on his body which the string meets should be carefully noted. For the ratio of that part of the body from the point noted up to the eyes, to the string pulled at right angles, will without doubt be the same as that of the whole body to the area to be measured. To make this understandable, let the flat area be BF, the cane, DE, the standing man, AB, the rod at right angles, DC. The ratio of CA to CD will be that of AB to BF.

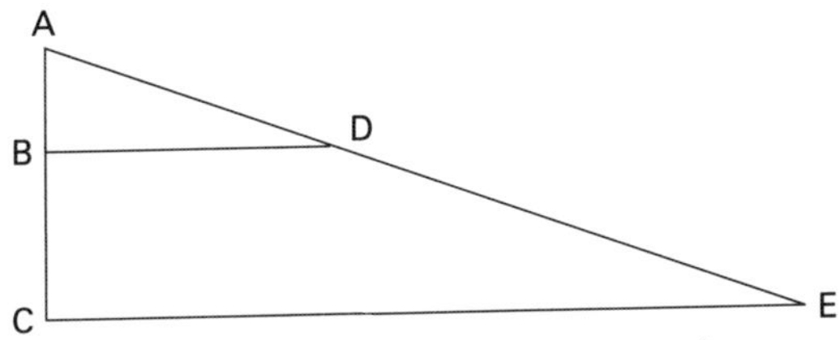

Another means. Let a cane equal in size to the measurer be placed on the edge of the flat area, to which another cane of any size you like should be joined at right angles. When these have been joined like this, the instrument should be moved up and down until the edge of the flat area can be seen in line with the top of each of them. When this has been noted, the joint of the canes themselves should be carefully considered. Then the part of the fixed cane above the join should be compared with the total length of that cane, and the same ratio will necessarily exist between the rod attached to the cane and the area. To make this clear, let the cane equal to

ficiei necessario habebitur. Evidentie gratia: sit arundo visui metientis equalis AC, sit planicies diiudicanda CE, virga orthogonaliter pendens BD. Sit AB medium AC. Erit igitur BD medium CE, quod experientem non latebit.

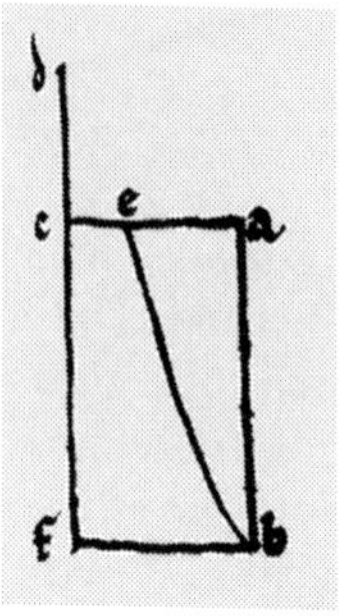

De putei vero cuiuslibet profunditate hec non minus subtiliter perdocuit. Primum enim ab artifice illud haut obscure inquiratur: ut videlicet circumductio putei circularis habeatur. Deinceps quantitas diametri colligatur. Postremo mensor super oram putei stans, pedibus suis lignum cuiuslibet quantitatis supponat, quod tamdiu ante et retro pedetemptim ducat, donec per summitatem ipsius ligni alterius partis putei profunditatem cernat. Deinde pars ligni que ab ora putei eminet cum statura metientis comparetur, eademque comparatio invenietur inter diametrum et profunditatem putei cum statura metientis.[114] Exempli gratia: sit putei profunditas CF, diametrum CA, statura metientis CD, lignum quod stature comparatur CE, altera pars putei AB. Sit DC triplum ad CE, erit DF[115] triplum ad CA.

Hec tibi de multis pauca, de gravibus levissima excerpsi, ut et eius precepta viderentur tibi credibilia et attendenti animo non minus intelligibilia. His itaque ceterisque eius preceptis si quis instruatur, per terram viam sibi in celum ingenio parabit et quantitatem superiorum differentiamque comprehendet. Et de hac hactenus; de illa vero que restat, altitudine[116] stature ceteris prelata, breviter perstringam.

114 profunditatem . . . metientis *Willner*, cum statura metientis et profunditatem putei P
115 DF *Willner*, CF P
116 altitudine *Willner*, -nem P

the height of the eyes of the measurer be AC. Let the flat area to be measured be CE, the rod attached at right angles, BD. Let AB be half AC. Therefore BD will be half CE, which will not escape the notice of anyone who tries it out.

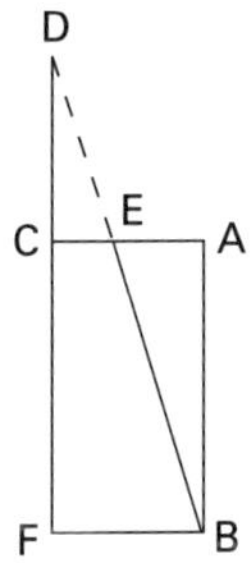

This maiden has just as subtly taught how to find out the depth of any well. First the geometer should enquire without obscurity whether the perimeter of the well is circular. Then the size of the diameter should be ascertained. Finally the measurer, standing on the brim of the well, should place under his feet a piece of wood of any size he likes which he should move slowly backwards and forwards until he can see the bottom of the well on the other side in line with the top of the piece of wood. Then the portion of the wood which overhangs the brim of the well should be compared with the height of the measurer, and the same ratio will be found between the diameter and the depth of the well plus the height of the measurer. For example. Let the depth of the well be CF, the diameter CA, the height of the measurer CD, the piece of wood which is compared with that height, CE, the other side of the well AB. Let DC be three times CE; DF will be three times CA.

I have excerpted for you a few examples from many, the easiest from the difficult, so that Geometry's precepts might seem both believable to you and no less intelligible to an attentive mind. If anyone is instructed in these and her other precepts, he will provide for himself a way through earth up to heaven, using his intelligence as his vehicle, and he will understand the size of, and the differences between, the higher bodies. But this is enough about this maiden. About her who remains, and whose stature is higher than the rest, I shall talk briefly.

ASTRONOMIA

Hec igitur quam vides splendore lucidam totumque corpus quodammodo oculeam, dextra radium, leva vero astrolabium gestantem, quicquid mobilis celestisque quantitatis infra applanon[117] continetur intelligentibus explicat. Hec enim sua disciplina comprehensam mundi formam, numerum quantitatemque circulorum, distantias orbium, cursus planetarum, situs signorum describit; paralellos colurosque depingit, zodiacum in duodecim partes rata ratione dispertit, stellarum magnitudinem, polorum oppositionem, axium extensionem non ignorat. Hanc si quis sibi privatam facere posset, non modo presentem rerum inferiorum statum, verum etiam preteritum vel futurum non diffiteretur. Superiora quippe illa divinaque animalia inferiorum naturarum et principium et cause sunt. Et de hac quidem multa non minus a supradictis desiderabilia proponerem, nisi si et brevi sermone non essent[118] comprehensibilia, expertique artis haut INTELLIGIBILIA.[119]

Tu vero, iuvenis, enitere pro viribus ut incepisti,[120] harumque quas diffinivi disciplinas amplectere! His namque imbutus et iuventutem rationis frenis honeste moderaberis, et senectutis onus[121] (ita enim quidam vocant) iocunde consolaberis. Quid enim pulcrius quam in iuventute, que vitiorum facilis imitatrix est, habere quo ipsam a vitiis temperes, virtuti applices; in senectute vero, que quodammodo naturaliter virtutis susceptrix est, quo alios – ac iuniores – a devio tramite divertas, vie reddas? Ita etenim et etas et mores concordi virtute honestata pulcrius elucescunt. Verumtamen quoniam hee ipse quas laudo non omnes apud eosdem facile reperiuntur, opere precium erit diversarum gentium doctores adire, quodque in singulis elegantius reperies, memorie mandare. Quod enim Gallica studia nesciunt, Transalpina[122] resera-

[117] applanō P
[118] essent *Willner*, sunt P
[119] INTELLIGIBILIA *capitals in* P
[120] incepisti *Willner*, intercepisti P
[121] honus P
[122] Transalapina P

This maiden whom you see shining with splendour, with her whole body in some way covered with eyes,[57] and wielding a rod in her right hand, an astrolabe in her left, explains to those who are intelligent whatever mobile and celestial quantity is contained beneath the *aplanos*.[58] For this maiden sketches the shape of the world, as contained in her teaching, the number and size of the circles, the distances of the orbs, the courses of the planets, the positions of the signs of the zodiac; she paints in the parallels and colures, she divides the zodiac into twelve parts with thoughtful reason, she is aware of the size of the stars, the opposite position of the two poles, the axis stretching between them. If anyone could make her his own,[59] he would be confident in declaring not only the present condition of lower things, but also their past or future conditions. For, those higher and divine animate beings are the principle and causes of the lower natures. I would bring forward many other points about this maiden which are no less desirable than those I have already mentioned, were it not for the fact that they cannot be contained in a short discussion, and are hardly intelligible to someone inexperienced in the art. Astronomy

But you, young man, strive with all your strength as you have begun! Embrace the teachings of these maidens whom I have defined! For, imbued with these teachings, you will honourably keep in check your youth with the reins of reason, and pleasantly console old age's burden (for this is what some call it). What is more beautiful than to have in youth, which is a ready imitator of vices, the means to restrain it from vices, to apply it to virtue? And in old age, which in a way is naturally susceptible to virtue, the means by which to divert others – and younger men – from a deviant path, to return them to the right way? For in this way both age and character shine out more beautifully, made honourable by a virtue which harmonizes with them. But since these very maidens whom I praise are not all easily found together amongst the same people, it will be worthwhile to approach teachers of different peoples, and to commit to memory what you may find is most finely expressed among each of them. For what French studies are ignorant of, those across the Alps

bunt; quod apud Latinos non addisces, Grecia facunda docebit. Ut enim anima, corporum dispensatrix, non in omnibus humanis partibus omnes exercere potest potentias, verum aliud in capite, aliud in corde, aliud item in epate operatur, ipsius etiam capitis partes diversas diversis officiis dedicavit – in prora[123] enim imaginatur, in medio ratione utitur, in puppi (id est occipitio) memoriam abscondit; cordis item sinistram procreandis arteriis, dextram vero venis suscipiendis accommodavit; in epate etiam ipso per choleram,[124] que calida[125] et sicca est, appetit, per melancoliam, que frigida et sicca est, continet, per sanguinem, qui calidus et humidus est, digerit, per flegma, quod frigidum et humidum est, expellit – sic eadem dispensatrix, cum non omnis regio omnis probitatis susceptibilium virorum fecunda sit, aliam in alia[126] disciplinam constituit, prout susceptibile corpus aliud alio eius divinitati convenientius est, ut, quod in una mundi parte efficere non valet, in uno mundo efficiat.'

His dictis, cum suis simul passibus moderatis discessit. Ego me rerum auditarum quodammodo stupidus domum contuli, exindeque aliquanto[127] studiosius singulis artibus animum applicavi, unaque lectione perlecta, aliam desideravi[128] maiori affectu, veluti si reliqua deessent, nichil michi lecta conferret, id ex hoc proposito expectans, ut iuventutem moderari, senectutem consolari possim. Et ego certe cum a Salerno veniens in Grecia Maiore quendam philosophum Grecum qui pre ceteris artem medicine naturasque rerum disserebat, sententiis pretemptarem, causam scilicet querens qua vi et natura magnetes ad se ferrum trahat, eiusque super hac re ceterisque similibus solutione audita, ipse michi vicissim proponeret, dicens: 'Si', inquit, 'per medium terre globum foraminis quam ampli hiatus pateret, rupesque non parvi ponderis vi magna iniceretur, quorsum, queso, evaderet, cum nec in celum pondera ferri possint, nec sit in precipiti illo a quo pondus sustineatur?' Mea item super hoc ratione audita,

[123] prora *scripsimus*, porta P
[124] choleram *Willner*, copulam P
[125] calida *Willner*, capula P
[126] alia *Willner*, alio P
[127] aliquanto *Willner*, aliquando P
[128] desideravi *Willner*, desiderare P

will unlock; what you will not learn amongst the Latins, eloquent Greece will teach you. For as the soul, the dispenser of bodies, cannot exercise all its powers in all the parts of the human body, but does one thing in the head, another in the heart, and yet another in the liver, and has also dedicated the different parts of the head itself to different duties – for in the prow it imagines, in the middle it uses reason and in the stern (that is the back part), it stores away the memory;[60] similarly she suited the left part of the heart to creating arteries, but the right to receiving the veins, and also in the liver itself, through yellow bile, which is hot and dry, she has appetite, through black bile, which is cold and dry, she retains, through blood, which is hot and moist, she digests, and through phlegm, which is cold and moist, she expels – so the same dispenser, since not every region is fertile with men who are capable of every kind of excellence, established different disciplines in different peoples, just as one receiving body is more suited to the soul's divinity than another, so that, what she is not able to effect in one part of the world, she might bring about within the compass of the one world.'[61]

Having said this, she left with her maidens, with measured steps. But I, quite stunned by the things I had heard, took myself home and thereupon I applied my mind somewhat more zealously to each of the arts. When I had thoroughly read one lesson, I desired the next with a greater passion, as if the one I had read would bring no benefit if what remained was lacking, hoping from this regime that I could keep in check my youth and console my old age. Well, this was the case for me, when, coming from Salerno, I put to the test with my opinions a certain Greek philosopher in Magna Graecia who, more than anything else, could talk about the art of medicine and the natures of things (*res*). I asked him by what force and nature do magnets attract iron to themselves, and when I heard his solution concerning this matter and other similar questions, he in turn put a question to me, saying: 'If there should open a very wide hole through the middle of the earth's globe and a rock of great weight were thrown in with great force, where, I ask you, would it end up, since neither can heavy things be carried in the direction of the sky, nor is there anything in the course of its fall by which its weight could be stayed?'[62] When, in turn, he heard my solution to

id in fine disputationis nostre eum asserentem audivi, lectiones scilicet disciplinares earumque labores penitentiam non consequi. Quod cum michi minus intelligenti exponeret, inquit, eum[129] enim qui se in secretum, rebus relictis, transfert, purgatoriis virtutibus roboratus, cum de his que inter mundanas illecebras gesserit peniteat, id certe quod didicerit numquam odisse – immo quod de tali fonte profundius non hauserit, id frequenter culpare.

Hactenus, carissime nepos, tibi causam itineris mei per diversarum regionum doctores flexi satagens explicavi, ut et me iniuste accusationis tue onere[130] alleviarem, et tibi eorundem studiorum affectum applicarem, ut, cum ceteri gazas suas multipliciter explicaverint, nos scientiam proponamus. Vale, et utrum recte disputaverim tecum diiudica.

[129] eum *Willner*, cum P
[130] honere P

this problem, I heard him asserting at the end of our discussion that one never regretted reading the liberal arts and working hard at them. In explaining this to me who was not quite understanding what he meant, he said: 'For he who, leaving his affairs, takes himself into a secret place, strengthened by cleansing virtues, while he repents of the things he has done enticed by worldly matters, certainly never hates what he has learnt. Rather, he frequently blames himself for not having drunk more deeply from that fount.'

Now, dearest nephew, I have sufficiently explained to you the cause of my winding journey to teachers in different regions, so that I might both lift from myself the burden of your unjust accusation, and urge the passion for the same studies on you, so that when the others display their riches in many ways, we may simply set forth knowledge. Good-bye, and judge for yourself whether I have disputed rightly.

Notes

1 The classical Latin word 'attrectationes' would seem more plausible than the manuscript-reading 'attrectiones' which is found as a variant for 'attritiones' (= 'bruises') in a Salernitan medical text of *ca.* 1250 (see *MlatWb*, s.v. 'attritio' I A 2).

2 The 'Prince of Philosophers' is Plato, and the reference is to the description of ingredients of the World Soul in Plato's *Timaeus* 35A ('Same, Different and Substance'). Later on, both Plato and Aristotle are called the 'Princes among Philosophers'; see pp. 11 and 19.

3 As Courcelle points out ('Adélard de Bath et la *Consolation* de Boèce', p. 572, n. 2), this term recalls the richly woven texture of Philosophy's dress described at the beginning of Boethius' *Consolation of Philosophy*.

4 The five serving-girls – riches, power, honour, fame, and pleasure – correspond exactly to the constituents of 'human happiness' listed by Boethius, *Consolation of Philosophy*, III, pr. 2: 'Habes igitur ante oculos propositam fere formam felicitatis humanae: opes, honores, potentiam, gloriam, voluptates.' ('You have before your eyes the epitome of human happiness: wealth, honours, power, fame, pleasures.') The scribe of MS P is inconsistent in his headings: 'fama' is written in the text, 'potentia', 'dignitas' and 'voluptas' are written in the margin, and 'divitie' is omitted altogether.

5 The phrase is that of Calcidius' translation of the *Timaeus*, 18B: 'auri argentique et supellectilis ceterae possessionem'.

6 This proverb is not listed in Walther.

7 Perhaps to avoid such a criticism William of Conches, in the *Dragmaticon*, I.1 (ed. Ronca, p. 9), calls himself 'philosophus sine nomine' ('a philosopher without a name'). 'False philosophers' are also referred to by John of Salisbury in the prologue to his *Metalogicon*: 'Potueram quidem scolarium et eorum qui philosophiae nomina profitentur utcumque in silentio cavere morsus' ('I could have completely avoided the bites of scholars and those who claim the name of philosophy, by being silent').

8 Virgil, *Georgics*, 2.490.

9 This could be Adelard's conclusion from comparing Cicero, *De inventione*, I.xxxvii.67 (the five parts of the syllogism) with Boethius, *De topicis differentiis*, bk. 2 or his *De syllogismo categorico*; see E. Stump, *Boethius's De topicis differentiis*, Ithaca and London, 1978, pp. 110–1. However, in William of Champeaux' commentary on the *De inventione* a different contrast between Boethius and Cicero is mentioned, but one which is

directly concerned with rhetoric: 'Nota enim quod in materia oratoris diversi sunt Boethius et Tullius. Dicit enim Boethius in IV libro Top. quodlibet negotium de quo intendit agere orator ipsius oratoris esse materiam, quae postea sub demonstrativo formatum vel sub deliberativo vel iudiciali cadat . . . Tullius vero nullam rem iudicat debere oratoris materiam apellari quamdiu consideratur non informatum aliqua illarum proprietatum, quae per demonstrativum et deliberativum et iudiciale significantur.' ('Note that concerning the material of the orator Boethius and Cicero disagree. For Boethius in the fourth book of the *Topics* says that whatever case the orator aims to discuss is his material, which afterwards falls into shape under a demonstrative or a deliberative or a judicial case . . . Tullius, however, judges that no matter should be called the material of the orator while it is considered in a form unshaped by one of those properties which are signified by the words "demonstrative", "deliberative" and "judicial"'); see Fredborg, 'The Commentaries', p. 7. Adelard could have been confusing the two instances of discrepancy.

10 The word for 'decaying' – 'carientem' from 'caries' – is a *hapax legomenon* in classical Latin, being found only in the β family of the manuscripts of Martianus Capella's *Marriage of Mercury and Philology*, 1, 10 (the α family gives 'cariantem'); see *TLL*, s.v.

11 Boethius, *Consolation of Philosophy*, III, pr. 2: 'Quae quidem sola considerans Epicurus consequenter sibi summum bonum voluptatem esse constituit.' ('Considering only these things, Epicurus consequently established pleasure as the highest good.') The whole of Adelard's passage here is replete with echoes from the *Consolation of Philosophy*, as has been shown with great sensitivity by Courcelle in his 'Adélard de Bath et la *Consolation* de Boèce'.

12 Adelard recalls the words of Boethius, *Consolation of Philosophy*, II, pr. 3: 'tum tantum cum audiuntur, oblectant' ('only then, when they are heard, do they delight').

13 The terms are those of Macrobius, *Commentary on the Dream of Scipio*, I.12.9: 'opinionis ortus est memorie defectus' ('the failure of memory is the birth of opinion').

14 The 'natural force' is the same as the 'soul'.

15 Adelard is recalling Philosophy's reference to the beguiling Muses at the beginning of Boethius' *Consolation of Philosophy*, I, pr. 1: 'Quis . . . has scenicas meretriculas ad hunc aegrum permisit accedere?' ('Who has allowed these little cabaret hussies to approach this patient?')

16 Fixing wings to the mind is an image used by Boethius in *Consolation of Philosophy*, IV, pr. 1 (Philosophy speaking): 'Pennas etiam tuae menti quibus se in altum tollere possit adfigam' ('I shall fix wings to your mind by which it can raise itself aloft').

17 This is perhaps a reference to Aristotle's authority in rhetoric (see below), especially considering the use of the word 'artificiosa' (= 'based on rules') in Cicero's definition of rhetoric, *De inventione*, I.v.6: 'artificiosa eloquentia

quam rhetoricam vocant'; cf. also Quintilian, *Institutio Oratoria*, II.17.42.

18 Adelard had written a whole treatise on how to use the abacus (*Regule abaci*). The 'digiti' are the single digits, 1 to 9, the 'articuli' are the numbers indicating a change in decimal place, such as 10, 100, 1000 etc. The words literally mean the 'fingers' (*digiti*) and 'joints' (*articuli*) of the fingers, which were used in finger-reckoning.

19 I.e., differentiation by number, as distinct from differentiation by species or genus.

20 This 'other disputation' has the same subject-matter as the continuation of the *QN*, promised at the end of the text, but not realised, or no longer extant; see p. xix above and p. 234, n. 95 below.

21 I. e., so that all apparently contradictory arguments can be made to seem to be compatible.

22 The phrase 'agamus pingui Minerva' is proverbial and occurs in this form in Cicero, *Laelius*, 19. A 'pinguis Minerva' would be an 'unsubtle' or 'rude intellect'. Adelard uses similar phrases ('vulgaris Minerva' and 'planior Minerva') in *QN*, Q37 and Q57, in each case as an excuse to introduce an example (as here). In the *commentum*-version of Euclid's *Elements* a distinction is made between 'pinguis Minerva' and 'philosophi': see *Robert of Chester's (?) Redaction*, p. 277: 'Res namque est subtilissima nec pingui Minerve satis perspicua solosque contemp<la>tores sui admittit, id est philosophos.' ('For the topic is most subtle and is not readily obvious to the rude intellect, and admits only its contemplators, that is, philosophers.')

23 Plato, *Timaeus*, trans. Calcidius, 44A.

24 The tone of this poem is similar to that of the poems in Boethius' *Consolation of Philosophy*, I, m. 2 and III, m. 10.

25 A recollection of Boethius' *Consolation of Philosophy*.

26 Plato, *Timaeus*, trans. Calcidius, 34C.

27 Plato, *Timaeus*, trans. Calcidius, 42A. Adelard's text includes two mistakes which are common to a large number of manuscripts of Calcidius' translation: 'voluntate' ('will') for 'voluptate' ('desire'), and 'promoventes' ('advancing') for 'permoventes' ('moving through'). These mistakes are, therefore, authorial, and should not be corrected.

28 'Insita', a noun from the verb 'inserere' (here 'to graft') is used for 'the graft' in *QN*, Q6, where Adelard upholds the different natures of the graft and the trunk into which it is grafted.

29 This argument is developed at greater length in Boethius, *Consolation of Philosophy*, II, pr. 5.

30 This sentence is a paraphrase of Cicero, *De inventione*, I.iv.5, just as Adelard's next phrase recalls *De inventione*, I.i.2.

31 A consistent translation of 'vox' is difficult to find. The sense is 'an audible sound made by an animate being, which has meaning'; compare Adelard's use of 'vox' in *QN*, Q13 and Q21.

32 For the 'material imposit/imposition' see Boethius, *In Categorias Aristotelis*, in *Patrologia Latina*, 64, col. 159, and Alison Drew's discussion in 'The *De*

eodem et diverso', p. 19; compare also John of Salisbury's discussion in *Metalogicon*, 1, 15 ('secunda impositio').

33 The term 'civilis disceptatio', like most of the terminology in the section on rhetoric, comes from Cicero's *De inventione*: see I.v.7.

34 In rhetorical terms, 'tropes and figures'.

35 This sentence summarises Cicero's paragraph on the civilization of man in *De inventione*, I.i.2.

36 In Cicero, Aristotle is the authority for the three genera of rhetoric: *De inventione*, I.v.7. The extra detail of the locations of the three genera is also found in Anselm of Laon, who criticizes this opinion on the grounds that all three genera could be held anywhere; Anselm's views are quoted in William of Champeaux' commentary on *De inventione*, see Fredborg, 'The Commentaries', p. 31.

37 Cicero attributes the five parts of the issue to 'most authorities': *De inventione*, I.vii.9.

38 Cf. Cicero, *De inventione*, I.v.6: 'eloquentia quam rhetoricam vocant'.

39 Adelard is referring to the subject-matter of Cicero's speeches *Pro Marcello*, *Pro Ligario* and *Pro rege Deiotaro*, the 'Caesarian speeches' which travelled together in the manuscripts and were the best-known of Cicero's forensic works in the Middle Ages.

40 The full hexameter occurs both in Cicero, *De officiis*, I.22.77, and in Quintilian, *Institutio Oratoria*, XI.1.24 (both reading 'concedat laurea linguae'); in Cicero, *In Pisonem* , 72–4 (referred to by Willner) only the first three words are quoted.

41 Adelard's authority is not Aristotle, but again Cicero; see *De inventione*, I.xv.20, where exactly the same terms are given. The mention of cases in *De inventione* occurs within the discussion of the appropriate exordium to a speech (I.xv.20). Different kinds of case require different kinds of introductions: hence Adelard's 'neque similiter in omnibus exoriendum sit'. In particular, the case called 'admirabilis' requires an indirect kind of approach which Cicero calls 'insinuatio'. 'Admirabilis' is Cicero's equivalent of the Greek παράδοξος and has a very specific meaning in rhetoric. Hubbell, the translator of *De inventione* in the Loeb series, explains it as meaning 'marvellous, apparently from the point of view of the juror, who thinks it strange that anyone should speak on behalf of such a defendant. It is therefore a "difficult" case to present'. 'Difficult' is the translation Hubbell uses, though 'discreditable' is generally preferred by historians of rhetoric.

42 The *Rhetorica ad Herennium*, I.iii.5, aside from only giving four kinds of cases, gives 'turpis' as equivalent to the 'admirabilis' kind. In the manuscript of *DED* we read 'in turpi admirabilis generis specie' which at first sight makes it appear that 'turpis' has been placed next to 'admirabilis' because Adelard or a scribe realized its equivalence. However, this does not work from the point of view of the syntax and must rather be explained by what follows in the *De inventione* (I.xvii.23): 'insinuatio' is used in the 'admirabilis' kind, but this case arises from three causes: (1) a certain

'turpitudo' ('scandal') in the case; (2) a prejudice introduced by those who have spoken before the orator; (3) the auditors' weariness. Adelard's 'turpi admirabilis generis specie' could therefore be simply the kind of 'difficult' case arising from the first of these causes. On the other hand one could emend 'turpi' into 'triplici' and include all three causes. The good will, good reception and attentiveness ('benivolentia, docilitas, attentio') of the auditors could apply respectively to the three causes. However, in *De inventione* these three aims are described separately from (and before) the discussion of the three causes of the 'difficult' case, along with 'odium', 'invidia' and 'contemptio' (I.xvi.22–3).

43 Cicero, *De inventione*, I.viii.10.

44 A serpent is in the right hand of Dialectica in the sculpture on the Portail Royal of Chartres Cathedral. The literary precedent (with a reversal of hands) is Martianus Capella, *The Marriage of Mercury and Philology*, IV.328: 'in laeva . . . serpens, in dextra formulae . . .' ('in the left hand . . . a serpent, in the right hand, rules').

45 Boethius, *Introductio ad syllogismos categoricos*, PL, 64, cols 761–94 (see cols 778–9). 'Univocation' is not mentioned here, but is included in Boethius' list of types of fallacy in his commentary on Aristotle's *De interpretatione*, editio secunda, 2, PL, 64, col. 460.

46 The means by which Aristotle's ten *praedicamenta* include genera, species and individuals is explained in Boethius' commentary on Porphyry's *Isagoge*, ed. S. Brandt and G. Schepss in *Corpus scriptorum ecclesiasticorum latinorum*, 48, Vienna and Leipzig, 1906, pp. 42–4, 180–1 and 211–6.

47 Adelard knew this definition from Nemesius, *Premnon phisicon*, where, however, it is attributed to Pythagoras, whom Xenocrates 'imitated and explained' (Nemesius, *Premnon physicon*, c. 2, pp. 40–1). Nemesius, apparently following Xenocrates, goes on to explain that the soul is not the same as number, using some arguments which are similar to those of Adelard: ' . . . non quod et numerus sit anima, sed quod in numerabilibus sit et in multiplicatis, et quia anima est quae discernit res, formas et figuras singulis apponendo . . . Amplius anima per se mobilis est; numerus vero descriptus est immobilis.' (' . . . not because number too is soul, but because it is in countable and multiple things, and because the soul is what discerns things, by applying forms and shapes to individuals . . . Also, the soul is mobile by itself, but the number referred to is immobile.')

48 The following sentences are a summary of the definitions in Boethius, *De arithmetica*, I.3–19 and 28. 'Officio stili' ('in writing') is a phrase used by Boethius, *Consolation of Philosophy*, I, pr. 1.

49 Boethius, *De arithmetica*, I.28.

50 Ibid., I.13 (with some paraphrasing).

51 Ibid., I.17.

52 Adelard's immediate source is Boethius, *De musica*, in which Pythagoras is described as having discovered the mathematics of music from comparing the sounds produced by anvils of different weights (I.10), as having

experimented with different lengths of strings and bores of pipes (I.11), and as stating that the tone consisted of two semitones and a comma (III.6) and that semitones are either major or minor (I.16).

53 Compare Walther, no. 15597 (= Peter Abelard, *Carmen ad Astralabium*, line 986): 'Quo pretiosior est res, ea rarior est.'

54 Adelard takes it for granted that the reader will understand the previous sentences to apply to Egypt.

55 Cf. *Geometria incerti auctoris*, ed. Bubnov, *Gerberti opera mathematica*, p. 324, line 1: 'a sapiente inventum', with 'sapiente' glossed as 'Pitagora'.

56 The examples Adelard gives all come from the *Geometria incerti auctoris*, Bk. II, x, ix, xiv, xi and xv, ed. Bubnov, pp. 323–7; Adelard uses other theorems from the same work in his *De opere astrolapsus*.

57 Martianus Capella, *The Marriage of Mercury and Philology*, VIII.811, where Astronomia is described as 'totis artubus decenter oculea' ('with all its limbs fittingly covered with eyes').

58 This is Adelard's term here and in the *QN* – taken from Macrobius, *Commentary on the Dream of Scipio* – for the sphere of the fixed stars; it is a corruption of the Greek 'aplanēs' (= 'unwandering').

59 Adelard uses 'privatum facere' as a technical term in hawking to mean 'to man' (i.e., to make the bird become used to being handled); see *DAT*, c. 6.

60 Cf. Nemesius, *Premnon phisicon*, c. 13, pp. 89–90. Adelard's terminology – 'prora', 'medium' and 'puppis' – is not found in Nemesius, but also occurs in Constantine the African, *Pantegni*, *Theorica* VI.11 and William of Conches, *Dragmaticon*, VI.18, ed. Ronca, p. 240.

61 Adelard's metaphor of the different functions of different parts of the human body is a secular equivalent to that of St Paul in Romans 12.4–5.

62 This question is solved in *QN*, Q49.

Questiones naturales

RICARDO BAIOCENSI EPISCOPO ADELARDUS BATHONIENSIS[1] SALUTEM.

Cum in Angliam nuper redierim,[2] Henrico Willelmi Anglis imperante, quoniam a patria causa studii diu me exceperam, occursus amicorum et iocundus michi fuit et commodus. Cum itaque in conventu nostro primo, ut fit, multum de nostra et amicorum salute quesitum esset, id quidem animo meo consequenter innotuit, ut gentis nostre[3] mores agnoscerem. Id igitur querens, violentos principes, vinolentos presules, mercennarios iudices, patronos inconstantes, privatos adulatores, mendaces promissores, invidiosos amicos, ambitiosos fere omnes[4] cum acceperim, nichil inquam michi inaccessius esse quam huic miserie meam intentionem subdere. Tum illi 'Quid autem,' inquiunt, 'agendum esse censes, cum hanc pravitatem moralem nec velis agere nec possis avertere?' Hic ego 'Oblivioni,' inquam, 'tradere. Unica enim malorum irrefragabilium medicina est oblivio. Qui enim quod odit retractat, quodammodo patitur quod non amat.' His itaque verbis hinc indeque habitis, cum adhuc diei non parva parte[5] restante dicendi aliquid[6] spatium non deesset, inter ceteros qui adventaverant nepos quidam meus in rerum causis magis implicans quam explicans, aliquid Arabicorum studiorum novum me proponere exhortatus est. Cui cum assentirent[7] ceteri, ego tractatum subscriptum recepi, quem quidem auditoribus suis utilem fore scio, iocundum nescio. Habet enim hec generatio ingenitum vitium,[8] ut nichil quod a modernis reperiatur putet esse recipiendum. Unde fit ut si quando inventum proprium publicare voluerim, persone id aliene imponens inquam: 'Quidam dixit, non ego'. Itaque ne omnino non audiar, omnes meas sententias dominus quidam invenit, non ego. Set hec hactenus. Nunc vero, quoniam me rogatu amicorum aliquid dicere convenit, utrum id recte dictum sit, tuo examine velim esse securior. Nichil enim in artibus liberalibus tam bene tractatur quod per te non possit luculentius efflorere. Ades igitur animo. Ego enim succincte agens in primis capitula preponam, deinde super rerum causis nepoti respondebo.

[1] BATHONICENSIS L (*for MSS headings see Introduction pp. xliii–xliv*)
[2] redirem L (M *omits prologue*)
[3] nostre LQV, mee G
[4] fere omnes LQV, G *transp.*
[5] non parva parte G *a.c.*LV, parte non parva G *p.c.*, Q *om.*
[6] aliquis L
[7] assentarent L
[8] ingenitum vitium G *a.c.*LQV, G *p.c. transp.*

ADELARD OF BATH GREETS RICHARD, BISHOP OF BAYEUX.

When I recently returned to England during the reign over the English of Henry,[1] son of William, since I had absented myself from my native land for a long time for the sake of study, meeting friends was both delightful and helpful for me. So when on our first meeting, as happens, the questions came thick and fast concerning our health and the health of friends, the fact impressed itself on my mind that I should get to know the character of our own people. Pursuing this plan, when I found the princes barbarous, the bishops bibulous, judges bribable, patrons unreliable, clients sycophants, promisers liars, friends envious, and almost everybody full of ambition, I said that nothing was further from my course than paying attention to this wretched situation. Then they said: 'So what do you think should be done, when you neither wish to join, nor are able to ward off, this moral depravity?' At this point I said: 'One should forget it. The only medicine of ills that cannot be rebuffed is forgetfulness. For whoever ponders what he hates, in some way suffers what he does not love.' When these words had been exchanged, since a considerable part of the day remained, so that there was time for saying something, among the others who were paying their calls was a certain nephew of mine, who, in investigating the causes of things, was tying them in knots rather than unravelling them. He urged me to put forward some new item of the studies of the Arabs. As the others agreed with him, I undertook the following treatise, which I know will be useful to its auditors, but whether it is pleasant, I do not know. For the present generation suffers from this ingrained fault, that it thinks that nothing should be accepted which is discovered by the 'moderns'. Hence it happens that, whenever I wish to publish my own discovery, I attribute it to another person saying: 'Someone else said it, not I!' Thus, lest I have no audience at all, some teacher came up with all my opinions, not I. So much for that! But now, since it is right for me to say something at the request of friends, I should like to be more confident, following your scrutiny, as to whether it is said correctly. For nothing is so well dealt with in the liberal arts that it cannot bloom more splendidly through your action. So, please lend me your attention. For I shall be succinct and start by giving the chapter-headings. Then I shall reply to my nephew's questions concerning the causes of things.

EXPLICIT PROLOGUS. INCIPIUNT CAPITULA[9]

.i. Qua ratione herbe sine preiacenti nascantur semine.
.ii. Quomodo alique herbe dicantur calide, cum omnes magis sint terrestres quam ignee.
.iii. Quomodo in eodem territorio herbe[10] nascantur contrarie.
.iv. Quare sicut a terra, non ita ab aqua vel aere vel etiam nascantur ab igne.
.v. Utrum ut in terra terree, in aqua aquatice, ita et in aere et igne, aerie et ignee nascantur nature.
.vi. Ut quid fructus insitam sequatur, non truncum.
.vii. Ut quid brutorum quedam ruminent, quedam vero minime.
.viii. Quare omnia[11] que ruminant a posteriori prius accumbunt,[12] et a priori posterius.
.ix. Quare eadem non ab anteriori prius surgunt.
.x. Quare non[13] omnia que potant mingunt.
.xi. Ut quid quedam animalia stomachum habeant, alia vero minime.
.xii. Ut quid quedam eorum nocte perspicatius videant.
.xiii. Utrum bruta animas habeant.
.xiv. Utrum opinio[14] in animato fundetur corpore.
.xv. Quare homines innata non habeant cornua vel arma alia.[15]
.xvi. Qua cautela nervorum et venarum contextio sit deprehensa.
.xvii. Quare bonum ingenium habentes memoria careant[16] et econverso.
.xviii. Qua ratione loca phantasie, rationis, et memorie[17] in cerebro deprehensa sint.
.xix. Ut quid nasus super os locatus sit.
.xx. Ut quid parte anteriore calvescant homines.
[Cur homines canescant in senectute][18]
.xxi. Qua natura voces prolatas audiamus.
.xxii. Qua natura dum vox ad aures venit quodlibet penetret[19] obstaculum.
.xxiii. Que de visu sententia tenenda sit.
.xxiv. Utrum visibilis spiritus substantia sit an accidens.

9 EXPLICIT PROLOGUS. INCIPIUNT CAPITULA G, EXPLICIT PROLOGUS L, QV *om.*
10 herbe L, GMV *om.* (Q *omits the capitula*)
11 plura L
12 accumbunt LMV, accumbant G
13 non *post* potant GM
14 opinio *post* corpore L
15 arma alia LV, GM *transp.*
16 careant GMV, sint carentes L
17 LV *add.* et
18 [Cur . . . senectute] *in* L *only*
19 penetret L, penetrat GMV

The prologue ends. The chapter headings begin.

1. The reason why plants grow without a seed being sown beforehand.
2. In what way some plants are to be called hot, when they are all more earthy than fiery.
3. How plants of contrary natures grow in the same spot.
4. Why, as they grow from earth, they do not grow in the same way from water, air or even fire.
5. Whether, as earthy natures grow in earth, and watery natures in water, so also, in air and fire, airy and fiery natures grow.
6. Why the fruit follows the graft, not the trunk.
7. Why some brute animals chew the cud, but others not.
8. Why all those that chew the cud lie down with their hindquarters first, and their forequarters afterwards.
9. Why the same animals do not stand up first from their forequarters.
10. Why not all those which drink urinate.
11. Why some animals have a stomach, others not.
12. Why some of them see more clearly by night.
13. Whether brute animals have souls.
14. Whether opinion is founded in an animate body.
15. Why men do not have innate horns or other armour.
16. By what observation the web of nerves and blood-vessels is detected.
17. Why men who have a fine intelligence are lacking in memory, and *vice versa*.
18. By what reasoning the seats of imagination, reason and memory in the brain were discovered.
19. Why the nose is placed above the mouth.
20. Why men go bald from the front part of their head.
    [Why men go grey in old age.][2]
21. By what nature voices are carried through the air and heard by us.
22. By what nature a voice, when it comes to the ears, can penetrate any obstacle.
23. What opinion should be held about sight.
24. Whether the visual spirit is a substance or an accident.

.xxv. Quomodo idem tam brevi spatio usque ad stellam eat et redeat.
.xxvi. Quomodo dum oculus clauditur visibilis spiritus non extra relinquitur.
.xxvii. Quomodo idem spiritus se ipsum non[20] impedit, si dum it redit.
.xxviii. Qua ratione anima ab illo spiritu formas accipiat.
.xxix. Quare spiritus ille sicut a speculo, ita ab albissimo non repercutitur vitro,[21] ut vultum reformet.
.xxx. Quare, ut ab obscuro in lucem videtur, non similiter a luce in obscurum videatur.
.xxxi. Qua natura odoremus, gustemus, et tangamus.
.xxxii. Qua natura gaudium sit causa fletus.
.xxxiii. Qua natura ex eodem ore nunc calidum nunc frigidum spiremus aerem.
.xxxiv. Ut quid flabellatio frigus,[22] si motus calorem generat.
.xxxv. Ut quid digiti inequales facti sint.
.xxxvi. Ut quid cava sit palma.
.xxxvii. Quare homines ut nati sunt[23] non ambulent, cum[24] hoc bruta faciunt.[25]
.xxxviii. Quare homines tam tenera habeant membra.
Quare erecti incedant.
.xxxix. Quare lacte magis nutriantur homines.
.xl. Quare eisdem iuvenibus vel senibus lac incongruum est.
.xli. Quare, si post leprosi coitum mulierem cognoscas, morbo afficieris, illa vero minime.
.xlii. Ut quid mulieres, si viris frigidiores, libidinis sunt petulantiores.
.xliii. Ut quid universaliter moriantur homines.
.xliv. Quare, semel equaliter commodeque humano constituto corpore, ab illa equalitate destituatur.
.xlv. Quare ciborum fiat vel aliqua vel tam frequens refectio.
.xlvi. Quare defunctorum corpora vivi pertimescamus.
.xlvii. Ut quid vivus, si in amnem ceciderit,[26] fundum tenet, cum post paucos dies idem mortuus natet.
.xlviii. Ut quid vel qua natura terre globus in medio sustineatur aere.
.xlix. Si perforatus foret terre globus, lapidi[27] iniecto quorsum foret casus.

20 se ipsum non LV, non se ipsum G, se ipsum M
21 repercutitur vitro GV, M *transp.*, repercutitur L
22 frigus M, GLV *om.*
23 L *om.* sunt
24 cum GV, cur L, sicut M
25 faciant G
26 deciderit L
27 lapidi L, lapide GMV

25. How that spirit goes to a star and returns in such a brief space of time.
26. How, when the eye is closed, the visual spirit is not left outside.
27. How the same spirit does not get in the way of itself if it returns at the same time as it goes out.
28. By what means the soul receives forms from that spirit.
29. Why that spirit is not reflected from the most transparent glass in the same way that it is reflected from a mirror, to shape again the face.
30. Why, when one can see from the dark into the light, one cannot similarly see from light into dark.
31. By what nature we smell, taste and touch.
32. By what nature joy is the cause of weeping.
33. By what nature we breathe out from the same mouth now hot air, now cold.[3]
34. Why fanning generates coldness, if movement generates heat.

35 (36).[4] Why the fingers are created uneven.

36 (37). Why the palm is hollow.

37 (38). Why men do not walk as soon as they are born, when brute animals do this.

38 (–). Why men have such soft limbs.
Why they walk upright.[5]

39. Why human beings are nourished rather by milk.
40. Why milk is not appropriate to the same human beings when they are youths or old.
41. Why, if you have intercourse with a woman after she has coupled with a man suffering from leprosy, you are affected with the disease, not the woman.
42. Why, if women are colder then men, they are more lustful.
43. Why all men die.
44. Why, when the human body has once been established in a balanced and appropriate way, it is deprived of this balance.
45. What the reason is for restoring the body with food, and why it is done so frequently.
46. Why we fear the bodies of the dead, when we are alive.
47. Why a live body, if it falls into a river, sinks to the bottom, while the same body, after a few days, now dead, 'swims'.
48. Why, or by what nature, the globe of the earth is held up in the middle of the air.
49. Where, if the globe of the earth were bored through, a rock thrown into the hole would end up.

.l. Qua de causa terre motus fiat.
.li. Quare marina aqua[28] sit salsa.
.lii. Unde accessus et recessus marini fiant.
.liii. Quomodo ex incursu fluminum non crescat Occeanus.
.liv. Quare flumina quedam salsa non sint.[29]
.lv. Quomodo fluminum cursus possit esse perpetuus.
.lvi. Qua de causa aque in excelso nascantur monte.
.lvii. Utrum sint aliqui veri fontes.
.lviii. Quare ex vase pleno inferius aperto aqua non exeat nisi superius foramen aperiatur.
.lix. Unde oriantur venti.
.lx. Unde primus procedat aerum motus an, si quicquid movetur ab alio movetur, nullus sit primus.
.lxi. Utrum uno athomo moto moveantur universi,[30] quoniam quicquid movetur movet aliud, sicque etiam motus semel inceptus sit infinitus.
.lxii. Quare circa terram non ad superiora movetur ventus.
.lxiii. Quare vel unde ei tantus accidat[31] impetus.
.lxiv. Unde procedant tonitrua.
.lxv. Unde procedant fulmina.
.lxvi. Quare non ex omnibus tonitruis elidantur fulmina.
.lxvii. Qua vi fulmen et lapideas et eneas penetret moles.
.lxviii. Quare nec simul nec semper cum videmus ignem talem audimus[32] fragorem.
.lxix. Ut quid Luna lumine curtata sit.
.lxx. Unde ei perpetua quedam[33] umbra incubuerit.
.lxxi. Qua de causa planete et Sol precipue per medium applanon sine anfractu cursus non teneant.
.lxxii. Qua necessitate vel ratione stelle contra applanon currant.
.lxxiii. Utrum cadant stelle cum videntur[34] cadere.
.lxxiv. Utrum animate sint stelle.
.lxxv. Quo cibo utantur stelle, si animalia sunt.
.lxxvi. Utrum applanon inanimatum corpus animalve an deum esse dicendum sit.

28 marina aqua LV, GM *transp.*
29 .liv. Quare flumina quedam salsa non sint L *om.*
30 uno athomo moto moveantur universi *scripsimus*, uno atomo moveantur universi G, una athomo moto (*sic*) moveantur universe L, una athomo mota moveantur universe M, nullo athomo moto moveantur universi V
31 accidit G
32 audimus LMV, audiamus G
33 perpetua quedam G *a.c.*LMV, quedam perpetua G *p.c.*
34 videntur LV, videantur G, existimantur M

50. From what cause an earthquake occurs.
51. Why sea water is salty.
52. Why the flows and ebbs of the tide occur.
53. How the Ocean is not increased by the flowing in of rivers.
54. Why some rivers are not salty.
55. How the flow of rivers can be perpetual.
56. From what cause waters arise on the tops of mountains.
57. Whether there are there any true springs.
58. Why water does not flow out from a full vessel which is open at the bottom, unless a higher opening is uncovered.
59. Where winds arise from.
60. From where the first movement of the air proceeds, or whether, if everything that is moved is moved by another, there is no first movement.
61. Whether, when one atom is moved, all atoms move, since whatever is moved moves another, and thus movement, once it has begun, is infinite.
62. Why wind moves around the earth and not to the higher realms.
63. Why and from where it has such force.
64. Where thunder comes from.
65. Where lightning comes from.
66. Why lightning is not struck from every clap of thunder.
67. By what force lightning penetrates stone and bronze structures.
68. Why it is that when we see the flashes, we do not hear such crashes at the same time, and sometimes not at all.
69. Why the Moon is deprived of light.
70. Why perpetual shadow enshrouds it.
71. From what cause the planets – and the Sun above all – do not keep to a course through the middle of the *aplanos* without winding about.
72. By what necessity or reason the planets run in the opposite direction to the *aplanos*.
73. Whether the stars fall when they seem to fall.
74. Whether the stars are animate.
75. What food the stars eat, if they are animate.
76. Whether one should say that the *aplanos* is an inanimate body, or an animate one, or a god.

EXPLICIUNT CAPITULA. INCIPIT LIBER.[35] SIC FACIUNT CAUSE RERUM.[36]

ADELARDUS: Meministi, nepos,[37] septennio iam transacto, cum te in Gallicis studiis pene puerum iuxta Laudisdunum una cum ceteris auditoribus meis dimiserim, id inter nos convenisse, ut Arabum studia ego pro posse meo scrutarer, tu vero Gallicarum sententiarum inconstantiam non minus adquireres.

NEPOS: Memini, eoque[38] magis quod tu discedens philosophie attentum futurum me fidei promissione astringeres. Id ego semper angebar, quare huic studio forem attentior. Quod utrum recte expleverim, re ipsa probari potest, hac precipue opportunitate, quia, cum Sarracenorum sententias te sepe exponentem auditor tantum notaverim, earumque non pauce satis futiles michi videantur, patientiam meam paulisper abrumpam, teque eas edisserente, ego sicubi michi videbitur obviabo. Quippe et illos impudice extollis, et nostros detractionis modo inscitia invidiose arguis. Erit itaque opere pretium te laboris tui assecutum esse fructum, si bene solveris, me item, si probabiliter opposuero, mea promissione non esse fraudatum.

ADELARDUS: Confidentius id quam valeas forsan presumis. Verumtamen quia cum[39] tibi, tum multis aliis hec disputatio non erit inutilis, protervitati tue morem geram, hoc tantum vitato incommodo, ne quis me ignota proferentem ex mea id sententia facere, verum Arabicorum studiorum sensa putet proponere. Nolo enim et si que dixero minus provectis displiceant, ego ipse etiam eis displicere. Novi enim quis casus veri professores apud vulgus sequatur. Quare causam Arabum, non meam, agam.

NEPOS: Sit sane, ne occasionem tacendi habeas.

ADELARDUS: A levioribus igitur incipiendum censeo; quod si de his non irrationabiliter egero, de maioribus eandem spem habeto. Itaque exordium ab imis, finem in summis habeamus.

[35] EXPLICIUNT CAPITULA. INCIPIT LIBER G, LMV *om.*, Incipit tractatus eiusdem ad nepotem Q

[36] SIC FACIUNT CAUSE RERUM GV, HIC FACIUNT CAUSE RERUM L, MQ *om.*

[37] Q, GLMV *add.* quod

[38] eoque LM, eo quoque GQV

[39] cum LQV, tum GM

The chapter headings end. The book begins. This is how the causes of things work.[6]

Adelard: You remember, dear nephew, that, seven years ago, when I dismissed you (still almost a boy) with my other students in French studies at Laon, we agreed amongst ourselves that I would investigate the studies of the Arabs according to my ability, but you would become no less proficient in the insecurity of French opinions.

Nephew: I remember, and all the more so because when you left me you bound me with a promise on my word that I would apply myself to philosophy. I was always anxious to know why I should be more attentive to this subject. This is an excellent opportunity to test whether I have been successful by putting my study into practice, because since, as a listener only, I took note of you when often you explained the opinions of the Saracens, and quite a few of them appeared to me to be quite useless, I shall for a brief while refuse to be patient and shall take you up as you expound these opinions, wherever it seems right to do so. For you both extol the Arabs shamelessly and invidiously accuse our people of ignorance in a disparaging way. It will therefore be worthwhile for you to reap the fruit of your labour, if you acquit yourself well, and likewise for me not to have been cheated in my promise, if I oppose you with probable arguments.

Adelard: Perhaps you are being more bold in your presumption than you are capable. But because this disputation will be useful both to you and to many others, I shall put up with your impudence, as long as this inconvenience is avoided: that no one should think that when I am putting forward unknown ideas, I am doing this out of my own head, but that I am giving the views of the studies of the Arabs. For I do not want it to happen that, even though what I say may displease those who are less advanced, I myself should also displease them. For I know what those who profess the truth suffer at the hands of the vulgar crowd. Therefore, I shall defend the cause of the Arabs, not my own.

Nephew: Agreed, so that you may have no occasion for silence.

Adelard: Well then, I think we should begin from easier subjects. For if I speak sensibly about these, you may have the same hope concerning greater things. So let us start from the lowest objects and end with the highest.

.i. NEPOS: Quoniam tibi diffidis, de minoribus agendo ingenium acuere vis. Ab herbis igitur que terre radicitus herent, radicem disputationis sumam. Quero itaque qua ratione herbe a terra nascantur. Que enim causa est vel edoceri potest, cum[40] primum plana et immobilis sit terre superficies; quid, inquam, est quod inde movetur, surgit, crescit, et ramos extendit? Licet enim si libet ut pulverem aridum colligas, subtiliterque cribratum in testeo vel eneo vase reponas, deinde accessione temporum cum herbas inde surgere videas, cui id nisi mirabilis divine voluntatis mirabili effectui imponas?[41]

ADELARDUS: Voluntas quidem Creatoris est ut a terra herbe nascantur. Set eadem sine ratione non est. Que ut aperte pateat, herbas a terra nasci concedo, set non a pura, immo commixta, ea certe commixtura ut in singulis suis particulis, que quidem sensui subiacent, omnia quatuor elementa cum eorum qualitatibus contineat. Ita etenim illa quatuor simplicia hoc mundi corpus unum componunt ut, cum in singulis compositis illa componentia existant, numquam tamen ita ut sunt sensibus appareant. Verum ipsum compositum sui simplicis nomine falsa assignatione vocamus. Nullus enim umquam terram vel aquam tetigit; nullus aerem vel ignem vidit. Composita enim hec que sensu percipimus non ipsa set ex ipsis sunt. Quare nec terra, nec aqua, nec aer, nec ignis, ut Philosophus ait, set terreum, aquaticum, aerium, igneum dicenda sunt. Ex eo tamen quod in singulis magis habundat, appellationem hanc sortita sunt. Itaque, cum in terreo isto licet subtiliter pulverizato, quatuor cause sint, necessario inde quiddam compositum surgit, terreum maxime, aquaticum parum, aerium minus, igneum minime, per terram coherens, per aquam diffundens, per aerem et ignem surgens. Nisi enim in eodem ignis esset, motum ad superiora non haberet. Nisi item aqua vel aer inessent, in latum diffundi non posset. Postremo nisi terra id prestaret, coherentiam non haberet. Quare illa que mira subtilitate in pulvere tuo inclusa erant, in publicum produnt. Ut autem id liquido intelligas, in exterioribus elementis suum simile irritantibus et extrahentibus suisque qualitatibus idem exigentibus causam huius proces-

40 cur L
41 effectui imponas GV, L *transp.*, effectui id imponas MQ

1. Nephew: Since you are not confident in yourself, you wish to sharpen your intellect by dealing with lesser things. I shall, then, take the root of the disputation from plants which are fixed in the ground by their roots. I wish to know, then, by what reason plants are born from the earth. For what is the cause or what cause can be expounded, when first the surface of the ground is flat and unmoved? What is there, I say, that stirs in it, rises up, grows and stretches out its branches? For you can if you like collect dry dust and sieve it finely into an earthenware or bronze container, and then, given the necessary length of time, when you see plants rise up from it, to what are you to attribute this unless to the wondrous effect of the wondrous divine will?

Adelard: It is indeed the will of the Creator that plants should be born from the earth. But that will is not without reason. To make this clear, I agree that plants are born from earth, but not from pure earth; rather, from mixed earth – in the kind of mixture that contains in each of its parts (those at least which lie open to the senses) all four elements with their qualities. For in such a way do those four simple elements compose this one body of the world that, although they exist as components in each composed object, they never appear to the senses as they are; but we assign to the composed thing the erroneous label of the name of one of its simple elements. No one has ever touched 'earth' or 'water', no one has seen 'air' or 'fire'. These composite things that we perceive with the senses are not the elements themselves, but are *from* the elements themselves. Therefore, as the Philosopher says,[7] they should not be called earth, water, air or fire, but the earthy, the watery, the airy and the fiery. Nevertheless, they have acquired this name from the particular element which is more prevalent in a particular composition. Thus, since in that earthy matter, even if it has been crumbled into fine dust, these four causes are present, from them a certain composed thing necessarily arises which is largely earthy, a little watery, less airy, and least of all fiery. Because of earth it sticks together; because of water it spreads out; because of air and fire it rises up. For unless fire were in it, it would not have movement upwards. Similarly, unless water or air were in it, it would not be possible for it to spread out sideways. Finally, unless earth provided it, it would not have coherence. Therefore, those elements which were enclosed in your dust in a marvellous fineness are revealed. But, that you may understand this clearly, I place the cause of this process in the outer elements stirring and drawing out what is similar to them, and in their own qualities which thrust out

sionis pono. Unde ista inferiora perpetua dissolutione ad sua similia recedunt. Et quoniam prima questio ratione soluta est, si quid de ceteris ambigis, propone.

.ii. NEPOS. Illi quidem qui nec nichil philosophie attigit, nec in eadem ad unguem nutritus est, soluta videretur; michi vero minime. Neque enim tu ita laqueos meos vitabis. Esto etenim ut ea causa exigente nascantur. Quero consequenter, cum omnes huiusmodi radices magis terree sint quam ignee, id est magis terra quam igne participant, ideo enim et locum suum et nativitatem iuxta terram occupant, cur quedam earum[42] calide, quedam frigide reperiantur nature. Sicut enim maior earum pars terra est, ita et magis in eis terre proprietatem habundare consequens est. Omnes itaque magis frigide dicende sunt quam calide. Si igitur, ut superius disseruisti, ex eo quod in compositis his maxime habundat, et proprietatem et nomen sortita sunt, omnia que a terra et circa eam nutriuntur, terrea magis et frigida esse dicendum est.

ADELARDUS: Id quidem negare nequeo, quin herbarum natura magis terrestris quam ignea sit; magis enim terra quam igne participat. Quod autem infers 'Si magis terrea, tunc magis frigida est', hic hereo. Magis enim et minus in rerum qualitatibus duobus se habent modis. Uno, secundum quantitatem, alio, secundum efficatiam. Est etenim hoc pallium meum totum viride, et hec[43] anuli mei smaragdus viridis. Est autem pallii viriditas ad quantitatem maior, smaragdi vero ad efficatiam. Unde si quantitatem respicis, dici potest istud illa[44] viridius; si vero ad efficatiam, illam[45] isto. Magis itaque et minus in qualitatibus secundum quantitatem quidem materiam sequitur, secundum efficatiam vero minime. Sint ergo omnes herbe quoniam magis terra participant, magis quidem quantitative frigide, earumdem tamen quedam magis efficaciter calide.

.iii. NEPOS: Quoniam id determinando exactum est, quero deinde cum ipse inter se radices contrarie sint – est enim una calida, alia frigida, hec sicca, illa humida – quomodo in eodem territorio vivant. Cum enim omnia huiusmodi a terra nutrimentum exigant, unde et ipsa a Grecis .ΓH. a Latinis mater corporum dicitur, constetque unumquodque suo simili nutriri, a suo vero contrario destrui, edissere,

42 earum LQV, eorum GM
43 hec LMQ, hic V, G *om.*
44 illa L, illo GMQV
45 illam L, illum G, istud M, illud QV

the same. The result is that those inferior elements[8] return to their similar elements by perpetual dissolution. And since the first question has been solved by reason, if you have any doubt about the rest, put it forward.

2. Nephew: To him who has a little experience of philosophy but has not been nourished in it completely, the question might seem to be solved; but not at all to me. You shall not avoid my snares in this way. For, granted that the plants are born because that cause thrusts them forth, I then want to know, since all roots of this kind are more earthy than fiery – that is, participate more in earth than in fire, for this is the reason why they take both their position in, and their birth from, the earth – why some of them are found to be of a hot, and some to be of a cold, nature. For just as the greater part of them is earth, so it also follows that the property of earth is more abundant in them. Thus they all should be called cold rather than hot. If, therefore, as you have argued before, these composed things take both their property and name from that which predominates in them, one should say that all things which are nourished from earth and in its vicinity should be rather called earthy and cold.

Adelard: Certainly, I cannot deny that the nature of plants is more earthy than fiery; for it participates more in earth than in fire. But when you infer 'If it is more earthy, then it is colder' here I stick. For 'more' and 'less' exist in two ways in respect to the qualities of things: one according to quantity, the other according to efficacy. For this cloak of mine is completely green, and this emerald in my ring is green. However, the greenness of the cloak is greater as far as quantity is concerned, but that of the emerald is greater in efficacy. Hence if you consider the quantity, it can be said that the cloak is greener, but if you consider the efficacy, the ring is greener. Thus 'more' and 'less' in qualities, when the quantity is considered, follow the matter, but when the efficacy is considered, by no means do so. Therefore, although all plants, since they share more in earth, are cold in regard more to quantity, nevertheless certain of them are hot with regard more to efficacy.

3. Nephew: Since that has been settled by defining terms, I then ask: given that the roots themselves are contrary to each other – for one is hot, another cold, this is dry, that is moist – how do they live in the same spot? For, since all things of this kind demand nourishment from the earth – hence it too is called 'Gē' by the Greeks and 'mother of bodies' by the Latins – and since it is established that each thing is nourished by what is similar to it, but destroyed by its contrary,

si potes, qua ratione herbe contrarie eodem in loco vivere possint, aut eas omnes eiusdem esse nature concede, quod item impossibile est.

ADELARDUS: Ut acumen ingenii tui experiar, eo diverticulo me protego: herbas a terra nullum habere nutrimentum.

NEPOS: Quod a terra nutriantur, id michi asserere pronum est. Quando enim ab ea avelluntur, tum virorem, tum incrementum, ipsumque etiam vivere amittunt. Unde palam fit ut aliquid ab ipsa matre acceperint, quo cum privantur simul et moriuntur. Item si nichil a terra traherent, frustra ab agricolis terre exercende tanta adhiberetur cura. Et quoniam diverticulum tuum opertum est, aliud quod firmius sit, si quid tamen firmum habes, profer.

ADELARDUS: Esto: a terra quidem nutriuntur, non simplici certe, set composita, immo non a terra, set a terreo, ut superius disputatum est, in quo cum quatuor illa elementa[46] sint, habent singule[47] nutrimentum quod sibi trahant, ut que calida est, calidum, que frigida est,[48] frigidum, sicca, siccum, humida, humidum affectet. Quod ex hoc tibi liquere potest. Nam cum aer exterior quicquid a terra traxerant[49] estivali[50] desiccatione eis auferat, omne quod extra terre superficiem est marcendo moritur. Quod vero intra sinum terre matris quodammodo contiguo lactantis radicitus latet, vitale reservatur. Et quoniam prolixitas fastidii mater est, ad cetera transeamus.

.iv. NEPOS: At non ita sine inconvenienti transibis. Tute ipse catenam qua stringaris nexuisti. Si enim, ut dicis, in singulis compositis corporeis illa quatuor simplicia elementa sunt, unde et aliis compositis nutrimentum prestare possunt, restat ut et aer quem videmus illud idem eis nutrimentum prebere valeat; quare a terra eradicate ab eo nutrimentum accipiant. Quod quia aer eis id quodammodo desiderantibus addere non potest, tota ratio tua debilitata est, et universorum effectus ad Deum magis referendus est.

ADELARDUS: Deo non detraho. Quicquid enim est, ab ipso et per ipsum est. Id ipsum tamen confuse et absque discretione non est. Que quantum scientia humana procedit audienda est; in quo vero univer-

[46] illa elementa LQV, elementa illa G, elementa M
[47] singule G *a.c.*LQV, singulare G *p.c.*M
[48] est LQV, GM *om.*
[49] traxerant LQ, traxerat G, trexerit (*sic*) M
[50] estivali MQ, estuali GLV

explain if you can how contrary plants can live in the same spot; or admit that they are all of the same nature – which again is impossible.

Adelard: To test the sharpness of your wit, I protect myself by this escape route: that plants have no nourishment from the earth.

Nephew: That they are nourished by the earth is easy for me to affirm. For when they are pulled from the earth they lose not only their greenness and power to grow, but even life itself. Hence it is obvious that they have taken something from their mother herself, and when they are deprived of it, they immediately die. Similarly, if they took no nourishment from the earth, in vain would farmers take such great care in working the ground. And since I have blocked your escape route, devise another one which is firmer, if you have any firm one.

Adelard: Very well: they are indeed nourished by the earth, but surely not by the simple earth, but the composite. Or rather: not by the earth, but by the earthy, as has been discussed before. Since those four elements are in it, each plant has the nourishment which it draws for itself, so that what is hot desires hot nourishment, what is cold, cold, what is dry, dry, and what is moist, moist. This is how it can become clear to you. In the drying heat of summer, when the outer air takes away whatever plants had drawn from the earth, everything which is above the surface of the earth dies by wilting, but what lies hidden with its roots within the bosom of the earth – a mother in some way giving milk by her contact – is kept alive. And since prolixity is the mother of boredom, let us pass to other topics.

4. Nephew: But you will not pass the problem by in this way without inconvenience. You yourself have linked the chain to bind yourself with. For if, as you say, in individual corporeal composite things those four simple elements are present and hence are able to provide nourishment for other composite things too, it remains that even the air which we see is able to provide that same nourishment for them; therefore, uprooted plants should receive nourishment from it. But because air is not able to supply it when they in a sense long for it, your whole line of reasoning is weakened and the execution of all things should rather be referred to God.

Adelard: I am not slighting God's role. For whatever exists is from him and through him. Nevertheless, that dependence <on God> is not <to be taken> in blanket fashion, without distinction. One should attend to this distinction, as far as human knowledge can go; but in the case where human knowledge completely fails, the matter should

saliter deficit, ad Deum res referenda est. Nos itaque quoniam nondum inscitia pallemus, ad rationem redeamus. Aer equidem, ut et terra, quatuor illa in se seminaria continet, nec tamen herbis eradicatis nutrimentum prebere potest, vel debet. Quod sic collige: omne sensibile – universum dico quod sentitur – licet ex illis causalibus principiis coniunctum sit, non tamen in illis compositis singula illa componentia equaliter assunt. In hoc enim corpore terra habundat, in illo aqua exuberat, hoc aere magis participat, illud igne maxime flagrat. Secundum hoc igitur quod unumquodque causis illis magis participat – secundum hoc, inquam, et illorum singulorum proprietates sequitur, et sensibus demonstrat, et illis que[51] in eo precipua sunt precipue[52] nutritur. Herbe itaque, si nescis, terrea natura magis participant, quare et a terra magis pascuntur. Quod si eradicentur, in aere quidem nutrimentum inveniunt, set aliter quam oportet. De eo enim quo magis egebant, minus ibi reperiunt, et de minori indigentia maius. Unde et eas dissolvi necesse est, hac videlicet necessitate, quia cum pars terrea in aere sufficientem refectionem non inveniat, a coniunctione tali solvi velit, ut ad suum simile, terram dico, redeat. Ea igitur remota, cetera componentia a pondere libera ad sua redeunt propria. Quam dissolutionem vulgus quidem, quod semper veris rerum vocabulis caret, mortem vocat, cum non mors, set mutatio dicenda sit. Unde Philosophus, de mundo loquens, ait:

> Neque quicquam ex eo recedit, nec est accedendi facultas, cunctis in se cohercitis, set corruptela partium senescentium intra se vicem quandam optinet cibatus.

Et meo certe iudicio in hoc sensili[53] mundo nichil omnino moritur, nec minor est hodie quam cum creatus est. Si qua enim pars ab una coniunctione solvitur, non perit, set ad aliam societatem transit.

Nunc vero ad rem redeamus, et quod predicta sic se habeant, ita assero. Quantum aqua a terra distat, tanto ipsa radicibus nutrimentum prestat, minus dico quam terra, maius tamen quam aer. Similiter aer minus eis quam aqua, maius quam ignis prebet. Et quia in his plus iusto stetimus, patet enim hoc et lippis et tonsoribus, ad altiora transeamus.

51 qui L
52 precipue L *om.*
53 sensili LQ, sensibili GMV

be referred to God. Thus, since we do not yet grow pale with lack of knowledge, let us return to reason. Certainly, air, like earth too, contains those four seed-beds in it, but nevertheless cannot and should not provide nourishment for uprooted plants. Understand it in this way: although every sensible object (I mean everything which is sensed) is put together from those causal principles, nevertheless those single components are not present in equal quantities in those composite things. For in this body earth predominates, in that, water is overflowing, this one has a greater part of air, that blazes up most with fire. According to the fact, therefore, that each body participates more in these causes, according to this, I say, it follows the properties of each of them, represents them to the senses, and is nourished especially by those which are dominant in it. Thus plants, of course, share more in earthy nature, and therefore are nourished more by earth. But if they are uprooted, they do indeed find nourishment in the air, but otherwise than is appropriate. For there they find less of what they need more, and more of what they need less. Hence it is also necessary that they are dissolved, the necessary cause of their dissolution being that, as the earthy part does not find sufficient food in the air, it wishes to be detached from such a joining, so that it may return to its like – namely, the earth. Therefore, when the earthy part is taken away, the other components, free from its weight, return to their proper places. The common people, who always lack the true names for things, call this dissolution death, although it should not be called death, but change. Hence the Philosopher, speaking of the world, says:

> Nothing at all departs from it, nor is there the capacity for anything to enter it, since everything is enclosed within it; but the corruption of parts as they grow old brings about a certain interchange of foods within it.[9]

And, in my judgement certainly, nothing at all dies in this sensible world,[10] nor is it smaller today than when it was created. For if any part is released from one conjunction, it does not perish but passes over to another association.

But now let us return to the matter in hand. Thus I assert that what I have described is just so. As much as water differs from earth, by so much does it provide less nourishment to the roots than earth, but more than air. Similarly air supplies for them less than water, but more than fire. And because we have spent more time than is right on these matters – for they are clear even to the bleary-eyed and barbers[11] – let us go on to more elevated topics.

.v. NEPOS: Revera hoc nulli nisi lippo vel ceco patet. Itane igitur illi Sarraceni subtilibus te nugis illaqueabant? Me vero numquam hodie falles, quin fucosas obscurasque rationes tuas subtili falsitate tectas aperiam.[54] Deprehendi enim quid agas: proponis quod nec michi nec tibi intelligibile sit, ut sic fastiditus falsa tibi concedam. Nescis cui loquaris;[55] sic autem experiere. Cogeris enim propria positione concedere, ut et in aqua quedam nature constent magis aquatice quam terree vel aerie vel ignee, et in aere quedam magis aerie quam terree vel aquatice vel ignee, et in igne quedam[56] magis ignee quam terree vel aquatice vel aerie. Hoc enim ad superiora sequitur. Quod quia non procedit, nec tu procedere debes.

ADELARDUS: Sane hoc quidem argumentum conicis; est namque ita ut conclusisti. Nam et in aqua nature nascuntur aquatice – id est pisces, qui nichil aliud sunt quam aqua concreta, et in aere aerie potestates; in igne etiam – id est superiori ethere – ignea animalia que videmus vivunt. Itaque et ratio procedit, et nos simul sentimus.

NEPOS: Licet parum michi sedeat, tamen ne iram tibi excitem de his taceo.

ADELARDUS: Dictum sit igitur et qua ratione radices nascantur,[57] et quomodo calide vocentur, et quomodo vivant, et cur magis in terra quam alibi durent, et postremo quare moriantur.

.vi. NEPOS: Nunc ergo ad arborum naturas veniamus. Quero itaque cum insita trunco inseritur, qua de causa totus fructus insite naturam sequatur. Cum enim truncus et insita diverse nature sint, constetque, ut dixi, crescentia ista nutrimentum a terra habere, aut idem nutrimentum ab ea trahent, aut dissimile et contrarium. Set de simili[58] vivere non possunt, cum diverse et contrarie sint nature.[59] Itaque diversum exigit insita quam truncus. Set illud aut per medium truncum trahit, aut aliunde. Aliunde autem non trahit. Itaque cum per medium truncum trahat, aut ipse truncus idem trahit aut non. Si non idem trahit, interruptus est tractus, nec potest ad insitam pervenire quod affectat. Si vero ipse truncus idem trahit, tunc suum contrarium trahit. Quod si est, suam dissolutionem et trahit et appetit. Quod nec esse potest, nec supradictis rationibus competens est.

54 aperiam G vl M, operiam GLV, comperiam Q
55 loquaris G *p.c.*, loqueris G *a.c.*(?)LMQV
56 quedam G *a.c.*LMV, G *p.c.*Q *om.*
57 nascantur LMQ, -untur GV
58 de simili LV, ex dissimili G *a.c.*M, de dissimili Q, ex simili G *p.c.*
59 sint nature LMQ, GV *transp.*

5. Nephew: The truth is that this is clear to nobody except to the bleary-eyed or the blind! Is this how those Saracens were tying you up in a net of subtle nonsenses? Today, however, you will never deceive me, but I shall reveal your sham and obscure arguments covered with subtle falsity. For I have found out what you are up to. You propose what is not intelligible either to me or to yourself, so that I, being wearied by this, may agree to your false words. You do not know what kind of man you are speaking to, but you will realize this. For you will be forced to agree by your own argument that in water too there are certain natures which are more watery than earthy, airy or fiery, in air certain things more airy than earthy, watery or fiery, and in fire certain things more fiery than earthy, watery or airy. For this follows from what has been said before. But, because the argument does not advance, neither should you advance.

Adelard: Certainly, you put forward a sensible argument; for it is just as you have concluded. In water are born watery natures, that is fish, which are nothing other than congealed water, and in air, airy powers; in fire too – that is in the upper ether – live the fiery animals which we see.[12] Therefore, both my reasoning advances and we think the same.

Nephew: Although the argument is hardly settled for me, I keep quiet about it, lest I make you angry.

Adelard: Let that, then, complete the discussion about the reason roots grow, how they are called hot, how they live, why they survive more in earth than elsewhere, and, finally, why they die.

6. Nephew: Now then let us turn to the nature of trees.[13] My question is: when a graft is inserted into a trunk, why does all the fruit follow the nature of the graft? For since the trunk and the graft are of a different nature and it is an established truth, as I have said, that such growing things take their nourishment from the earth, either they will draw the same nourishment from it, or a different and contrasting one. But they are not able to live from similar nourishment, since they are of a different and contrary nature. Thus the graft needs a different nourishment from the trunk. But it draws that either through the trunk as an intermediary, or from elsewhere. Yet it does not draw it from elsewhere. Therefore, since it draws it through the trunk as an intermediary, the trunk itself either draws the same nourishment or not. If it does not draw the same nourishment, the draught is interrupted, and what the graft desires is not able to reach it. But if the trunk draws the same nourishment, then it draws its own contrary. If this is the case, it draws and desires its own dissolution,

ADELARDUS: Quia sophistice agis, determinanda est importunitas. Et insita quidem a terra nutrimentum trahit, et cum aliunde non, per truncum id facit. Concedo. Inde procedis ut et[60] ipse truncus trahat. Esto. Exin descendis dicens: 'Itaque suum contrarium trahit'. Set et hoc libens accipio. Postremo infers: 'Suam igitur dissolutionem sponte appetit'. Hoc vero non descensus set precipitatio est. Quod brevi absolvam. Trahit truncus et sibi et alii. Quod sibi trahit, retinet; quod alii, eidem transfert. Verbi gratia: stomachus quidem quicquid corporis partibus necessarium est, et[61] appetit et trahit. Itaque quod sibi est, retinet; quod aliis, eisdem prebet. Sic demum et truncum in propria dico remanere[62] natura et insitam. Unde et si post triennium vel aliud spacium insitam abscidas, truncus eiusdem invenietur nature cuius et prius fuerat.

NEPOS: De istis, que puerilia sunt, verisimilia magis quam necessaria dixisti. Quare ad ipsam animalium naturam ascendamus. Ibi enim, ut animus michi presagit, scrupulum tibi iniciam.[63]

ADELARDUS: De animalibus difficilis est mea tecum dissertio. Ego enim aliud a magistris Arabicis ratione duce didici; tu vero aliud, auctoritatis pictura captus, capistrum sequeris. Quid enim aliud auctoritas dicenda est quam capistrum? Ut bruta quippe animalia capistro quolibet ducuntur, nec quo aut quare ducantur discernunt, restemque qua[64] tenentur solum sequuntur, sic non paucos vestrum bestiali credulitate captos ligatosque auctoritas scriptorum in periculum ducit. Unde et quidam sibi nomen[65] auctoritatis usurpantes nimia scribendi licentia usi sunt, adeo ut pro veris falsa bestialibus viris insinuare non dubitaverint.[66] Cur enim cartas non impleas, cur et a tergo non scribas, cum tales fere huius temporis auditores habeas, qui nullam[67] iudicii rationem exigant, tituli nomine tantum vetusti confidant? Non enim intelligunt ideo rationem singulis datam esse, ut inter verum et falsum ea prima[68] iudice discernatur. Nisi enim ratio iudex universalis esse deberet, frustra singulis data esset. Sufficeret enim preceptorum scriptori datam[69] esse, uni dico vel pluribus; ceteri

60 L *om.*
61 et LV, GMQ *om.*
62 in propria dico remanere LV, dico in propria remanere G, dico in propria remanere (*supra* vel remeare) Q, dico remanere in propria M
63 iniciam V, innectam GLMQ
64 qua Q, quo GLMV
65 sibi nomen LV *p.c.*, nomen sibi GMQV *a.c.*
66 dubitaverint Q, -erunt GLMV
67 G *add.* sibi
68 prima GMQ, primum LV
69 datam *scripsimus (following Be)*, data GLMQV

which is impossible and not in accordance with the earlier reasoning.

Adelard: Because you argue with sophisms, the fallacy of your reasoning should be checked. The graft indeed draws nourishment from the earth and, since it is not able to take it from elsewhere, it does this through the trunk. I agree. Then you proceed by saying that the trunk itself draws the nourishment. Very well. From there you go downhill when you say: 'Thus it draws its contrary.' But even this I am willing to accept. Finally you infer: 'Therefore it freely seeks its own dissolution.' This is not going downhill, but falling headlong! I shall briefly resolve the question. The trunk draws both for itself and for another. What it draws for itself it keeps; what it draws for another, it passes on to it. For example: the stomach seeks and draws whatever is necessary for the parts of the body. Thus it keeps what is for itself; what is for others, it provides for them. Thus, finally, I say that both the trunk and the graft remain in their proper natures. Hence even if you cut the graft after three years or another interval, the trunk will be found to be of the same nature as it was before.

Nephew: In answer to these questions, which are appropriate to children, you have said what is likely rather than what is necessary. Therefore, let us ascend to the nature of animals. For there, my heart tells me, I shall throw a stumbling-block in your path.

Adelard: About animals my conversation with you is difficult. For I have learnt one thing from my Arab masters, with reason as guide, but you another: you follow a halter, being enthralled by the picture of authority.[14] For what else can authority be called other than a halter? As brute animals are led wherever one pleases by a halter, but do not know where or why they are led, and only follow the rope by which they are held, so the authority of written words[15] leads not a few of you into danger, since you are enthralled and bound by brutish credulity. Hence too, certain people, usurping the name of 'an authority' for themselves, have used too great a licence to write, to such an extent that they have not hesitated to trick brutish men with false words instead of true. For why should you not fill pages, why not write on the back too,[16] when these days you generally have the kind of listeners that demand no argument based on judgement, but trust only in the name of an ancient authority? For they do not understand that reason has been given to each single individual in order to discern between true and false with reason as the prime judge. For unless it were the duty of reason to be everybody's judge, she would have been given to each person in vain. It would have sufficed for reason to have

eorum institutis et auctoritatibus essent contenti. Amplius: ipsi qui auctores vocantur, non aliunde primam fidem apud minores adepti sunt, nisi quia rationem secuti sunt, quam quicumque nesciunt vel negligunt, merito ceci habendi sunt. Neque tamen[70] id ad vivum reseco, ut auctoritas me iudice spernenda sit. Id autem assero, quod prius ratio inquirenda sit, ea inventa, auctoritas si adiacet demum subdenda. Ipsa vero sola nec fidem philosopho facere potest, nec ad hoc adducenda est. Unde et logici[71] locum ab auctoritate probabilem non necessarium esse consenserunt. Quare si quid amplius a me[72] audire desideras, rationem refer et recipe. Non enim ego ille sum quem pellis pictura pascere possit. Omnis quippe littera meretrix est, nunc ad hos nunc ad illos affectus exposita.

.vii. NEPOS: Sit sane ut postulas, cum michi rationabiliter opponere facile[73] sit, neque Arabum tuorum auctoritates sequi tutum sit. Stet igitur inter me et te ratio sola iudex ut sit. Et quoniam de brutis agendum est, quero quare eorum quedam ruminent, quedam minime.

ADELARDUS: Animalium ut et hominum, diverse nature sunt. Sunt etenim quedam naturaliter calida, alia frigida, quedam humida, alia sicca. Que calida sunt, acceptos cibos melius digerunt, et in sanguinem facilius convertunt. Que vero frigida, peius. Omne enim quod mutatur, calore quam frigore levius convertitur. Habet enim ignis quasi proprium, ut coniuncta dissolvat. Illa itaque animalia que calidum habent stomachum, cibos facile coquunt. Alia vero que frigide nature sunt, dum cibos vertere ex defectu caloris nequeunt, ad dentes eos revocant, ut ibi secundo triti,[74] emolliri facilius queant, ut boves, cervi, capre, et similia id genus que Greca appellatione phisici melancolica vocant. Quod autem hec omnia frigide nature sint, licet phisicis patens sit, tibi tamen sic innui potest. Ob hoc enim et pinguedinem suam duriorem et solidiorem habent, quam vulgus sepum vocat. Alia vero ut calidiora, pinguedinem habent molliorem, utpote magis decoctam, quod communi usu unctum dicitur.

70 tamen LV, enim G *p.c.*, MQ *om.*
71 loyci G
72 Quare si quid amplius a me LQV, Quere siquidem aliud a me amplius G *p.c.*, Quare siquidem amplius a me M
73 facile GMQV, difficile L *p.c.*, ficile L *a.c.*
74 triti Q, tritis GLMV

been given to a legislator – to one, I say, or several legislators: all the others would have been content with their decrees and authoritative statements. Moreover, those who are called 'authorities' did not obtain their initial trust among lesser men, except in that they followed reason, and whoever ignores or neglects reason, should worthily be thought blind. However, I do not state categorically that in my judgement authority should be spurned. Rather I assert that first, reason should be sought, and when it is found, an authority, if one is at hand, should be added later. But authority alone cannot win credibility for a philosopher, nor should it be adduced for this purpose. Hence the logicians too have agreed that an argument from authority is probable, not necessary. Therefore, if you wish to hear anything more from me, give and receive reason. For I am not the kind of man whom the painting of the skin can satisfy. Every letter is a prostitute, open now to these affections, now to those.[17]

7. Nephew: By all means let us do as you demand, since it is easy for me to oppose with reasonable arguments, nor is it safe to follow the authorities of your Arabs. Therefore, let us keep to this rule: between you and me reason alone should be the judge. And since we should talk about brute animals, I ask why some chew the cud, some not at all.

Adelard: The natures of animals, like those of men, vary. For some of them are naturally hot, others cold, some moist, others dry. Those which are hot digest the food they receive better and convert it more easily into blood; but those which are cold are worse at this. For everything which is changed is converted more easily by heat than by cold, since fire has, as a property so to speak, the ability to break up things which are joined. Thus those animals which have a hot stomach easily digest their food. But others which are of a cold nature, since they are unable to convert their food because of a lack of heat, return it to their teeth so that the food, ground down there a second time, can more easily be softened – such animals are oxen, deer, goats, and animals of this kind which the natural scientists, using a Greek expression, call 'melancholic'. But, that all these are of a cold nature, although it is obvious to natural scientists, can be shown to you in this way: because of this they have both a harder and a denser fattiness which the common people call 'tallow', whereas other animals, being hotter, have a fattiness which is softer because it is more cooked; this is called in common usage 'grease'.

NEPOS: Si igitur propter frigidam naturam ruminant, cur item equus et asinus non ruminant, que, ut a phisicis didici, melancolica sunt?

ADELARDUS: Set non adeo: habent enim et stomachum aliquanto calidiorem, et dentes ad molendum efficatiores.

NEPOS: At non sic extorquebis. Si enim ex frigiditate, ut dicis, hoc eis accidit, cur inquam oves, cum calide complexionis sint, idem tamen exigunt?[75]

ADELARDUS: Licet ipse quidem[76] calide nature sint, tamen eundem eis laborem instrumentalis penuria inferre potest. Carent enim dentium efficatia; unde et bidentes appellantur. Preterea, tam modici caloris sunt, ut etiam sepum ad modum melancolicorum habeant.

NEPOS: Quid igitur? Si quia melancolica sunt ideo ruminant, tunc et homines,[77] quia melancolici et frigide nature quidam[78] sunt, ruminare contingeret.

ADELARDUS: Licet ad comparationem hominum homines frigidi sint, tamen quantum ad cetera muta[79] animalia calidos omnes esse iudicamus. Verumtamen ut te omnino a conclusione tua deiciam, vidi quosdam qui tante frigiditatis stomachum habebant, ut cum cibos sumptos digerere non possent, male eos retinebant. Unde et hoc eis accidebat, ut quicquid grossius et indigestibilius sumpserant, haut multum post prandium reicerent.

.viii. NEPOS: De his hactenus satis dictum sit. Nunc vero illud querere destino, quare omnia[80] que ruminant et sepum habent, a posteriori prius accumbant parte et a priori posterius.

ADELARDUS: Discrete quesivisti. Hec enim tria coaccidentia sunt: ruminare, sepum habere, a posteriori accumbere; et ex eadem causa – id est frigiditate – oriuntur. Frigiditas autem non equalis omnibus membris adest: maior enim posterioribus est et a corde remotis; igitur et eisdem gravitatem maiorem inesse necesse est. Iure itaque[81] a graviori prius accumbunt.

.ix. NEPOS: Itaque teneo quod volo. Si enim quia posterius frigidiora sunt, a posteriori prius descendunt, et quia in anteriori calidiora sunt,

75 exigunt LV, non exigant GM, non exigunt Q
76 quidem L *om.*
77 G *p.c.*MQ *add.* quidam
78 quidam LMV, GQ *om.*
79 muta GM, multa LQ, V *om.*
80 omnia GMQ, plurima LV
81 igitur L

Nephew: If, then, they chew the cud because of their cold nature, why do the horse and the ass not likewise chew the cud, since, as I have learnt from the natural scientists, they are melancholic animals?

Adelard: But not so much. For they have both a slightly hotter stomach and teeth which are more efficient at grinding.

Nephew: But you will not wriggle out in this way. For if this happens to them because of coldness, as you say, why, I ask, do sheep, though they have a hot complexion, have to do the same?

Adelard: Although they are of a hot nature, nevertheless the deficiency in the instrument can occasion for them the same chore; for they lack efficiency in their teeth; hence they are called 'bidents'. Moreover, they are of such a modest heat, that even they have tallow like melancholic animals.

Nephew: So, what about this? If they chew the cud because they are melancholy, then men also, because certain of them are of a melancholic and cold nature, should chew the cud.

Adelard: Although, in comparison with other men, certain men are cold, in comparison with the other, mute, animals, we judge them all to be hot. Nevertheless, so that I do not deprive you entirely of your conclusion, I have seen some men who had a stomach of such coldness that, since they could not digest the food they ate, they had difficulty in keeping it down. Hence this also happened to them, that whatever they had taken which was rather coarse and indigestible, they regurgitated quite soon after the meal.

8. Nephew: Enough has now been said about these matters. Next I am resolved to ask this: why do all animals which chew the cud and have tallow lie down with their hindquarters first, and their forequarters afterwards?

Adelard: You have asked a clever question. For these three accidents occur together: chewing the cud, having tallow, and lying down back first. They all arise from the same cause – that is, from coldness. But coldness is not equally present in all limbs. It is more marked in the hindquarters and those far from the heart. So it is also necessary for a greater heaviness to be in those parts. Rightly then they first lie down with the heavier part.

9. Nephew: So I have what I want. For if, because they are colder at the back, they lower themselves first from the hindquarters, then, because they are hotter at the front, they ought also to stand up first

ex anteriori prius surgere deberent. Quod quia consequenter non accidit, et antecedens absurdum est.

ADELARDUS: Quicquid quiescit, ideo id facit ut vires resumat, quas ei labor minuerat. Animalia igitur de quibus agimus per quietem reparatis viribus, dum vires ille recentes et efficatiores sunt, illam aggrediuntur molis sue partem prius levare que ponderosior est. Ut et tu operarius cum fueris recens et viriosus,[82] primum gravia (nisi fallor) adis pondera; deinde, fatigatus cum fueris,[83] leviora.

.X. NEPOS: Ita esse nec necesse nec impossibile est. Nunc vero illud dissolvas volo: cur non omnia que potant mingunt? Cum enim superfluitas potus ad renes descendens confluat et[84] exinde per virgam effluat, mirum est, cum volucres quedam ut columbe multum potent, quare etiam conveniente causa non mingant.

ADELARDUS: Animalia alia in primitiis nativitatis sue molli quadam et virosa materia educantur, lacte dico, cuius maxima pars aquatica est, minor vero terrestris; alia dura quadam materia, lapillis dico et harenis, ceterisque talibus implentur ut volucres. Illa igitur que aquosa degunt materia, et vesicam habent et mingunt. Cetera vero ut superfluam humiditatem non recipiunt, ita nec vesicam habent nec mingunt. Ingruente tamen in eis[85] cibaria siccitate, eadem aliquantulum potare necesse est, ut tantilla saltem humiditate et cibo facilior via[86] paretur, et membro per quod aspera descendunt vel sic parcatur. Neque enim putandum est discretas esse cibi et potus vias ingrediendi stomachum, ut Aristotili imponunt. Simul enim in stomachum veniunt, deinde in epar, siquidem[87] illud habent, descendunt. Cum igitur humor ille non nutrimentum[88] sit corpori, set tantummodo ducatus cibi, ob modicam sui quantitatem singularem non exigit exitum.

.xi. NEPOS: Quoniam in decursu sermonis de stomacho mentionem fecisti, quero quare quibusdam animalibus stomachum habere

[82] viriosus *scripsimus*, virosus GLMQV
[83] cum fueris MQG *i.m.* V, cum eris G, curris L
[84] et LQV, GM *om.*
[85] eius L
[86] facilior via LV, GQ *transp.*, via familior M
[87] siquidem GMQV, si quod G *i.m.*L
[88] non nutrimentum LQV, G *transp.*, M nutrimentum

from their forequarters. But since this does not happen to follow, the first premiss also is without sense.

Adelard: Whatever rests, does so that it might recover the strength which hard work had taken away from it. The animals, therefore, about which we are talking, having recovered their strength through resting, while that strength is fresh and more effective, start to raise that part of their bulk which is heavier first. Just as you also, when you have just started physical work and are still full of strength, tackle heavy weights first – unless I am mistaken – and then, when you are tired, tackle the lighter ones.

10. Nephew: That this should be the case is neither necessary nor impossible. But now I want you to solve this question. Why don't all those animals that drink urinate? For since the excess of their drink flows into one place, descending into the kidneys, and flows out from there through the penis, it is surprising that certain birds like doves, although they drink a lot, do not also urinate, since the cause to do so is there.

Adelard: Some animals in the early days after their birth are brought up on a certain soft and viscous material – I mean milk – whose greater part is watery, and lesser part earthy; others, like birds, are satisfied with a hard material – I mean gravel, sand and other such things. Those, then, which live off a watery material both have a bladder and urinate. But the others, as they do not take in excess moisture, thus neither have a bladder, nor urinate. However, because of the danger to them from the excessive dryness of their food, it is necessary for them to drink a little, so that at least by such a small amount of moisture an easier passage may be prepared for the food, and the bodily part through which rough foods descend may thus be spared. For it should not be thought that the paths of food and drink entering the stomach are separate – an opinion people impute to Aristotle. They enter the stomach together, then they descend into the liver – if indeed they have that organ. Therefore, since that moisture is not the nourishment for the body, but only the lubricant of food, because of its modest quantity it does not require a separate exit.

11. Nephew: Since in the course of our talk you have mentioned the stomach, I ask why it happens that some animals have a stomach,

contingat,[89] quibusdam vero minime.

ADELARDUS: Quoniam rationabiliter procedis, rationem audi et recipe. Illa quidem animalia que lapillis harenisque vivunt, stomachum non habent. At quiddam[90] durius habere necesse est, iecur dico, qui ipso stomacho ad digerendum fortior est, harenarumque asperitatem lapillorumque fracturas ferre potest. Nam et interius et exterius fortissimo nervo munitum est, in medio vero carne solidiuscula aptum. Per interiorem enim substantiam contenta patitur; per carneam certe, que calida est, eadem digerit; per exteriorem vero, si quid interiori claustro rupto exire nititur, ne ad vitalia penetret patrocinatur.

.xii. NEPOS: His ita discussis, adhuc michi de eisdem animalibus brutis nonnulla dubitatio obstat. Quero enim quare quedam illorum nocte quam luce perspicatius videant. Scimus enim lucem visui amicam, tenebras vero contrarias.

ADELARDUS: In oculo, qui visus instrumentum est, plures humores esse necesse est. Est itaque unus albus visui amicus, per quem visibilis spiritus et exit et diffunditur. Est et alius niger, per quem ne nimium diffundatur cavetur; per eum enim quia tenebrosus est, et coalescit et constringitur. Illa igitur animalia que die quam nocte minus vident, albugineum humorem habundanter habent, minus vero nigrum. Unde et die necessario magis exit spiritus, et nimietate diffusionis videndorum discretionem amittit. Nocte vero propter tenebras exteriores in se colligitur, constantiorque exit et melius discernit. Quod et sic asserere licet: vidisti enim quosdam qui plus iusto albos habebant[91] oculos, unde et parum discernebant, quos si per nivis albicantem late superficiem deduxeras, aut parum aut nichil discernere poterant. Set et tu ipse cum aliquid directe videre volueris, supercilia stringis, ut collectus non vagetur visus, set percipienda discretive pertingat. Quod in sequentibus, si querere calles, patebit.

.xiii. NEPOS: His dissertis, postremo utrum bruta animalia animas habeant necne, solvendum suscipe. Id enim nostre quidem etatis hominibus ambiguum est.

[89] contingat Q, continget GM, contigeret L, contigit V
[90] quiddam L, quoddam GMQV
[91] habebant LQV, habeant GM

but others do not.

Adelard: Since you proceed in a rational way, listen to and accept reason. Those animals which live on gravel and sand do not have a stomach, but have to have something harder – I mean the gizzard[18] – which is stronger than the stomach itself for digesting and is able to bear the roughness of the sand and the crushing of the gravel. For both on the inside and the outside it is protected by a very strong muscle, and in the middle it is fitted with rather solid flesh. Because of its interior substance, it can bear its contents; because of the fleshy substance which is hot, it digests them securely; and through its exterior substance it makes sure that, if anything threatens to come out having broken the internal seal, it should not penetrate as far as the vital organs.

12. Nephew: Having discussed these things in this way, I am still worried by a doubt about these same brute animals. My query is why some of them see more clearly by night than by day. For we know that light is friendly to sight, but darkness is opposed to it.

Adelard: In the eye, which is the instrument of sight, there have to be many humours. Thus there is a white one which is friendly to sight; through this the visual spirit both goes out and is diffused. There is also another one which is black; through this it is ensured that the spirit is not diffused too much. For through it, because it is dark, the spirit both coalesces and is restricted. Therefore those animals which see less by day than by night have the albugineous humour in abundance,[19] but less of the black. Hence by day the spirit both necessarily goes out more, and, because of the excess of its diffusion, loses its power of distinguishing between what is to be seen. But by night, because of the external darkness, it is collected into itself, it goes out in a more focused way and it distinguishes more clearly. This can also be confirmed in this way. You have seen some people who had whiter eyes than normal, hence they could distinguish little. Whenever you took them over snowy ground which was white in all directions, they could distinguish either very little or nothing at all. But even you yourself, when you wish to look at something straight in front of you, screw up your eyes so that the visual spirit, collected together, does not wander about, but reaches what is to be seen in a discerning way. This will be clear in what follows,[20] if you ask the right questions.

13. Nephew: Having discussed these things, finally take up solving the question of whether or not brute animals have souls. For this is uncertain, at least to men of our time.

ADELARDUS: Ut vulgus de negatione non dubitat, ita philosophis affirmatio certa est. Habent enim, et eas habere sic assero. Bruta sensus habent. Set omnem sensum aut iudicii motus sequitur aut non. Sensum igitur brutorum aut iudicium sequitur aliquod aut nullum. Verbi gratia: cum oculos intendo, aliquando super eis que video iudicium habeo, ut scilicet intelligam 'Album est'[92] quod vidi, vel 'nigrum' vel aliquid huiusmodi; aliquando vero rem aliquam video, et tamen de ea visa nullum iudicium habeo, ut cum est in aliis occupatus animus.

NEPOS: Sensus quidem habent, iudicium vero nullum, neque intellectum de his que sentiunt.

ADELARDUS: Si nullum de sensibilibus iudicium habent, nullum de eisdem propositum appetendi haberent vel fugiendi.[93] Verbi gratia: quicquid vides de quo nullum intellectum habes, id ex illo sensu nec appetis nec fugis. Quod singulis horis et tibi et cuilibet alii accidit. Quod si bruta nullum de exterioribus propositum habent, dicat qui potest, cur canis quo maxime properabat,[94] velocissime avertitur, si quid in ipso itinere non quod noceat dico, set quod nocere possit aspiciat. Quid,[95] inquam, illum affectum tanta celeritate in contrariam convertit actionem? Habent nimirum, ut manifestum est, appetendi et fugiendi propositum. Habent igitur de eisdem quare ea appetunt vel fugiunt aliquod iudicium. Set illud iudicium quia corporis non est, non nisi in anima constare potest. Quare et illa animas habere necesse est.

Amplius: omnis vox audita ab aliquo aut aliquem intellectum in eodem aut nullum generat. Si nullum in auditore intellectum generat, nullum propter vocem illam auditor momentum[96] actionis facit, quippe nichil ei significarat. Itaque cum brutis aliquod significativum dicitur, aut in eis intellectum generat aut non. Si nullum in eis[97] generat, nullum in eis actionis motum suaderet, aut efficeret. Itaque si sic esset, nullum[98] animalia propter voces auditas propositum vel inciperent vel vitarent. Edissere igitur, si potes, quomodo canis voce audita inceptam sponte actionem statim finiat, contrariam incipiat.

92 est L, esse GMQV
93 haberent vel fugiendi L, habent vel fugiendi V, vel fugiendi haberent GQ, vel fugiendi habent M
94 properabat GMQ, probat L, p(ro)abat V
95 Quid MQV, Quod GL
96 momentum LQV, movimentum (*or* monimentum) G, M *om.*
97 in eis LV, GMQ *om.*
98 nullum GMV, nulla L, Q *om.*

Adelard: As the common people do not hesitate to say 'No', so the philosophers say a clear 'Yes'. They have souls, and I justify that they have them for this reason: brute animals have sensation. But either a movement of judgement follows every sensation or not. So either some judgement follows the sensation of a brute animal or none. For example, when I use my eyes, sometimes I make a judgement on what I see, so that, for instance, I understand what I have seen to be white or black *et cetera*; but at other times I see something and yet make no judgement about what I have seen, as when my mind is occupied with other things.

Nephew: Indeed, brute animals have sensation, but no judgement nor understanding about what they perceive with their senses.

Adelard: If they have no judgement about what they sense, they would have no intention to seek or avoid anything as a result of what they sense. For example, whatever you see about which you have no understanding, you neither seek nor avoid as a result of that sensation. This happens all the time both to you and to anyone else. But if brute animals have no intention concerning things outside themselves, let him who can tell me why a dog is diverted very quickly from the place to which he was making in great haste, if he sees anything in his path – I do not say, which actually harms him, but which *could* harm him. What, I ask, turns that desire so quickly into an opposite action? Brute animals clearly must have intentions of seeking and avoiding. Therefore they have some judgement about why they seek or avoid something. But because that judgement does not belong to the body, it can only exist in the soul. So it is necessary for them also to have souls.

Moreover, every spoken word heard by anyone either gives rise to understanding in him or does not. If it does not give rise to any understanding in the listener, he makes no movement of activity because of that word, because it has not meant anything to him. So, when something meaningful is spoken to brute animals, it either gives rise to understanding in them or does not. If it gives rise to no understanding in them, then it would neither persuade them into any movement of activity nor cause that movement. So if this were the case, animals would not begin or avoid anything that they intended to do because of the words they heard. Tell me then, if you can, how a dog, when he hears a spoken word, spontaneously and immediately stops what he is doing and starts doing the opposite.

Amplius: sensus et sensuum discretio non idem sunt. Sensus enim in corpore et circa corpora[99] sunt; discretio vero quarumlibet rerum, precipue similium, non nisi in anima esse potest. Set bruta animalia[100] discretione sensuum utuntur. Nam et canis bestie quam sequitur odorem sentit, set si alia ei eiusdem generis occurrat, tunc cum utriusque odorem occupet, inter eosdem discernit, et uno spreto, alterum[101] magistri memor insequitur. Habet enim in animo quodammodo pictum: 'Hic ille est quem sequi debeo, ille vero alius ab eo diversus'. In illa igitur differentia notanda non parva iudicii discretivi subtilitas est. Quare et bruta fundamentum discretionis habere necesse est.[102]

NEPOS: Numquam hodie a me aliud extorquebis, nisi quod vulgari clamore didici, ut scilicet bruta sensum tantum, non intellectum habeant.

ADELARDUS: Depuduit igitur te dum dissimulas quod superius concessisti, ut ratione iudice disputatio nostra constaret. Nunc autem, quoniam sensus obscurum nomen est, ex ipsa eius significatione sententia sumenda est. Et est sensus diffinitio, ut michi quidem videtur, 'Animati corporis exteriorum applicatione non levis mutatio'. Que diffinitio ut fidem facere possit, prius ei fides astruatur. Et 'mutatio' merito sensus dicitur, quia in omni sensuali motu qualitas corporis commutatur, aut accedendo ad aliud – ut in visu et tactu – aut alterius ad ipsum accessu – ut odoratu, gustu, auditu. Omnium enim instrumentorum officialium qualitates offensa exteriorum corporum immutantur. Est igitur omni sensui commune genus mutatio. Set quia fiunt quedam frequenter circa corpus mutationes que ob nimiam sui subtilitatem instrumentorum qualitates parum convertunt, quibus numquam anima ad[103] iudicium irritatur, ut athomi offensione vel minutissimi odoris accessione, ad horum differentiam, addimus 'non levis'. Item quia et circa inanimata corpora ex aliorum et sua coniunctione non leves frequenter conversiones in suis qualitatibus fiunt ut in contrarios mutentur effectus, ad eorum separationem

99 corpora G *a.c.*L, corporea G *p.c.*M(?)Q, corpus V
100 animalia LV, GMQ *om.*
101 alterum LV, alium GMQ
102 *This is the point where L ends and P begins.*
103 ad PQV, GM *om.*

Moreover, sensations and the discernment about sensations are not the same. For sensations occur in the body and around bodies, but discernment concerning anything – but especially in identifying similar things – cannot occur except in the soul. But brute animals make use of discernment of sensations. For, again, a dog senses the scent of the wild animal that he is following, but if he comes across another animal of the same kind, then, when he picks up the scent of both animals, he discerns between them, and mindful of his master's command he spurns the one and follows the other. For he has somehow pictured in his mind 'This is the one which I must follow, but that is another scent, different from this.' In noticing that difference there is considerable subtlety in discriminating judgement. Therefore it is necessary that brute animals too have the fundamentals of discernment.[21]

Nephew: You will never prise out of me today anything other than what I have learnt through the proclamation of the common people, that is, that brute animals have only sensation, and not understanding.

Adelard: So you have lost any sense of decency by just pretending, when you agreed earlier on that our disputation should be established with reason as guide! But now, since 'sensation' is an unclear term, we should form a clear concept from its very meaning. The definition of sensation, as it seems to me, is 'a significant mutation in an animate body caused by contact with outside objects'.[22] So that this definition can inspire confidence, first confidence should be built into it. Rightly sensation is called a 'mutation', because in every movement of sensation, the quality of the body is changed, either by approaching another object, as in sight or touch, or by the approach of that other object to it, as in smell, taste and hearing. For the qualities of all the organs appointed to the task are changed by their impact with outside bodies. 'Mutation' is the genus common to each sensation. But, because certain changes frequently occur in the body which because of their excessive fineness hardly alter the qualities of the organs, and by which the soul is never stimulated into making a judgement – such as in the impact of an atom or the approach of a very faint smell – , to differentiate from these, we add 'significant'. Similarly, because in inanimate bodies too, as a result of their contact with other bodies and with each other, significant alterations in their qualities frequently occur, so that they are changed into the opposite conditions, to

'animatum corpus' expressimus. Postremo, quia et ipsius animati corporis quedam non levis mutatio fit ex ipsius et solius anime motu[104] – ut ex ira rubeum, ex tristicia pallidum – ut hec separemus, adiunximus 'exteriorum applicatione'. Est itaque plena diffinitionis nota terminique convertibilis paritas, quam si nec superfluam nec diminutam dicere potes, recte assignatam esse non dubium est. Quod si hec ratio recta est, sicut sensus non nisi circa animatum corpus est, ita et sine anima esse non potest, quoniam et sine anima animatio non est.[105]

Amplius: ut de sensibus sententiam transeam, de motibus rationem sumam. Cum enim plures motus sint species, quidam proprie ad corpus, alii principaliter referuntur ad animam. Habent enim corpora sursum moveri ex igne, deorsum ex terra, dextrorsum vero et sinistrorsum et ante et retro ex aere et aqua. Orbicularis vero motus ad solam primo refertur animam. Hic itaque quoniam brutis convenit – moventur enim motu voluntario – et animas habere ea consequenter conveniat.

.xiv. NEPOS: Ne omnino rationi repugnare videar, animalia non intellectum, non discretionem, solam vero opinionem[106] habere concedam, quam non in anima set in corpore fundari satis sane asseram. Rationis enim iudicium in anima fundari eadem ratio probat. Opinionem vero que in brutis est, non animale set corporale subiectum exigere certum est.

ADELARDUS: Utrum vel ipsa opinio in animato fundari possit, sic exigo. Nos ipsi quidem cum rationales simus, et iudicii discretionem et sensualitatis[107] opinionem habemus, set nec hec nec illa in nostro corpore ablata anima adesse potest. Quod si in humano corpore quod omnibus[108] terrestribus ficticiis elementari coniunctione compositius[109] est,[110] illud quod ex sensibus oritur, quoquomodo vocetur, anima ablata residere non potest, multo minus id in bruto expectare, nisi insanias, potes. Aut si ea sine anima id posse ex sola coniunctione elementari[111] non pudet dicere, incurris quod nolles – humana scilicet corpora anima spoliata digniori iure idem exequi posse. Quod falsitate non caret.

[104] motus P
[105] quoniam . . . est PV, GMQ *om.*
[106] opinionem Q, opinionem ea GMPV
[107] sensualitatis PV, sensibilitatis GMQ
[108] omnibus G *a.c.*PV, G *p.c.*MQ *om.*
[109] compositius PV, compositum GMQ
[110] G *add.* vel
[111] coniunctione elementari PV, GMQ *transp.*

differentiate from them, we have specified 'animate body'. Finally, because even in the animate body itself a certain significant change occurs as a result of the movement of the soul itself and the soul alone, such as redness from anger, and paleness from sadness, to separate these we have added 'by contact with outside objects'. Thus the annotation of the definition is complete, and there is parity between the definition and what it defines; for there is no doubt that the definition has been correctly set out, if you cannot add more to it or take anything away from it. But if this reasoning is correct, just as sensation does not occur except in animate bodies, so it cannot exist without a soul, since animation too does not exist without a soul.

Moreover, leaving aside our opinion about sensation, I shall take a reasoned argument from movements. While there are many species of movement, some properly belong to bodies, others principally to the soul. Bodies have their ability to move upwards because of fire, downwards because of earth, right and left and backwards and forwards because of air and water. But movement in a circle belongs to the soul first and foremost. Thus, since this movement is appropriate to brute animals – for they move with a voluntary motion – it follows that it should be appropriate for them also to have souls.

14. Nephew: Lest I seem to oppose reason completely, I shall concede that brute animals do not have intellect or discernment, but only opinion, which I shall assert sensibly enough is founded not in the soul but in the body. For that the judgement of reason is founded in the soul is proved by that same reason. But it is certain that the opinion found in brute animals demands a substrate which is not ensouled but corporeal.

Adelard: Whether even opinion can be founded in something animate, I find out in this way. We ourselves, since we are rational, have both discernment, which belongs to judgement, and opinion, which belongs to sensation, but neither the one nor the other can be present in our bodies when the soul has been taken out. But if in the human body, which is better composed by the conjunction of the elements than all the fabrications on the earth, that which arises from the senses, whatever it should be called, cannot reside once the soul has been taken out, much less can you expect it to do so in a brute animal, unless you are mad. Or if you are not ashamed to say that that is possible without the soul and only as a result of the conjunction of the elements, you incur an unwelcome consequence: namely, that human bodies deprived of the soul can achieve the same thing by a more worthy right.[23] Which is not devoid of falsity.

NEPOS: Esto. Animata sint bruta ut elaboras. Set hee ipse eorum anime, simul ut a corpore separantur, pereunt.[112] Neque enim penam vel premium meritorum expetunt.

ADELARDUS: O perversa rerum conversio! Corpus, quod vilis ac mutabilis essentie est, generationi et corruptioni subiacens, ex diversitatibus compositum, non existens certe, set continua inconstantia ab hoc in illud, nec hoc nec illud transiens – hoc tale, inquam, non perire dicis, animam vero ducem ac dominam[113] incorpoream agentemque naturam, totalem, non particularem, perire deliras? Immo potius tu, dum talem essentiam tam male diffinis, peris.

NEPOS: Audienda sunt que dicis, non credenda. Ego vero ad altiora accingar, ut quantum scientiola mea[114] potest ex fumo lucerna prodeat. Licet enim nec Grecas iactantias noverim, nec antrum Vulcani viderim, tamen cum[115] vera cognoscere, tum falsa improbare et didici et expertus sum. Age igitur de humana natura quid sentias experiri affecto. Licet enim que superius asseruisti magni facias, tamen, si te ipsum non novisti, supradicta parvi pendenda iudico. De homine enim homines disputare convenit. Nempe sicut tractatus futilis a subtilitate inquisitionis animum deicit, sic materia sublimis ad inquisitionem mentis oculum acuit.

ADELARDUS: Quoniam ad hoc conveni ut questioni solutio subiceretur, tuum erit quamcumque agi[116] volueris proponere, meum discutere.

.XV. NEPOS: De compositione hominis sermo michi habendus est. Unde in primis istud ambiguum dissolve: quare homines cornua non habeant.

ADELARDUS. Ut intentio tua locum questionis habeat, primo aliqua ratio vel vera vel verisimilis a te trahenda[117] est, unde debere ea habere videantur. Alioquin talis disputatio inter[118] philosophos locum habere non meretur. Neque enim ego sum ille qui omnium que sunt causas dare opere precium iudicem, set eorum tantum que ratione monstrante aliter ac sunt esse debere videntur.[119]

NEPOS: Equum postulas. Cum igitur omnia que Conditor ex

112 separantur, pereunt GMQV, vel separantur exeunt pereunt P (*probably a gloss* 'vel exeunt' *incorrectly incorporated into the text*)
113 ac dominam G *om.*
114 mea G *a.c.*PV, G *p.c.*MQ *om.*
115 cum P, tum GMV, tum tunc Q
126 agi PQV, agere G, M *om.*
117 trahenda G *p.c.*MQ, tradenda PV
118 intra P
119 viderentur P

Nephew: So be it. Let brute animals possess a soul, as you strive to prove. But these very souls of theirs perish as soon as they are separated from the body. For they do not seek punishment or reward for their deeds.

Adelard: Oh, perverse reversal of things! That the body which is of a base and mutable essence, being subject to generation and corruption, composed of discordant parts, and not existing with certainty, but in continual inconstancy passing from this to that and being neither this nor that – that such a thing, you say, does not perish, but the soul, its guide and incorporeal mistress, and the active nature,[24] as you crazily say, perishes – and perishes completely, not partially? Rather you perish, when you so wrongly define such an essence![25]

Nephew: One should listen to what you say, not believe it. I shall gird myself for higher matters, so that as far as my meagre knowledge can manage, a flame should be produced from this smoke. For though I neither know what the Greeks boast of, nor have seen the Cave of Vulcan,[26] nevertheless I have learnt and experienced both how to recognize what is true and reject what is false. Come then, I desire to find out what you think about human nature. For although you make much of what you asserted before, nevertheless, if you do not know yourself, I judge that the preceding statements should be held of little value. For it is appropriate for men to dispute about mankind. Just as dealing with useless things dislodges the mind from the subtlety of enquiry, so the sublimity of the matter sharpens the eye of the mind for enquiry.

Adelard: Since I have agreed that the solution should come after the question, it will be your task to propose any question you wish to discuss, and mine to discuss it.

15. Nephew: I must talk about the make-up of man. So first, resolve this problem: why do men not have horns?

Adelard: For your intention to be ranked as a real question, first I must get from you some reason – either a true one or a likely one – why men seem as if they ought to have them. Otherwise such a disputation does not deserve to have a place among philosophers. For I am not the kind of man who judges that it is worthwhile to give reasons for everything that exists, but only for those things which, in the light of reason, seem as if they ought to be something other than they are.

Nephew: Your request is fair. Reason shows that, since all things which the Creator led forth from the formlessness of nothingness into

informitate nichili in formam duxit essentie, ut ab optimo facta sunt, ita optime facta esse ratio monstret, omnibusque a se factis facultatem existendi ipse Auctor dederit; nam et esse amant et ut sint substantive munita sunt: habent enim quedam animalia – de his nempe agitur – robur substantivum sibi innatum, ut aper dentes, leo ungues, taurus cornua, quibus si quid eis periculi imminet, esse suum defendant; alia vero que minus gravia sunt, ut mole sua tueri se nequeant, ipsum esse suum fuge levitate conservant, ut capreoli, damme, lepores, et cetera id genus – cum, inquam, ea que indigniora sunt, defensionem sibi innatam habeant, cur homo, qui omnibus inferioribus naturis dignior est, nec innata habet arma[120] ut cornua vel letiferos dentes, nec fuge levitate hostem imminentem vitare potest? Unde, quia sibi a Conditore data non sunt, accidit ut extrinsecus petita arma sibi laboriose adaptet et, cum necesse est, pedibus animalis alterius magis quam suis confidat. Illud igitur quod Summe Bonitati carius est, supremo auxilio destitutius est.

ADELARDUS: Vulgari me prius responsione liberabo. Cedo[121] enim ut homo ceteris animalibus Conditori carior sit, tamen nec arma sibi innasci convenit, nec levissima fuga aptari; habet enim id quod his longe melius digniusque est: rationem dico, qua[122] etiam adeo ipsa bruta excellit, ut per eandem domentur, domitis frena adiungantur, frenatis diversa exercicia adhibeantur. Vides igitur[123] rationis donum corporeis instrumentis quantum precellat.

NEPOS: Insipienti satis dictum esset, michi vero nequaquam. Neque enim quod ambigebam dissolvis. Illud nempe non dubito quin ratio omnem molestiam corporalem precellat. Set cum homines ipsi, etiam qui inter eos rationabiliores sunt, quandoque contra imminentiora pericula tutela egeant, eo usque ut necesse sit extrinsecus petere et arma et vehicula, ipsa illa ratione hoc idem suadente, ut quid ab hoc auxilio divinitus elongati sunt? Honestius enim erat ut cum ipsa ratione in se haberent quo indigent, quam simul cum ea extrinsecus petere quo non possunt carere.

[120] habet arma PMV, GQ *transp.*
[121] Credo P
[122] qua M, que GPQV
[123] ergo P

the form of being, as they have been made by the Best, so they have been made in the best way, and the Author himself has given to all things made by himself the capacity to exist. For they both love to exist and they are protected in a substantive way in order to exist. Now certain animals – because it is about these that we are, of course, speaking – have a substantive strength born in them by nature, such as the boar, which has tusks, the lion, which has claws, the bull, horns – by which they defend their being if any danger threatens them. But others, which are less heavy, so that they are not able to protect themselves by their sheer weight, preserve their being by the swiftness of flight, such as wild goats, deer, hares and others of this kind. Why, I ask, when those which are less worthy have a means of defence inborn in them, does man, who is more worthy than all these lesser natures, not have inborn weapons, like horns or deadly tusks, and cannot avoid a threatening foe by the lightness of flight? The result is that, because he has not been given these things by the Creator, he laboriously puts together for himself weapons which he has looked for outside himself, and, when it is necessary, he puts more confidence in the legs of another animal than in his own. That creature, therefore, which is dearer to the Supreme Good is the more deprived of the supreme help.

Adelard: I shall free myself first by the common answer. For I agree that man is dearer to the Creator than the other animals. Nevertheless, it is not appropriate either for arms to be innate in him, or for very swift flight to be attached to him. For he has that which is much better and more worthy than these – I mean reason, by which he excels the very brute animals so much that they are tamed by it, and, once tamed, bridles are put on them, and, once bridled, they are put to various tasks. Thus you see how much the gift of reason is superior to bodily instruments.[27]

Nephew: That would be an adequate answer for a fool, but not at all for me. For you do not solve my problem. Of course, I do not doubt that reason is superior to every bodily encumbrance. But since men themselves – even the more reasonable among them – sometimes need protection against the more threatening dangers to such an extent that they must seek weapons and means of conveyance outside themselves, this being dictated by that very reason itself, why have they been kept away from such a resource by divine will in the first place? For it would have been a mark of greater honour to have what they needed in themselves, alongside reason, than to seek outside what they could not do without (while still having reason).

ADELARDUS: Quoniam pro ratione irrationabilia non recipis, quid inde michi probabilius videatur accipe. Homo quidem animal rationale atque ideo sociale est aptum ad duas operationes – actionem dico et consilium – que aliis appellare placet bellum et pacem. In bellicis igitur negotiis armorum usum desiderari conversatio cotidiana docet. In pacis vero quiete eadem deponere, ac longe a secreto consilii removere veritas astruit. Quippe quorum unum ira movet, alterum ratio mitigat. Si itaque arma innata sibi homo haberet, cum pacis federa tractaret, ea deponere non posset. Si item levitate fuge munitus esset, et constantiam vagus amitteret et gracili membrorum ineptitudine, cum in bellicis[124] adesset, debilis reperiretur. Nunc vero et quando necesse est recipit, et pace postulante eadem deponit, et, cum eget velox esse, industria potest, et necessitate sublata ad constantiam redire non nequit.

.xvi. NEPOS. Satis commode dicta sunt[125] ista. Nunc vero quandoquidem de compositione hominis disputare pergimus,[126] illud enucleari libet quod nobis a[127] phisicis dicitur, qua ratione scilicet contextio nervorum[128] et venarum in corporibus nostris deprehensa sit. Cum etenim adeo implicita sint ut ex similitudine rete[129] a philosophis[130] dicatur, si mortuorum divisio aperitiva facta est, accideret ut ibidem et nervi et vene secando rumperentur.

ADELARDUS: Ingenium michi abesse non timeo solvendi ea que postulas. Set multitudo lectionum memoriam forsan hebetavit. Audivi enim et hec et alia multa quendam senem apud Tharsum Cilicie disserentem, quod tamen quam breve potero faciam: cum opere precium philosophis[131] videretur extensiones nervorum et venarum cognoscere, hoc modo id eos assecutos esse arbitror. Cadaver quidem humanum puto eos in fluminis impetu ligasse, ibique donec tota cutis et caro deciderit dimissum esse, nervis tamen et venis, que tenatioris substantie sunt, ibidem remanentibus. Itaque eis talis contextio retexta est.

124 in bellicis Q, in bellis G *a.c.*, inbecillis G *p.c.*PMV
125 sint P
126 pergimus PQV *a.c.*, peregimus GMV *p.c.*
127 nobis a *Müller*, a nobis GPM, a vobis QV
128 contexio nervorum PV, GQ *transp.*
129 rete PMQ, recte GV
130 physicis M
131 philosophis PQV, ph(. . .)is G, physicis M

Adelard: Since you do not accept irrational arguments in place of reason, take what seems to me most probable about this. Man is a rational animal, and hence a sociable animal, well-fitted for two modes of operation – action and deliberation, which others like to call war and peace. Everyday experience teaches that the use of weapons is needed in the business of war.[28] But in the quietness of peace truth requires the same weapons to be laid aside and removed far from the intimacy of deliberation. For while anger motivates the one course, reason soothes the other. If, then, man had weapons innate in him, when he came to make peace treaties, he would not be able to lay them down. Similarly, if he had been armed with the lightness of flight, he would both lose constancy in flightiness, and be found weak in the face of battle, because of the slender ineptitude of his limbs. But now he both takes up arms when it is necessary, and lays them down when peace demands it, and when he needs to be swift, he can work out a way do this, and when the need is no longer there, he is equally able to return to firmness.

16. Nephew: That's well enough said. But now, since we are going on with our disputation on the fabric of man, I should like a clarification of what the natural scientists say to us:[29] namely, how the web of nerves and blood-vessels in our bodies can be detected. For since they are so intertwined that philosophers call them a 'net' because of their likeness to one, if corpses are opened up it would happen that both the nerves and the blood-vessels would be broken on the spot in the process of cutting.

Adelard: I am not afraid that I shall lack the talent for solving what you ask. But perhaps the number of lectures I have heard has blunted my memory. For I heard an old man in Tarsus in Cilicia discussing both this topic and many others. Nevertheless, I shall repeat what he said as briefly as I can. When it seemed worthwhile to philosophers to get to know the spread of the nerves and the blood vessels, I think that they achieved this in this way. I believe they bound a human corpse in the current of a river, and it was left there until all the skin and flesh fell away, while the nerves and the blood vessels, because they were of a tougher material, stayed in place. In this way such a web was unravelled for them.

.XVII. NEPOS: Licet ita non fuerit, tamen quia ita esse potuit, procedamus. Et quoniam in agmine verborum tuorum de ingenio et memoria mentionem fecisti, illud quod michi inde enodandum occurrit, in commune ponam. Querendum namque est, cum memoria nichil aliud sit nisi eorum que ingenio capiuntur certa repositio, cur quidam qui bonum habent ingenium[132] memoriam omnino non habeant, et ea que facile capiunt, eadem facilitate, cum necesse est, rememorare non possunt. Nam, et quod magis miror, illi qui difficilius intelligunt, firmius sepenumero memorant.

ADELARDUS: Recte queris, si quesita benivole recipis.

NEPOS: Nichil libentius.

ADELARDUS. Quicquid anima in corpore agit quod huiusmodi sit, per aliquod corporis auxilium agit. Ita nempe aliud in cerebro, aliud in corde, aliud in membris aliis. In cerebro enim utitur fantastico motu, id est ingeniali; rationali etiam, id est iudicio; set et memoriali, id est recordatione. Prius enim intelligit; deinde quod intellectum est iudicat; tertio ipsum iudicium constanti memorie[133] commendat. Set et hoc et illud per aliud et per aliud est. Ingenium quippe per humiditatem viget; memoria vero per siccitatem. Quicquid enim humidum est, cuiuslibet sigilli inpressione facile signatur, set eiusdem humiditatis inconstantia facilius deletur. Quod vero siccius est, difficile quidem inpressionem forme recipit, set ut semel eam admittit, non leviter immutatur. Itaque qui humidum habent cerebrum, ingenio quidem pollent, set memoria fatiscunt. Qui vero siccum habent, hii memoria vigentes ingenio privati sunt.

.xviii. NEPOS: Si de sequentibus non minus probabiliter dixeris, ista concessu digna non diffiteor. Consequens enim est ut et istud quod a multis dubitatum est discutias. Cum enim de comitantibus cerebrum nobis sermo sit, elice, si vales, qua ratione loca fantasie, rationis et memorie a philosophis deprehensa sint. Nam Aristotiles in Phisicis et alii in tractatibus aliis sic discernunt, ut fantasiam exerceri dicant in parte cerebri anteriore, rationem in medio, memoriam in occipitio.[134] Unde et tribus inde cellis nomina imposuerunt, fantasticam, rationa-

[132] habent ingenium PV, GMQ *transp.*
[133] constanti memorie V, constantie GMPQ
[134] occipitio GMV, occipio P, occipite Q

17. Nephew: Even if that did not happen in that way, nevertheless, because it could have happened like that, let us go on. Since in the full flood of your speech you mentioned intelligence and memory, I shall bring out into the open something that has occurred to me concerning these which needs explanation. For one must ask: since memory is nothing other than a reliable repository of those things which are grasped hold of by the intelligence, why is it that some people who have a fine intelligence are completely lacking in memory, and cannot, when necessary, recall with the same facility those things which they easily grasp? For – and this surprises me even more – those who find it more difficult to understand things very often have a more accurate memory.

Adelard: You seek in the right way, if you are well-disposed to accept what you have sought.

Nephew: Nothing would please me better.

Adelard: Whatever activity of this kind the soul performs in the body, she does through some assistance from the body.[30] Thus she does one thing in the brain, another in the heart, and another in other members of the body. In the brain she uses the movement of imagination, that is of the intelligence; rational movement, namely, judgement; and also the memorizing movement, namely, recollection. For first she understands, then she judges what has been understood, and thirdly she consigns the judgement itself to the firm memory. But these different parts operate through different things. Intelligence thrives on moisture, memory on dryness. For whatever is moist can easily be marked with the impression of any seal, but because of the lack of firmness of this same moisture, it can more easily be erased. But what is drier receives the impression of a shape with difficulty, but, once it has received it, does not change easily. Thus, whoever has a moist brain, has a lively intelligence, but lags behind in his memory. But, as for those who have a dry brain, while their memories flourish, they fall short in intelligence.

18. Nephew: If you speak no less convincingly about what follows, I shall not deny that this is worth conceding. The consequence is that you should discuss also what many people are uncertain about: since our discussion is about the attributes of the brain, work out, if you can, the way by which the positions of imagination, reason and memory were discovered by philosophers. For both Aristotle in his *Physics* and others in other works divide them in such a way that they say that imagination operates in the front part of the brain, reason in the middle, memory in the back. Hence they have given to those three

lem et[135] memorialem. Qua vero astutia id eis iudicatum est, ut operationes anime ita localiter distinguerent, ut tam parvam presertim cerebri aream diversis eius actibus dedicarent, cum nec ipsa nec talis eius operatio sensu aliquo discerni possit?

ADELARDUS: Non intelligenti omnia[136] impossibilia esse videntur. Intellecta vero eadem facultate perspicua esse iudicantur. Quicumque igitur primus de cellis illis discretive egit, id ipsum sensuali experimento didicisse auguror. Erat quippe aliquis qui cum fantastica formarum recollectione bene uteretur, parte illa capitis anteriore lesus est, adeo ut inde virtutem fantasticam amiserit, ratione tamen et memoria non privatus. Unde factum est ut id a Philosopho notaretur. Non dissimiliter etiam si lesione aliarum[137] partium alie anime actiones impedite sunt, constare potuit pro certo ut in singulis cellis singule exerceantur, quippe cum et ipse celle quibusdam licet tenuibus interstitiis distincte sint.[138] Ita igitur ex eis que sensus extrinsecus notant insensibilis et intellectualis anime operatio patefacta est. Nam et ipsa anima, cum incorporea essentia sit nullique sensui subiacens, ex his tamen que sensualiter in corpore agit, ibidem esse non dubitatur. Ex motu enim quem corpori dare videtur, quem a se habere nequiret, in corpore quiddam incorporeum esse arguitur.

.xix. NEPOS. Quoniam eorum que aliter atque sunt esse debere videntur[139] rationem recipis, quanto magis audio, tanto magis dubito. Cum enim, ut supradictum est, omnia que a tali Conditore facta sunt optima esse ratio doceat, est quiddam quod in humana non modicum[140] miror compositione, quare scilicet super os nasus formatus sit; nichil quippe in tali compage quam os honestius, set nec aliquid mundius servari ipsa conversatio docet. Naso autem quid humidius – ne inmundius dicam – esse potest? Adeo ut multe inde oris honestati importunitates immineant. Quare ipsum, si dicere fas est, aptius videri posset, si super nasum os elatum esset.

ADELARDUS: Ita percutit qui non videt. Non enim intelligis quia

[135] et GMQ, PV *om.*
[136] omnino P
[137] aliorum P
[138] sint PQ, sunt GMV
[139] videntur PMV, viderentur GQ
[140] modicum PMV, modice GQ

cells the names 'imagining', 'rational' and 'memorizing'.[31] But by what cleverness did they make that judgement, so distinguishing the operations of the soul spatially that they could dedicate even such a small area as that of the brain to the different actions of the soul, when neither the soul nor such an operation of hers can be discerned by any of the senses?

Adelard: For a person who does not use understanding, all things seem to be impossible, but, once understood, they are judged by the same faculty to be evident. Whoever first treated these three cells separately, I guess, learnt this very thing by experiencing it with his senses. I suppose there was a man who was well able to use his imagination to gather the forms of things, but he was injured in that front part of his head to such an extent that he lost his power of imagination, but was not deprived of reason and memory. Hence it happened that this was noticed by the Philosopher.[32] In a similar way if, by the injury of other parts, other actions of the soul were impeded, it could be stated for certain that single actions operated in single cells, seeing that the cells themselves are separated by some, albeit narrow, divisions. This then is the way that the insensible and intellectual operation of the soul has been revealed from those things which the senses note on the outside. For even the soul herself, although she is an incorporeal essence and lies open to none of the senses, nevertheless, from those things which she does in a sensible way in the body, is known for certain to exist in it. For, from the movement which she is seen to give to the body and which the body cannot have from itself, she is proved to be a certain incorporeal essence existing in the body.

19. Nephew: Since you are adopting reason for those things that seem to have to be other than they are, the more I hear, the more I doubt. For since, as was said before, reason teaches that all things that have been made by such a Creator are the very best,[33] there is something which I am not a little surprised about in the composition of man: namely, why the nose has been formed above the mouth. For in the fabric of man there is nothing more honourable than the mouth. But daily experience teaches us that nothing should be kept cleaner than the mouth. Yet, can anything be more moist, not to say more unclean, than the nose? To such an extent that many nuisances threaten the honour of the mouth from that source. Therefore it would seem more suitable, if one may say so, if the mouth had been raised above the nose.

Adelard: So strikes the man who cannot see! For you do not

nullum naturale inmundum vel indecens est. Quicquid autem contra nature rationem est, id licet quantum ad visum picturaliter decoratum sit, in se tamen et inmundum et iniquum esse iure dicitur. Ad rem igitur redeundum est. Et nasus quidem et os non[141] principalia membra set principalibus deservientia sunt.[142] Servit enim et nasus cerebro et os stomacho. Cum enim cerebrum quadam actionis sue necessitate humidum sit, itemque ad idem alieni humoris a stomacho in fumo quodam ex cibis orto non parum ascendat, necessarius ei canalis fuit, per quem humoris superfluitas purgaretur. Item cum a stomacho ex decoctione innumerorum ciborum quedam levis spumositas naturaliter nascatur, opportunum fuit eandem per os purgari. Unde et nasum cerebro, et os stomacho vicinius esse necesse est, presertim cum non aliunde nisi per os quicquam cibarii in stomachum descendere queat.

.XX. NEPOS: His que proposita sunt solutis, ad alia inquirenda affector. Et quia de humana proprietate sermo nos detinet, id item a te exigo: ut quid parte anteriore non[143] posteriore calvescant homines. Si enim pars anterior, ut ratio docet, quodammodo dignior est, unde et sensibus magis illustrata est, consequenter quidem et florali capillorum decore eam privari[144] minus competens est, presertim cum et inter homines obprobrii nomine calvities vicissim opponatur, et ipsa oppositio posterius quam anterius galeari operimento fieri posset absconditior.

ADELARDUS. Ex predictis ea que causaris procedunt. Quoniam enim ea que in stomachum descendunt per os intrare necesse est, eademque in ipso stomacho digeri et decoqui patens est. Ex tali item decoctione fumum ascendere per eundem meatum ad superiora consequens est. Fit demum ut, sicut meatus ille anteriori quam occipitio propinquior est, ita et in ea poros ipsius cranei – id est teste – apertiores esse necesse sit. Quod et in testa defuncti, si quis notare velit, perspicuum est. Exalat enim cerebrum calorem, qui, si in eo obturaretur, frenesim proculdubio singulis generaret. Causali igitur illius caloris egressione cum incendio habundante nimia pororum apertio fit; capilli quidem nasci intendentes, dum glebe sue continuita-

141 P *add.* sunt
142 sunt GQV, PM *om.*
143 non *scripsimus* (cf. V: parte anteriore calvescant homines non posteriore), quam GPQ, M *om.*
144 privari GMQ, privare PV

understand that nothing natural is unclean or unfitting. Whatever is against the reason of nature, although to the sight it is decorated with surface beauty, is rightly said to be both unclean and flawed in itself. But we must return to the matter at issue. Both the nose and the mouth are not principal members, but serve the principal members. For the nose serves the brain, the mouth the stomach. Since the brain is moist because its activity in a certain way necessitates this, and also a considerable amount of foreign moisture rises from the stomach in fumes of a sort generated from food, a channel was necessary for it, through which the excess moisture could be purged. Similarly, since a certain light foaminess arises naturally from the stomach as the result of the cooking of innumerable foods, it was appropriate for it to be purged through the mouth. Hence it was necessary both for the nose to be closer to the brain, and for the mouth to be closer to the stomach, especially since whatever one eats cannot descend to the stomach by any other route than through the mouth.

20. Nephew: Since you have solved the questions I have posed, I am minded to ask others. And because we are involved in a discussion concerning what is special to man, I also demand this from you: why do men go bald from the front part of their head, not from the back part? For if the front part, as reason teaches, is in a way more worthy (hence it is also more illustriously graced by the senses), it follows that it is certainly less appropriate for it to be deprived of the flowering ornament of hair, especially since among men 'baldness' is bandied around as a term of reproach, and the shame of baldness can be hidden more easily at the back than at the front, by covering it with a hat.

Adelard:[34] The cause you plead can be defended from what we have said before. Since those things which descend into the stomach must enter through the mouth, it is clear that they are digested and cooked in the stomach itself. As a consequence of such cooking fumes ascend along the same route to the higher parts. The eventual result is that, just as that route is closer to the front part than the back part of the head, so it is necessary that in that very part the pores of the cranium itself (that is, the skull)[35] should be more open. This is in fact clear in the skull of a dead man, if any one cares to check. For the brain exhales heat, which, if it is blocked up, without doubt generates madness in every case. Because that heat goes out of the body with excessive burning, the pores are excessively wide; the hairs intending to grow there lose the unbroken surface of their soil and

tem amittunt, id quod ceperant fieri non possunt. Licet enim ex fumo nasci noscantur, tamen sine aree sue continuitate radicari non possunt, neque sine humiditate aliquam extensionis coherentiam mereri queunt. Unde fit ut iuvenibus et eorum magis calidis natura hec penuria frequentius accidat.

NEPOS: Hactenus ista sufficienter quidem tractata sunt. Illud nempe quare homines in senectute precipue canescant, quidam etiam eorum raro et in aliis etatibus, quamobrem item soli ipsi barbescant – numquam enim aut raro idipsum mulieribus et eunuchis contingit[145] – quoniam ab aliis dictum est, a te querere supersedendum est. Quoniam vero de his que circa vultum sunt disserere incepimus, de sensibus ipsis enodandum esse iudico. Nichil enim in corporea[146] hominis compositione, meo iudicio, vel intellectu difficilius vel mutua allocutione inexplicabilius est. Et merito. Dum enim vulgares homines rerum scientias vel falsa ascriptione sibi arrogent, vel parvi pendendas esse iudicent, fit demum ut in se ipsis inveniant quiddam quod[147] disserere nequeant. Unde et Boetius in Musica de eisdem sensibus ait: 'Inter philosophos quidem dubitabile est; vulgus vero dubitatio preterit'. In primis igitur de auditu quid sentias audiamus.

.xxi. ADELARDUS: Audio quod rogas, non intelligo quid interrogas.[148]

NEPOS: Id certe interrogo: utrum in sententia de auditu Boetio consentias necne.

ADELARDUS: A tanto philosopho non dissentio.

NEPOS: Set hec eius sententia in Musica ab eodem sic exposita est ut cum vox aer sit, aer autem corpus, corpus vero individualiter idem et unum in diversis locis eodem tempore totum esse non convenit, sic eadem vox in diversis auribus hominum simul esse non possit. Subdit igitur quomodo esse possit, dicens:

> Aer quidem ore loquentis formatus linguaque impulsus proximum sibi eadem forma afficit aerem, et secundus tercium, tercius quartum, itaque usque ad aures astantium. Sicque similes non eidem audiuntur aeres.

[145] contingit P *p.c.*, contingat GP *a.c.*QV, contingatur M
[146] in corporea G *p.c.*M *p.c.*QV, corporea G *a.c.*, incorporea PM *a.c.*
[147] quiddam quod V, quedam que G *p.c.*, quoddam quod PMQ, quedam quod G *a.c.*
[148] non intelligo quid interrogas PMQ, quid interrogas nescio G, non intelligo quod interrogas V

are not able to complete what they had begun to do. For although they are acknowledged to grow from the fumes, nevertheless without an unbroken surface of their field, they are not able to put down roots, nor can they acquire any cohesion in extension without some moisture. Hence it happens that this lack more frequently occurs to young men and those among them who are more hot by nature.

Nephew: That is enough of these subjects. We must pass over the question why men go grey especially in old age, why some of them go grey rarely, or at other times of their lives; why, too, only men grow beards – for this never, or rarely, happens to women or eunuchs – since others have dealt with this question. But since we have begun to talk about those things which are around the face, I judge that we should solve the problem concerning the senses themselves. For nothing in the bodily composition of man – in my opinion – is either more difficult to understand or more impossible to unravel by mutual discussion. And rightly so. For while common people either falsely claim for themselves knowledge of things, or judge that such knowledge is worth little, it happens that they finally stumble upon something in their very own bodies which they are unable to explain. Hence Boethius too in his *Music* says about these senses: 'Among philosophers it is a problematic matter, but the problems escape the notice of common people altogether.'[36] So first let us hear what you think about hearing.

21. Adelard: I hear the question you pose, but I do not understand what you are asking.

Nephew: I am asking just this: do you agree with Boethius's opinion on hearing or not?

Adelard: With such a great philosopher I do not disagree.

Nephew: But Boethius explains his opinion in the *Music* in this way: since sound is made up of air, but air is a body, and one and the same body taken on its own cannot be complete in different places at the same time, thus the same sound cannot be simultaneously in the ears of different men. To explain how this can happen, he therefore has to add that:

> The air is formed by the mouth of the speaker and propelled by the tongue. It then shapes the air closest to itself with the same form, and the second <parcel of> air shapes the third, and the third the fourth, and so on until it reaches the ears of the bystanders. Thus they hear similar <parcels of> air, but not the same.[37]

Unde et idem Boetius exempli similitudinem in iactu lapidis circulorumque formatione[149] subdit.

ADELARDUS: Satis sane sensum eius exponis.

NEPOS: Verumtamen qui idem sentit, non satis sanus esse videtur. Si enim aeris inpulsiva informatione, dum ad diversos diversi veniunt, auditus fit, quomodo si non a me aerem pellam, set illum qui in ore meo est, expresse intus ducendo dum ad me traho, sibilum faciam – quomodo, inquam, hic a diversis audietur, et tamen audietur?

ADELARDUS: Ex similibus similia perpendi fas est. Illud nempe quod Boetius de aeris informatione diversa et auditus natura disserit, ut a philosopho dictum est, ita et[150] philosophis concedendum est. Ex eo autem quod de impulsiva formatione fit cum diversi ad diversos transeunt aeres, similitudo effectus etiam in attractiva formatione intelligenda est. Sicut enim, dum a me aerem impellendo formo, ad diversos diversi venientes sentiuntur, sic, dum ad me trahendo infra labia eum illido, a diversis per attractionem venientes ab eis a quibus veniunt[151] sentiuntur. Ut enim anima natura vigil essentia cum ad aures suas aliquid extraneum accedit, de eadem re ad iudicium excitatur, ita et cum ab auribus suis aliquid exit et ad aliud ducitur, anima inde ad discernendum provocatur. Boetius vero de illa sola auditus natura disserit, que ad voces significativas pertinere potest.

.xxii. NEPOS: Nec nichil penitus dixisti, nec omnino errorem removisti. Quid enim si inter me et auditores paries eneus, aut alio modo solidus, interpositus sit, numquid, si aerem trahendo sibilem, audiri non potero? Neque tamen ab auribus eorum per tam solidum corpus duci potuit. Quid igitur dici convenit? Fingere si potes hic, fies michi magnus Appollo.

ADELARDUS: Non fingendo quod non est, set naturam rei exprimendo que occulta est, Appollo dici merear. Nescis enim, nepos, neque id in verboso sophistarum agmine didicisti, omne corpus, metallicum etiam aut si quid eo solidius est, quoniam porosis interstitiis subiectum

149 formatione GQV, formationem PM

150 et G *p.c.*PV, G *a.c.*MQ *om.*

151 venient P

Hence the same Boethius adds as an illustration the throwing of a stone into water and the formation of circles.

Adelard: You explain the sense of Boethius's opinion soundly enough.

Nephew: Well, whoever thinks that, does not seem to be sound enough! For, if hearing is caused by air being shaped into forms by propulsion, while different <parcels of> air approach different ears, if I do not propel a <parcel of> air from me, but, by drawing inwards under pressure that air which is in my mouth and while sucking it in, I make a hissing sound, how, I ask you, will this sound be heard by different people – indeed be heard at all?

Adelard: One should proceed by analogy. That which Boethius has to say about the different shaping of the air and the nature of hearing, as it was said by a philosopher, so it should also be granted by philosophers. From the fact that it happens by a shaping caused by propulsion, when different <parcels of> air pass into different ears, a similarity of effect should be understood also in shaping caused by attraction. For just as, when I shape the air in propelling it away from me, and different <parcels of> air passing into different ears are 'heard', so, while by sucking air in through my lips I crush the breath, the <parcels of> air coming away from different ears by attraction are heard by those from whom they come. For since the soul, being by nature a wide-awake essence, when something outside comes to her ears, is roused to make a judgement about that same thing, so also, when something goes out from the ears and is drawn towards something else, the soul is provoked into noticing that. In fact, Boethius discusses only that kind of hearing which can pertain to significant sounds.

22. Nephew: There is something in what you have said, but you haven't removed the problem altogether. For what if a wall of bronze or some other solid material is placed between me and my auditors? Can I not be heard if I hiss by sucking in the air? Yet the air could not have been drawn away from their ears through such a solid body. What is it appropriate to say? If you can invent a way of answering *that* question, you will be to me Great Apollo himself![38]

Adelard: I should deserve to be called Apollo, not by inventing what does not exist, but by explaining the nature of something that is hidden. For you are ignorant, dear nephew, and among the host of garrulous sophists you have not learnt even this simple fact: that every body – even a metallic one or something yet more solid (if there is one) – is subject to porous openings, and therefore can be penetrated

est. Quare et subtilissimo aeri pervium est. Minime igitur illud ita esse ut dixi impossibile est. Itaque ad sequentia transeundum est.

.xxiii. NEPOS: Nichil michi hoc transitu dulcius, nichil hoc tractatu iocundius. Ad illud enim quod michi obscuris omnibus obscurius est, ipso rerum ordine id exigente accedendum est. Audita enim de auditu sentencia, quid de visu dicendum sit videamus.

ADELARDUS: Quoniam de visu multi multas protulerunt sententias, in tali disceptatione eas proponere conveniens est, et cui earum ratio potius accedat, discutiendum est.[152]

NEPOS: Meum igitur sit eas, si ita laudas, in communi ponere, tuum vero si quid eis oppugnat nichilominus positis opponere.

ADELARDUS: Equum postulas.

NEPOS: Quantum apud diversos colligere potui, de visu sententia quadripertita est.

.i. Quidam enim[153] animam in cerebro utpote principali sedentem sede perque fenestras patulas – oculos dico – exteriora intuentem formas rerum comprehendere dicunt, comprehensas iudicare, ita quidem ut nec ab ea quicquam ad exteriora exeat, nec ab exterioribus ad eam formarum quicquam accedat.

.ii. Alii vero formarum accessione visum fieri autumant, dicentes corporum formas aerem usque ad oculum interiacentem informare, sicque ad animam transire quod possit iudicare.

.iii. Quamplures item et aliquid ab anima mitti – visibilem dico spiritum – et formas videndorum ei occurrere in medio aere asserunt, quibus ipse informatus spiritus ad sedem redeat, representetque anime formam quam iudicet.

.iv. Quartam quorumdam sancit auctoritas, nullas penitus a suis subiectis ad oculum formas accedere, immo quiddam quod 'vim igneam' vocant in cerebro natum per nervos concavos primum, per oculos demum ad videnda usque traduci, inpressamque sibi more sigilli formam eadem levitate ad ortum redeundo anime referre.

Prima itaque primo tractanda sit.

[152] est GM, PQV *om.*
[153] enim G *om.*

by the finest air. Therefore it is not at all impossible that this should be as I have said. So let's go on to the next question.

23. Nephew: Nothing is more sweet to me than this transition, nothing more delightful than this disquisition. To approach the following question, which to me is more obscure than all obscure questions, is demanded by the very order of things. For having heard your opinion about hearing, let us see what should be said about sight.

Adelard: Since many people have put forward many ideas about sight, it is appropriate to set them out in such a debate and discuss which of them reason approaches more closely.

Nephew: Let it be my job then (if you approve) to bring them out into the open, and yours, no less to object to whatever is objectionable in them, once they have been put forward.

Adelard: That's fair.

Nephew: As far as I have been able to gather from various authorities, the opinions about sight are divided into four.

i) Some people say that the soul, seated in the brain, this being her principal seat, and looking at things outside through open windows (I mean, the eyes), takes in the forms of things and judges what she takes in, in such a way that neither does anything go to the outside things from her, nor do any of the forms come from outside things to her.

ii) Others declare that sight happens as a result of the approach of forms, saying that the forms of bodies inform the air lying in between and right up to the eyes, and thus there passes to the soul that which she can judge.

iii) Similarly, many assert that something also is sent forth from the soul – I mean the visual spirit – and the forms of the objects to be seen meet it in mid-air; the spirit itself then, having been informed by these forms, returns to its seat and presents the form to the soul for her to judge.

iv) The authority of some people sanctions a fourth opinion, that no forms at all come from their objects to the eye; rather, something which they call a 'fiery force', arising in the brain, passes first through the hollow nerves, then through the eyes to the objects of vision, and when the form has been impressed on it like that of a seal, returning to its source with the same swiftness, it relays the form to the soul.

One must deal first with the first opinion.

ADELARDUS: Omnium quidem habitudo sensuum, ut Boetius in Musica testatur, omnibus animalibus presto est, set que eorumdem sit vis quisve modus, non nisi intellectui philosophantis perspicuum est. Preclare. Nam et rerum effectus nexu subtilissimo a causis precedentibus procedunt et ipse inter se cause cum suis effectibus subtilissimis quibusdam intersticiis differunt. Unde et ipsis philosophis rerum natura sepe se subtrahit. Si enim, ut hii de quibus hic sermo est profitentur, anima in cerebro ut in sede locata per oculorum fenestras exteriora contemplatur, aut id dicentes vim aliquam in fenestris oculorum facient, aut nullum in talibus fenestris exteriora[154] contemplandi auxilium ponent. Set si[155] in oculis dicendo eos fenestras aptitudinem videndi exteriora faciunt, utrumne animam rem corpoream intelligunt, cui necesse sit locum pervium exitumque non impeditum, ne egressum obturatum offendat, neve articulos ledat? Utrumne item ipsam formarum conceptionem, quam visum vocamus, corpoream vim in anima putant, que sibi vereatur ne obstaculi offensione impediatur, unde sibi fenestre exigantur? Atqui nec anima nec ipsa eius de exteris consideratio corporea sunt, quare nec corporalem offensionem metuunt. Itaque quantum ad eorum sensa nec corporalem apertionem exigunt. Quod si nullam vim verba eorum claudunt, neque in fenestris oculorum viam aliquam anime id non postulantis ponunt, quid est quod eam per oculos exteriora videre dicunt, precipue cum nec ipsi oculi perforati sint, immo ipso naso ceterisque sensualibus instrumentis solidiores? Unde si quam corporalem patentiam anima exigeret, magis ei aures vel nares que cave sunt id prestarent. Item si ita ut aiunt anima exteriora tuendo formas rerum videret, qua ratione humana anima speculum contemplando sue faciei formam videret? Speculum enim suis fenestris oppositum videre posset, facies vero propria, ut opposita non est, ita hac intentione videri non potest. Itaque eorum sententia vacillans est. Quare secunda quam superius posuimus opinio agenda est.

NEPOS: Agenda certe. Neque enim video superioris sententie assertores, qua ratione hec offendicula vitent.

[154] P *add.* et
[155] P *add.* et

Adelard: The possession of all the senses, as Boethius attests in his *Music*,[39] is available to all animals. But what their force is (or their mode of operation) is clear only to the intellect of someone who practises philosophy. I'll explain. The effects of things follow from the preceding causes by a most subtle kind of chain, and the causes themselves with their effects differ among themselves by some most subtle distinctions. Hence the nature of things often slips out of the grasp even of the philosophers themselves. For if, as those whom we are talking about profess, the soul, placed in the brain as if on her seat, contemplates outside things through the windows of her eyes, in saying this either they create a certain force in the windows of the eyes, or they do not locate any aid for contemplating outside things in such windows. But if they place the aptitude for seeing outside things in the eyes, saying they are windows, do they understand the soul to be a corporeal thing for which it is necessary to have an opening and an unimpeded exit, lest it bump against a closed-off outlet and hurt its limbs? Do they think, too, that the very conception of forms which we call sight is a corporeal force in the soul, which fears that she might be hindered by the blockage of some obstacle, and hence demands windows for itself? But neither the soul nor her very consideration of outside things is corporeal, and therefore they do not fear a corporeal hindrance. Thus, in respect to what is sensed by them, they do not require a corporeal opening. But if their words carry no weight, and they do not place in the windows of the eyes any route for the soul (which does not require such a route), why is it that they say that she sees outside things through the eyes, especially when the eyes themselves are not perforated – indeed they are more solid than the nose and the other instruments of the senses? Hence, if the soul demanded some corporeal opening, the ears or the nostrils, which are hollow, would provide that better. Similarly, if the soul, by looking at outside things in the way they say, were to see the forms of things, how would the human soul see the form of her own face by looking into a mirror? For she could see the mirror opposite her windows, but her own face, as it is not opposite, cannot be seen by this means. Thus their opinion totters. Therefore the second opinion which you put forward before should be dealt with.

Nephew: It should be dealt with, certainly. For I do not see how the supporters of the first opinion can avoid these little obstacles.

ADELARDUS: Secunda igitur de visu secta a Stoicis primum relata est, formas scilicet exteriores ad ipsam usque animam accedere, eique more cere insigniri. Unde Boetius in Philosophica Consolatione:

Quondam Porticus attulit
obscuros nimium senes,
qui sensus et imagines
e corporibus extimis
credant mentibus inprimi,
ut quondam celeri stilo
mos est equore pagine
inpressas figere notas.

Que licet in eodem loco ab eodem ut futilis inpugnetur, a nobis quoque quomodo improbetur sic habeto. Si forme videndorum ad animam videntis accedunt, aut in suo subiecto et cum eo, aut in alio et cum alio; etenim sine omni impossibile est. Set in suo et cum suo non: neque enim corpus visum ad me accedit. Si in alio et cum alio, dicant cum quo.

NEPOS: In aere et cum aere. Informat enim, ut aiunt, corpus aerem, aer alium, usque ad illum qui in cerebro est, qui et ipsam inficit animam.

ADELARDUS: Si igitur tali formarum traditione ad animam usque accessus fit,[156] quomodo si vitreum corpus interponatur? Utrumne ille formarum progressus per tale obstaculum protundetur?[157] Utrumne aer vitro circumiacens a vitro non inficietur, a forma vero longinquioris corporis minusque splendidi informabitur? Atqui hoc absurdum est. Absurda itaque et eorum opinio est.

Item cedamus, ut ait Boetius in Topicis 'argumentandi gratia', ut quid inde sequatur videamus. Concedamus, inquam, formarum progressionem usque ad oculum ferri, ab oculo etiam ad cerebrum, vel ad cerebri aerem. Illum vero transitum, quem quasi sigilli inpressionem vocant, notemus. Si enim aer cerebri formam suam anime inprimet, aut eandem essentialiter insigniet, aut aliam ei similem. Set eandem inprimere non potest. Nulla enim individua forma ab uno subiecto ad aliud transire potest. In ipso enim transitu sine subiecto esset, quod non convenit. Si vero similem ingerit, iam in anima partes locales ut

156 sit P
157 protundetur P, protendetur GMQV

Adelard: The second doctrine about sight was advocated first by the Stoics: that outside forms come as far as the soul herself and they are imprinted on her as on wax. Hence the words of Boethius in the *Consolation of Philosophy*:

> Once the Portico brought forth
> exceedingly obscure old men.
> They believe that the sensations and images
> from the most remote bodies
> are imprinted on minds
> as at times it is the custom
> to inscribe with a swift-moving stilus
> the marks of letters on the flat surface of a page.[40]

Although this opinion is impugned as idle by the same author in the same place, this is the way we also prove it wrong. If the forms of the object of vision come to the soul of the one who sees, either they are in their subject and with it, or in something else and with it. For it is impossible for them to be without any subject at all. But they are not in their subject and with it, for the object of vision does not come to me. If they are in another subject and with it, let them say with what!

Nephew: In the air and with the air. For, as they say, the object gives form to the air, and that parcel of air to another, and so on until it reachcs that air which is in the brain and affects the soul itself.

Adelard: Well, if the arrival in the soul happens by such a relay of forms, what happens if a solid glass body is put in between? Will that progression of forms be forced through such an obstacle? Will the air surrounding the glass not be affected by the glass, rather than be shaped by the form of an object further away and less bright? But this is absurd. Therefore their opinion too is absurd.

Another thing: let us accept this 'for the sake of argument', as Boethius says in the *Topics*,[41] and see what the consequence is. That is, let us accept that a progression of forms is carried as far as the eye, and from the eye also to the brain, or to the air in the brain. Let us take note of that transition, which they call an 'impression', as if of a seal. For if the air of the brain impresses its form on the soul, it will impress either a form which is the same in essence or another form similar to it. But it is not possible to impress the same form. For no individual form can pass from one subject to another. For in its very transition it would be without a subject, which doesn't make sense. But if it brings in a similar form, it will now need spatially distinct

in aere obtinuerat exiget, quas quia in ea non invenit, nec ita sigillando ei formam dare poterit. Quoniam igitur ista eradicanda esse apparent, ad cetera traducamur.

Et[158] est tercia de visu sententia eorum qui exire quidem concedunt a cerebro spiritum visibilem, set nequaquam usque ad corpus quippe[159] longinquissimum protendi, eo quod in medio aere formam quam querit inveniat. Unde non ulterius aciem elaborare necesse est. Itaque inpressus redit, formamque anime a qua missus est refert. Id autem dicentes non attendunt quid ex eorum dictis consequenter procedat. Si enim, ut aiunt, inter videntem et rem videndam aer ex progressione formarum similium visibili spiritui quod querebat offert, si quis ex orientali parte in occidentalem aspiciens ad formam albi corporis discernendam intendat, eam in medio, ut aiunt,[160] aere reperiet. At vero sit ut alius eadem hora ab occidentali plaga in orientalem ad nigrum corpus aspiciendum visum dirigat. Sequetur itaque ut nigredinis formam in eodem medio aere eadem progressione reperiat. Ita ille idem medius[161] aer albedini simul et nigredini subiectus erit. Sicque duo contraria in eodem subiecto in vi contrarietatis invenientur. Quod quia stare non potest, nec eorum qui id dicunt secta sequenda est.

Postremo quiddam esse non dubium est quod ceteris etiam oppositionibus supersessis supradictas omnes sententias satis debilitare potest: illud scilicet quod in speculis formam propriam communiter[162] concipimus. Id enim cum usu rerum cotidiano verum sit, cum supradictis non convenit.

Conveniat igitur ut illam quam Philosophus approbat tractemus sententiam, aliisque velut debilibus spretis, huic Achademice veritati fidem faciamus. Est igitur quam testamur talis sententia: in cerebro quidem quendam aerem subtilissimum nasci, ignee eundem nature, ideoque[163] levissimum utpote qui ab anima, quociens libet[164] atque necesse est, ad exteriora videnda per nervos emittitur. Unde et a phisicis 'spiritus visibilis' nuncupatur. Hunc itaque, quia corpus est, exitum localem exigere conveniens est. Habet autem egressum per duos nervos concavos, quos Greci vocant opticos, a cerebro usque ad oculos protensos; perinde exiens ad corpus usque videndum mira

[158] Et P *om.*
[159] quippe GMQ, PV *om.*
[160] in medio ut aiunt P, ut aiunt in medio GMQ
[161] medius G *p.c.*PV, individuus G *a.c.*MQ
[162] communiter GQV, PM *om.*
[163] ideo P
[164] libet GMQV, voluerit (*supra* libet) P

parts in the soul, as it had acquired in the air. Because it does not find them in soul, it cannot give the soul a form by means of this kind of sealing. Since, then, these views clearly have to be pulled out by the roots, let us pass on to the others.

The third opinion concerning sight is that of those who agree at least that a visual spirit goes out from the brain, but say that by no means does it extend as far as the object, being very far away, because it finds the form which it is seeking in the air in between. Hence it is not necessary for the glance of the spirit to make an effort to go any further. So, having received the impression, the visual spirit returns and brings back the form to the soul from whom it had been sent out. Those who say this do not pay attention to the consequences of their words. For if, as they say, the air between the one who sees and the object of vision, as a result of the progression of forms, yields up to the visual spirit what it was seeking, if anyone looking from east to west aims to make out the form of a white object, he will find it, as they say, in the air in between. But say there is another person who, at the same time, directs his sight from west to east, to look at a black object. It will follow that he finds the form of blackness in the same air in between, where it had arrived by the same progression. Thus that same air in between will be subject to whiteness and blackness at the same time. So two contraries having a contrary force will be found in the same subject. Since this cannot stand, the doctrine of those who say this should not be followed either.

Finally there is undoubtedly something which, leaving aside every other objection, is able to weaken all the views mentioned before – the fact that we all see our own forms in mirrors. Since daily experience of reality proves that this is true, it does not fit with the previous opinions.

It would be fitting, then, for us to deal with that view which the Philosopher approves, and, having discarded the others as weak, to fix our trust in this truth of the Academy. The view that we defend is as follows. In the brain a kind of very fine air comes into being, which has a fiery nature and is consequently very light, being that which is sent through the nerves by the soul for seeing outside things, whenever the soul wishes and when necessary. Hence it is called by the natural scientists too 'the visual spirit'. Because this is corporeal, it is appropriate that it needs an exit which has 'place'. So it has an egress through the two hollow nerves which the Greeks call the 'optic nerves', extended from the brain as far as the eyes. Leaving by this route it travels to the object of vision at a remarkable speed, and,

celeritate evadit, eique inpressus formam eius et recipit et retinet, rediensque ad ortum suum, formam retentam remonstrat. Hic autem spiritus a Philosopho 'vis ignea' nominatur.[165] Que vis cum speculum sibi oppositum vel tale aliquid elucidatum corpus reperit,[166] ab eo repercutitur, ex repercussione ad vultum proprium redit, eiusque formam sibi retinens, introiens anime renuntiat. Nec vero putes vim[167] hanc igneam formam vultus in speculo repperisse, set a tam levi planitie in qua herere non potest, reverberata,[168] in vultu illam redeundo recipit, receptam refert. Et hec est Achademicorum divina quam sanxit Plato sententia in Timeo inter cetera his verbis:

> Due sunt, opinor, virtutes ignis: altera edax et peremptoria, altera mulcebris innoxio lumine. Huic igitur ex qua lux diem invehens panditur, domesticum et familiare corpus oculorum divine potestates commente sunt, intimum siquidem nostri corporis ignem utpote germanum ignis perlucidi, sereni et defecati[169] liquoris, per oculos fluere ac demanare voluerunt, ut[170] serenus ignis per eandem deflueret medietatem. Itaque cum diurnum iubar applicat se visus fusioni, tunc nimirum incurrentia semet invicem duo lumina similia in unius corporis[171] speciem coherent,[172] quo concurrunt oculorum acies emicantes, fluentis quoque intima fusionis acies contigue imaginis occursu repercutitur. Totum igitur hoc similem eandemque sortitum passionem eiusdemque passionis effectum, cum quid aliud tangit vel ipsum ab alio tangitur, tactum motu diffundens se per omne corpus, perque corpus ad animam porrigens, sensum efficit qui visus vocatur.

De speculis item et aquis paulo post ait:

> Siquidem utriusque ignis tam intimi quam extra positi concursu incidente in tersam levemque superficiem, formatique in multas et varias figuras, simulachra ex levigati corporis conspectu resultant.

165 nuncupatur P
166 reperit P, repperit GQ, reppererit M
167 vim GMQ, PV *om.*
168 reverberata GMQ, -tam PV
169 defecati G *p.c.*PQV, defricati G *a.c.*M
170 Q *adds the missing phrases from Calcidius' translation*: per leves congestosque et tanquam firmiore soliditate pavidos orbes luminum quorum tamen esset angusta medietas subtilior.
171 corporis Q, GPMV*om.*
172 Q, GPMV *add.* et

having received an impression from the object, it both takes in and retains the object's form, and, returning to its origin, shows again that form which it has retained. This spirit is called by the Philosopher the 'fiery force'.[42] This force, when it finds a mirror or any such bright object in front of it, is propelled back from it, and as result of this propulsion returns to its own face and, retaining the form of the face, it announces that to the soul as it enters. But you should not think that this fiery force has found the form of the face in the mirror. Rather, since the surface is so smooth that it is not able to stick to it, it is thrown back from it and, in returning, it receives that form in the face, and brings back the form it has received. And this is the divine opinion of the Academicians, to which Plato gave his approval in the *Timaeus*, in these words (among others):

> In my opinion, the qualities of fire are two. One is corrosive and destructive, the other mild with harmless light. For the second, then, from which the light bringing in the day spreads out, the divine powers contrived the homely and familiar substance of the eyes. For they wished the inner fire of our body, being the brother of a very bright fire – a cloudless and spotless liquid – to spring from and flow out through our eyes, as the cloudless fire might flow down through the same medium.[43] Thus when the brightness of day joins itself to the flowing of vision, then, not surprisingly, the two similar lights running into each other cohere into the species of one body, and wherever the flashing glances of the eyes run to, the inner glance of the outpouring flow too is thrown back by meeting the neighbouring image. So all this, having acquired a similar and the same feeling and the effect of the same feeling, when it touches something else or is itself touched by another, having been touched, pouring itself over the whole body and stretching through the body up to the soul, it brings into being the sense which is called sight.

Similarly, about mirrors and water he says a little later:

> By the meeting of each fire, both the inner and the outer, hitting a polished and smooth surface and formed into many different figures, the images jump back from looking at the smooth body.

Intelligis demum ex his Philosophi verbis, quem sicut numquam quesita veritas latebat, ita et aliquando falsitas non fallebat – intelligis, inquam, quomodo igneum spiritum per oculos fluere commemorat, eundemque etiam vidende rei occursu repercuti non tacet, similem etiam sortiri passionem, id est similem formam non diffitetur, perque oculi corpus redeundo se diffundere, deinde ad animam porrigi visumque sic effici tandem autumat, idemque in speculis esse consequenter affirmat.

.xxiv. NEPOS: Intelligo inquam et nichil hoc intellectu michi gratius. Cum enim et in philosophicis contemplationibus et in phisicis causarum effectibus, ethicisque etiam consultibus Platoni te penitus consentire perceperim, inveni inpresentiarum quo tibi et Platoni imponam; habeo quo utriusque[173] bilem moveam. Si enim, ut superius dissertum est, aliquid a cerebro exit – sive illud visibilis dicatur spiritus, ut phisicis placet, sive nostri corporis ignis, ut Platoni sedet – id ipsum aut[174] substantiam esse necesse est aut accidens.

ADELARDUS: Substantia quidem corporea. Est enim,[175] ut ait, ignis subtilissime substantie ex quatuor elementis compositus.

.xxv. NEPOS: Satis sane. Neque[176] enim, si accidens esset, sine subiecto ad exteriora transire posset. Cum itaque corpus id esse intelligas, quod inde incommodum sequatur intelligere aptum est. Cum enim firmamentum, quod applanos dicitur, a quibusdam etiam ultima spera vocatur, omne corpus omnemque contineat locum, stelle autem superiores – ille dico que non moveri videntur – in applano fixe[177] dicantur, igneus autem ille spiritus uno oculi ut ita dicam ictu stellam videat, quoddam corpus tantam aeris densitatem que infra Lunam est infinitumque spatium quod a Luna ad Solem, a Sole usque ad ipsum applanon continetur, tam brevi tempore transire, iterum rediens remetiri concedendum est? Qua insania[178] nichil impossibilius est, cum ipsa tocius terre latitudo nullam vel minimi puncti proportionem quantum ad celi infinitum fere diametrum optinere possit.

ADELARDUS: Sic sic equidem pugnat, qui rerum naturas ignorat. Tu autem ades animo et quod sequitur memorie trade. Sicut enim inter ipsa spatia quedam ampla, quedam ampliora, quedam amplissi-

173 utriusque PV, utrisque GMQ
174 aut P *om.*
175 enim PM, etenim GQV
176 Neque MQ, ne GPV
177 fixe M, fixa GPQV
178 Que insanius P

You understand at last, from these words of the Philosopher (just as the truth once sought never eluded him, so falsity did not deceive him at any time) – you understand, I say, how he relates that the fiery spirit flows through the eyes, and is not silent about the fact that the same spirit is thrown back by hitting the object of vision. He does not deny either that it acquires a similar suffering – that is, a similar form – and, on its return, it pours itself through the substance of the eye. Finally he claims that it then reaches as far as the soul and thus sight is effected, and he affirms that the same happens as a consequence in the case of mirrors.

24. Nephew: I understand, I say, and nothing is more pleasing to me than this understanding. But now that I have perceived that you completely agree with Plato in philosophical speculations, in the physical effects of causes and even in ethical deliberations, I have found in the present circumstances a problem by which I may trick both Plato and you; I have something by which I may annoy both of you. For if, as was discussed before, something comes out of the brain – whether that should be called a visual spirit, as the natural scientists like to call it, or the fire of our body, as pleases Plato – that thing must be a substance or an accident.

Adelard: It is of course a corporeal substance. For it is, as he says, fire of a most subtle substance, composed from the four elements.

25. Nephew: Very well. For if it had been an accident, it would not be able to pass to exterior objects without a subject. So, since you understand it to be a body, it is appropriate to understand what inconvenience follows from that. Since the firmament which is called the *aplanos*, and by some is also named the ultimate sphere, contains every body and every place, and the higher stars – I mean those which seem not to move – are said to be fixed in the *aplanos*, but that fiery spirit with one glance of the eye, so to speak, sees a star, must one concede that a certain body can pass through such density of air below the Moon, and the infinite space which stretches from the Moon to the Sun and from the Sun as far as the *aplanos* itself, in such a short time, and returning again, measures out the same space once more? Nothing is more impossible than this madness, when the very breadth of the whole earth can occupy not even a fraction of the tiniest point in comparison with the almost infinite diameter of the heavens.

Adelard: Thus fights the man who does not know the natures of things! But you must lend me your attention, and commit to memory what follows. I think that just as among spaces themselves some are large, some larger, some most large, so among bodies themselves some

ma sunt, ita ipsorum corporum quedam velocia, quedam velociora, quedam velocissima esse arbitror. Id autem non omnibus perspicuum est. Sicut enim hec celi amplitudo eiusque forma, que a geometra describitur, vulgari animo vix aut numquam patet, sic ei tanta velocitas tam brevi spatio patere non potest, quippe cum sit eorum oculus velocior quam animus. Metitur enim, immo ut verius dicam mentitur, omnia secundum hoc quod iuxta terram sibi fallaces promittunt sensus, estimatque celi ambitum terre undique versum insidere, tanteque nec maioris quantitatis quam lippientibus oculis eius videtur formas Lune et Solis ceterorumque potest iudicare, cum ipsa vera ratione terre ambitum excedant. Illos[179] vero qui magis in huiusmodi rebus incorporeo animi oculo ratione duce utuntur, sicut amplitudo extime continentie eiusque diametri fere infinita spatiositas non latet, sic et ipsius uranice conversionis visibilisque spiritus velocissimus motus latere non potest. Quod utrisque rationis merito accidere puto. Sicut enim illa extime spere anima omnibus conditorum animabus[180] virtutum exsecutione prestantior est, sic et hic visibilis spiritus omnibus elementaliter[181] compositis mirabili creatricis virtutis opera subtilius elimatus est. Illius itaque tam velocis conversionis causa perfectio anime est; huius vero tam velocis itus et reditus ratio perfectio corporis est. Quod qui nescit, simul et effectus nescire et ammirari non mirum est.

Illud tamen preterea tibi audiendum est quod non omnibus id philosophis placet quod superius posuisti, scilicet superiores stellas in applano fixas esse. Aiunt enim et eas longe inferiori ambitu infra celum in aere volvi. Quod suo in loco si vitam vixero diffiniens tractabo. Dum itaque et impossibilitas et spacium decrescit, decrescat et dubitatio.

.xxvi. NEPOS: Verumtamen magis magisque crescit, priusque ego ipse ut reor penitus decrescam, quam ab hac dubitatione decidam. Si enim quocumque modo visibilis spiritus stellam tangit, dum id facit, oculus claudi potest, itaque cum hoc frequenter accidat, spiritum illum extra relinqui necesse est.

ADELARDUS: Si ea que superius disserta sunt mente teneres, minime michi id opponeres. Ipse etenim spiritus velocissimus est, ut supradictum est; eundem autem ab anima et per animam mitti constat. Ipsum item oculum non nisi per animam claudi, a qua omnis

[179] Illud P
[180] animalibus G
[181] elementaliter PMQ *p.c.*, elementabiliter GQ *a.c.*(?), elementorum V

are swift, some swifter, some are most swift. But that is not obvious to all people. For just as this largeness of the heavens and their shape, which is described by the geometer, is scarcely or never clear to the mind of the common people, so that enormous speed occurring in such a brief space of time cannot be clear to them, since their eyes are swifter than their brains. For the common person measures – I should rather say, mis-shares – all things according to what the deceiving senses suggest on the terrestrial level, and he considers the circuit of the heavens to sit on top of the earth on all sides, and can even judge the forms of the Moon, the Sun and the other planets to be of the same size and no larger than it seems to his bleary eyes, although they are larger than the circumference of the earth, in the light of true reason. But as for those who rather use the incorporeal eye of the mind in matters of this kind, with reason as guide, just as they are not unaware of the size of the outermost container and the almost infinite extension of its diameter, so the most swift movement of the turning of the heavens themselves and of the visual spirit cannot escape their notice either. This, I think, happens to each of these by the right of reason. For, just as that soul of the outermost sphere is more excellent than all the souls of created things in the execution of its powers, so this visual spirit also has been refined more subtly than all bodies composed of elements by the wonderful work of the creating power. Thus the cause of the very fast rotation of the former is the perfection of the soul, but the reason for such a swift going to and fro is the perfection of the body. It is not surprising that the one who does not know this is at the same time ignorant of the effects and amazed.

However, you must also bear this in mind: that not all philosophers agree with what you posited before – that the higher stars are fixed in the *aplanos*. For some say that they too turn about in the air below the heavens in a much lower orbit. I shall define and deal with this subject in its right place if I live long enough. Thus while both the impossibility and the space become less, the doubt also should diminish.

26. Nephew: No, rather it grows more and more, and I myself, as I think, will fade into nothing before I fall away from this doubt.[44] For if the visual spirit touches the star in any way at all, while it does this, the eye can close and thus (as this in fact often happens) it is necessary for that spirit to be left outside.

Adelard: If you were to hold in your mind what was discussed before, you would not raise this objection against me at all. For the spirit itself is very fast, as was said; and it is agreed that it is sent by the soul and through the soul. Similarly, it is clear that the eye itself is

motus voluntarius corpori datur, patet. Pateat igitur et eius consequens, ut que mittit et recipit, portamque quando vult claudit, nullum sibi incommodum intercludat, nichil suum extra remaneat, precipue cum ipsum tante velocitatis sit[182] ut se statim in suo recipiat. [Set etiam ut tue protervitati satagam, licet intercludi concedam, non tamen inde videndi sensus magis perit. Sicut nec si tangentis manus abscidatur, ideo ipse tangendi sensus minor efficitur.][183] Frustra igitur hoc incommodum quis doleat, cum mittenti[184] nature et missum pareat et parenti porta non resistat, [etiamque si intercludatur, non ideo magis quam iam visum est ipsam animam latet].[185]

.xxvii. NEPOS: At tu ne doleas quod, dum Platonem sequeris, in foveam incidis. Licet enim quod esse potest quocumque modo in hoc conventu recipiamus, sequitur tamen aliquid unde tibi nodum innectam.[186] Si enim ita ut dicis hic spiritus vel ignis usque ad videnda porrigitur, aut antequam redierit aut post reditum visus efficitur.

ADELARDUS: Nichil antequam redeat nuntiat.

NEPOS: Si igitur postquam redierit visus constat,[187] cum directis oculis aliquid diu intuemur, statim cum illud videmus, illum ivisse et rediisse necesse est. Set continue et sine intervallo id spectamus. Itaque contiguum et sine intervallo habebit itum et reditum. Quare et dum it[188] redit et se ipsum impedit.

ADELARDUS: Immo tu te ipsum[189] impedis. Nam cum prave intelligis, tum[190] oblique opponis. Non enim ex eo quod continuo obtuitu aliquid spectamus, sequitur ut ille spiritus dum it[191] redeat, vel se ipsum impediat. Nos enim eum tam velociter ire et redire dicimus, ut, uno reditu finito, altera progressio tam cito fiat quod nulla ibi mora esse aliquo sensu percipi possit, cum sit tamen aliqua, non que sensui set intellectui pateat. Itaque non iam necesse est ut dum it redeat, set ut ea velocitate qua post itum redit, eadem post reditum

182 velocitatis sit PQ, GV *transp.*
183 [Set etiam . . . efficit] *in* P *only.*
184 cum mittenti P, committenti GMQV
185 [etiamque . . . latet] *in* P *only.*
186 innectam GMQ, nectamus PV
187 constat PQV, constet GM
188 id P
189 ipsum GMV, PQ *om.*
190 tunc G
191 id P

only closed through the soul, by whom every voluntary movement is given to the body. The consequence, then, should also be clear – that what sends and receives, and closes the door when it wishes, should be hindered by no inconvenience, and nothing belonging to it should remain outside, especially since it is of such velocity that it immediately receives itself into its own. [But also, to satisfy your audacity, even if I should agree that it is shut out, the sense of sight still does not perish because of this, just as, if the hand of the man who touches something is cut off, the sense of touching itself is not made less thereby].[45] So there is no point in anyone's fretting about this objection, since both the thing sent obeys the nature of the sending nature, and the door does not resist what obeys, [and even if it is shut out, that does not escape the notice of the soul any more than what has already been seen.][46]

27. Nephew: But it is you who should be warned not to fret, because, while you are following Plato, you are falling into a pit. Although we accept in this meeting, in some way or other, what *can* be, nevertheless something follows from which I can tie a knotty problem for you. For if this spirit or fire reaches as far as the objects of vision in the way you say, either the act of sight occurs before it returns or after its return.

Adelard: The spirit can announce nothing before it returns.

Nephew: If then sight is established after it has returned, when we look at something directly for a long time, it is necessary for that spirit to have gone and come back as soon as we catch sight of that thing. But we are looking at it all the time and without a break. Thus the spirit will have to go to and fro continuously and without a break. Therefore it returns at the same time as it goes out and gets in the way of itself.

Adelard: Rather, it is you who are getting in the way of yourself. For when your understanding is warped, your attack is askew. It does not follow from the fact that we look at something continuously, that that spirit returns at the same time as it goes out or gets in its own way. For we say that it goes and returns so fast that when one round trip has been finished another journey takes place so quickly that no delay can be perceived to be there by any of the senses, although there is some delay which is not obvious to the senses but only to the intellect. Thus it is not necessary that it returns at the same time as it goes out, but that with the same velocity with which it returns after going out, it should go out after returning. Because this interval is

eat. Quod intervallum quia sensu non discernitur, continuum iter esse falso videtur.

.xxviii. NEPOS: O illigabilem verborum responsionem. Dum enim meam oppositionem meditor, nichil ea michi ratione firmius, nichil verborum consequentia coherentius videtur. Cum vero ad responsa[192] mentem applico, nichil in eis irritum video. Ita me indefessa solutionum tuarum scaturigine invitum quidem ad sententiam tuam reducis. At vero per propria verba tibi imponere ulterius satagam. Inexhausto enim verborum tuorum fonte cum superius Stoicorum sensa reprimeres, non posse ab aere in anima sigillari more formam inprimi, disseruisti. Quod si eis rationabiliter obiecisti, tu ipse idem recipias iustum est. Esto igitur ut Platonicus ignis cum forma sine omni offensione ad animam redeat, sitque impossibile ut eam illi induat. Itaque totus iste labor vacillat.

ADELARDUS: Tu vero vacillas magis, dum, quod discipuli forma addiscere deberes acclinis, ita cum impudenter tum inefficaciter arguis. Neque enim ego iuxta obscuram portam set medium forum habito; etiam si id erras, non Stoicum me set Bathoniensem dico. Quare non Stoicorum errata, set meum intellectum docere debeo.

NEPOS: Ego itaque taceo.

ADELARDUS: Primum id igitur[193] tacens intellige, quia, cum Conditor immensitatis huius mundum decore illustraret, precellenti nature quam animam vocamus intellectuales formas omnium creaturarum induit. Unde sibi presto est et preterita refigurare, et absentia representare, et futura augurari. Illum itaque formarum intellectualium thesaurum non semper set cum necesse est atque vult explicat. Cum itaque nova ad presens forma visibilis insignitur spiritus, anima statim non illam set similem explicans, non corpoream set intelligibilem ostentans, sue habundantie exprimit qualitatem, indeque ad iudicium irritatur. Cum igitur et visibili igni itus et reditus presto sit, set nec etiam formarum traditio obsit, auctor huius divine rationis Plato simul cum suis celebretur, ametur.

192 responsam P
193 igitur PV, GMQ *om.*

not discerned by the senses, the movement seems falsely to be continuous.

28. Nephew: What an untieable string of words in reply! For when I try to work out my opposition tactics, nothing seems firmer than that reasoning, nothing to hang together more than that sequence of words. Yet when I apply my mind to your replies, I see nothing invalid in them. Thus, by the untiring torrent of your solutions, you bring me – unwillingly indeed – round to your opinion. But I shall endeavour to trick you with your own words. Within the inexhaustible fountain of your words, when earlier you found fault with the views of the Stoics, was your claim that it was not possible for a form to be impressed on the soul by the air, like a seal. If you raised this objection against them in accordance with reason, it is just that you yourself should accept the same reason. Let it be that the Platonic fire with its form returns to the soul without any hindrance, and that it is impossible for it to clothe the soul with that form. And so this whole structure of yours totters.

Adelard: But you totter more, while you both impudently and ineffectually argue in this way against what you ought to learn docilely as a pupil. I do not lurk under the shade of some dark doorway, but occupy the centre of the square. Though you also err in this: I call myself not a Stoic, but a man of Bath.[47] Therefore I should not be expected to teach the errors of the Stoics, but my own views.

Nephew: So I am silent.

Adelard: Keeping silent, then, understand this above all: that when the Creator of this huge expanse illuminated the world with beauty, he clothed the pre-eminent nature which we call the soul with the intellectual forms of all creatures. Hence it is possible for the soul to reshape past forms, to picture absent things, and to predict the future. She opens that treasury of intellectual forms not all the time, but when necessary and when she wishes. Thus when the visual spirit is impressed with a form which is new at the time, the soul, by immediately revealing not that form but a similar one and showing not a corporeal but an intelligible form, expresses a quality of her own wealth, and hence she is spurred into making a judgement. Since, therefore, both going out and returning are possible for the visual fire, and the handing over of forms is not an objection, the author of this divine reasoning, Plato, and his followers should be both praised and loved.

.xxix. NEPOS: At nondum ita[194] nobis festivus habetur, nisi prius de speculorum formis explicetur. Si enim, ut Plato tuus velle videtur, dum igneus ille spiritus in detersam ac defecatam inciderit materiam, inde repercuti ac reverti necesse est, ut quid,[195] cum album vitrum ac detersissimum intuemur, simili causa vultus non refertur?[196]

ADELARDUS: In causa quidem talis reciprocationis plana superficies recipienda est, set non solitarie. Addenda est etenim eiusdem corporis aut densitas aut obscuritas, unde reverti aut necesse sit aut voluntarium. Inde etiam si eidem albo vitro quod, quia rarum et lucidum est, penetrari a tali spiritu potest, si illi inquam nigrum aliquid subponatur,[197] iam repercussus spiritus ad suam redibit faciem. Odit enim tenebras, amatque sibi similem lucem. Unde et in aquarum levi superficie si simul obscuritate obumbrentur, eedem[198] repercussiones fiunt, quas Philosophus umbras vocat. Itaque in tali reditu tam corporis levitas[199] quam eiusdem densitas, quandoque etiam et umbrositas notanda est.

[Hic deest cur in occipitio careatur oculis][200]

.xxx. NEPOS: His usque[201] evolutis, adhuc michi de visus natura id obstat, quod quivis in obscuro existens quicquid in luce agitur discernere potest; qui vero in luce fuerit, quod in obscuro agetur minime videbit. In quo visus natura mira videtur. Cum enim spacium equum sit, tantumdem enim est a luce ad obscurum, quantum ab obscuro ad lucem, mirum est quod sicut visus hominis in obscuro existentis usque ad lucem pervenit, non similiter existentis in luce visus in obscurum usque perveniat, precipue cum fortior exire videatur qui in luce exit, quam qui in tenebris, unde et longius procedere debere videri potest.

ADELARDUS: Ego vero de tua ammiratione non miror. Qui enim cecus est, ita de luce loquitur. Nunc vero intelligas quia quanto visibilis spiritus a sua nativitate elongatur, tanto debilior efficitur; quanto vero debilior est, tanto facilius impeditur. Unde fit ut ille qui in luce emittitur, licet ibi vi magna pollens exeat, tamen qui in itinere a cerebro elongatur, necessario in via debilitatur. Debilis vero factus,

194 P *add.* a
195 quid GQV, quicquid etiam P, M *om.*
196 refertur MQV, referetur G, referuntur P
197 subponatur GQV, superponatur PM
198 heedem P
199 lenitas V
200 [Hic deest . . . oculis] *in* P *only*
201 itaque G

29. Nephew: But Plato is not yet thus fêted by us, unless first an explanation of the forms in mirrors is given. For if, as your Plato seems to wish, when that fiery spirit falls on a polished and spotless object, it must be thrown back from it and return, how is it that when we look at transparent glass of the most polished kind, the face is not reflected for a similar reason?

Adelard: In the case of such reflection a smooth surface must be provided, but not that alone. For one must add either the thickness or the opacity of that object, so that the spirit either has to or wishes to return. Hence also if behind that same transparent glass, which can be penetrated by such a spirit because it is thin and light, if, as I say, behind that glass something black is placed, the spirit will now be thrown back and return to its face. For it hates darkness and loves the light in that it is similar to itself. Hence also in the smooth surface of water, if it is at the same time overshadowed by darkness, the same reflections occur, which the Philosopher calls 'shadows'.[48] Thus, in such a reflection, both the smoothness of the object and its density, and sometimes even its shadiness too, should be taken note of.

[Here is missing 'Why there are no eyes at the back of the head'][49]

30. Nephew: Having got this far, I still have this problem concerning the nature of sight, that someone in the dark can discern whatever is happening in the light, but whoever is in the light cannot see at all what is going on in the dark. This appears to be a strange characteristic of vision. For when the distance is equal – for as much space separates the light from the dark as the dark from the light – it is strange that the sight of the man in the light does not penetrate the dark in the same way as the sight of the one in the dark reaches the light, especially since it would seem that the sight that goes out in the light would be stronger than the one that goes out in the dark, and could seem therefore to have to penetrate further.

Adelard: I, at any rate, do not find strange the fact that you find it strange. For the one who is blind speaks about light in this way. But now you should understand that the more the visual spirit is removed from its birthplace, the weaker it becomes, and the weaker it is, the easier it is for it to be impeded. Hence it happens that the spirit that is sent out in the light, although it goes out thriving there with great force, nevertheless, becoming distant from the brain as it travels, it necessarily becomes weak on the route. Then, having been weakened,

cum iam tenebris offenditur, omnino alienatus frustratusque nichil efficit nisi si ad sedem suam sine efficatia redit. Qui vero in obscuro procedit, dum iuxta cerebrum suum est, facile suum contrarium patitur, nec ab eo vincitur. Est enim presto iuxta se unde vires continuo reparet. Item si in ipsa extensionis elongatione aliqua sui parte deteritur, quia non iam tenebras set lucem repperit, ipsa lucis cognatione vires resumit quodque in luce repperit, non diffidens discernit. Sicque quicquid superius obieceras ipso rationis merito in contrarium effectum cedit.

.xxxi. NEPOS: Quia de his ulterius non ambigo, de ceteris que ad sensus pertinent audire desidero. Licet enim ante tuas assertiones dum dubitatio animum pregravat, livor michi oculum perstringat, eadem tamen supplosa in benivolentiam redit animus. Neque enim ego ille sum qui ob invidiam disciplinam relinquam.

ADELARDUS: De illis quidem breviter perstringendum puto, presertim cum ex non[202] dissimili causa procedant, eandemque explicationem non fugiant. Quod sic habeto: odoratus quidem et gustus et tactus, instrumentorum suorum cum rebus exteris coniunctione[203] fiunt. Ad odorem enim efficiendum aer odore affectus quibusdam mammulis a cerebro suspendulis, que huius sensus instrumenta sunt, coniungitur. Lingue item et palato, que gustus instrumenta sunt, res gustande adhibentur. Applicatur item alias rei tangende manus, alias vero ipsi manui res tangenda adiungitur. Unde et ipsorum instrumentorum qualitates necesse est immutari, hac videlicet necessitate. Hec enim ipsa instrumenta, ut communiter agam, pre ceteris corporis partibus qualitates habent temperatas, mammule videlicet ad odorem, lingua et palatum ad gustum, manus vero ad tactum. Unde fit ut cum aliquid extrinsecus adhibetur, cuius aliqua qualitas instrumentorum temperantiam excedit, similis subiecti passio ad cognationem excrescens, in distemperantiam convertit instrumentum. Quod in tactu figuraliter videri licet. Cum enim vola manus calorem frigoremque ceterasque confamulas passiones habeat temperatas, cum quid magis calidum quam frigidum offendit, calor eius prius equalis, modo fit inequalis.[204] Similiter etiam de ceteris figurari licet. Hanc itaque instrumenti sui inequalitatem anima provida percipiens, ex uno simili

[202] ex non PQV, GM *transp.*
[203] coniunctione GQV, -nem PM
[204] fit inequalis PQV, GM *transp.*

when it is struck by darkness, it is completely alienated and frustrated, and achieves nothing except the return to its seat without result. But the spirit which goes out in the dark, when it is close to its brain, easily endures what is opposite to it and is not defeated by it. For it has readily available to it the means to recover its strength immediately. Similarly, if, when it extends some distance away, it is partially weakened, because it does not now find darkness but light, it recovers its strength through its very relationship to the light, and confidently discerns what it finds in the light. Thus, whatever objection you made before yields to the contrary result by the right of reason.

31. Nephew: Because I am no longer unsure about these things, I want to hear about everything else which pertains to the senses. For although, while doubt oppressed my mind in the face of your assertions, ill-will touched my eyes, now that that same doubt has been stamped out, my mind returns to a spirit of goodwill. For I am not the kind of man who refuses to follow instruction out of spite.

Adelard: I think that one must touch briefly upon these other senses, especially since they proceed from a not dissimilar cause, and do not refuse the same explanation. Look at it like this: smell, taste and touch come about through the contact of their instruments with outside things. For, to create a smell, the air which is affected by the smell comes into contact with certain little teats dangling a little from the brain which are the instruments of this sense. Again, things that are to be tasted are brought into contact with the tongue and palate, which are the instruments of taste. Again, sometimes the hand is applied to the thing that is to be touched, but at other times the thing to be touched is brought into contact with the hand itself. As a result of these contacts the qualities of the instruments too are bound to change, for this necessary reason: these instruments (to take them all together) – namely, the little teats for smell, the tongue and palate for taste, the hand for touch – more than any other parts of the body have temperate qualities. Hence it comes about that when something outside is presented, some quality of which exceeds the temperateness of the instruments, a corresponding affect in the subject, exceeding the temperateness in the same ratio, changes the instrument into a state of distemper. This can be shown using the example of touch. Since the palm of the hand has heat and cold and the other related affects in a temperate state, when something more hot than cold meets the hand, its heat, which before was balanced, now becomes unbalanced. One can give similar examples concerning the other senses. The ever-watchful soul, thus perceiving the imbalance of her instrument, and

aliud auctum esse augurans, calidum esse quod tactum est proponit. Similique ratione odoratus et gustus sensuales fiunt passiones. Dictum itaque sit et ibidem de eisdem.

NEPOS: Dictum sane. Et quoniam de sensibus haut incommode enodatum est, quid de reliquo humano statu intelligas intelligere velim. Si enim de his que[205] circa hec constantia sunt dissolveris, nichil deinceps dubito quin ad altiora animum habeas expeditum.

ADELARDUS: Ascendere cum voles, alatum me reperies.

.xxxii. NEPOS: Erit quidem cum ascendam; nunc vero circa hec leviora versari oportet. Cum enim nuper a parte Orientali venires qua causa studii diutine steteras, ego qui nichil magis quam te desiderabam pre gaudio lacrimas fudi. Itaque quod michi risus causa esse debuit in fletum deiecit. Quod cum non soli michi accidere dicatur, magis rerum causas quam me mutatas fore credo.[206]

ADELARDUS: Cause rerum sane intelligenti sane reperiuntur. Et gaudium quidem causa risus esse solet cum anima pre gaudio motu quodam hilariori exultat. Accidit item cum et eiusdem gaudii causa calorem naturalem per corpus extendens ad cerebrum usque eum[207] traducit, qui, cum intensior est, si ibidem humorem talem reperit, naturaliter eliquat ac per exitum suum expellit. Itaque rerum cause cum ratione constant.

.xxxiii. NEPOS: Nunc vero id quod nobis cotidie usu accidit in commune ducam: qua ratione scilicet ex eodem ore cum vult aliquis[208] calidum proferat aerem, quando vero libet, contrarium. Cum enim, ut phisici perhibent, ex uno eodemque pulmone talis omnis aer exeat, perque eandem viam prodeat, mirum est unde tam subito in contrarias mutetur affectiones.

ADELARDUS: Hoc etiam, quoniam edubitari potest, querendum est. Scire itaque potes, ut dicis, quoniam a pulmone et calidi et frigidi huiusmodi[209] procedunt aeres. At non confuse. Cum enim inde procedit, si statim patulo emittatur ore, utpote a calore veniens neque vim ullam interpatiens, calidus sentietur. Cum vero non statim emittitur,[210] set in ore tenetur, quod pulmone non dico verum ipso

[205] qui P
[206] credo fore P
[207] eum PV, GMQ *om.*
[208] vult aliquis PQV, aliquis vult G, vult M
[209] huiuscemodi P
[210] emittitur PQV, emittetur G, mittitur M

guessing from one similar thing that another has been increased, proposes that what has been touched is hot. By a similar reasoning occur the sensible affects of smell and taste. So let us take them as described also in the same place <as where I described the others>.

Nephew: Certainly, let us take them as described. And since the problems concerning the senses have been conveniently solved, I should like to understand what you understand about the rest of the human condition. For if you resolve the problems that remain concerning these matters, I shall have no doubt afterwards but that you will have a mind ready for higher topics.

Adelard: When you wish to take off, you will find me with my wings attached!

32. Nephew: The time will come when I take off. But at present we should deal with these more down-to-earth matters. When you came from the Orient recently, where you had spent a long time for the sake of study, I, who desired nothing more than to see you, wept tears of joy. Thus, what should have been the cause of laughter for me, cast me down into weeping. Since this is said to happen not just to me, I believe that the causes of things went awry rather than that I did.

Adelard: The causes of things are correctly discovered by one who correctly understands them. Joy is usually the cause of laughter when the soul leaps up with a certain happier movement because of the joy. It also happens when, because of that same joy, the soul, diffusing the natural heat through the body, brings it as far as the brain, and when the heat is more concentrated, if it finds such moisture there, it naturally causes it to flow and expels it through its own exit. Thus the causes of things conform with reason.

33. Nephew: But now I shall bring into the open what happens to us in practice every day: by what reason from the same mouth, when a person wishes, he can blow hot breath, but when it pleases him, the opposite. For since, as the natural scientists say, all such breath comes out from one and the same pair of lungs, and goes along the same passageway, it is strange that it can be changed into opposite affects so quickly.

Adelard: One should enquire about this too, since it can be a matter of doubt. You can know, then, as you say, that both hot and cold breath of this sort come from the lungs – but not in a confused manner. For when the breath is coming from them, if it is immediately breathed out with an open mouth, since it comes from a hot source and suffers no other influence on its journey, it will be felt as hot. But when it is not immediately breathed out, but held in the mouth, which

meatu minus calidum est, in teporem[211] a calore descendit. Huc accedit ut cum eum ad frigiditatis usum producere volueris, inter labia illisum in frigiditatem ipsa sibili ventilatione convertas. Nullum enim elementum[212] in diversas qualitates aere mutabilius est.

.xxxiv. NEPOS: Quoniam de ventilatione mentionem fecisti, est ab ea quiddam quod inter nos ventilari vellem. Cum enim motum caloris esse causam indubitabile sit, cur ad frigiditatem faciendam aer flabellando ventiletur, ventilando moveatur, ratione nutante, dubitabile fit.

ADELARDUS: Motum quidem calorem augere non dubito, set id quod opponis aliorsum respicit. Cum enim flabella admovemus, id ideo certe facimus, ut calorem nostrum interiorem admixtione exterioris aeris, qui non adeo calidus est, refrigeremus. Modis enim omnibus interior noster aer, cum sic sudamus, exteriore calidior est. Unde et exteriorem motu adiungimus, ut sic applicatus per poros intrans interiori commisceatur, eumque moderetur.

NEPOS: Tu vero modum omnem respuis dum hoc modo loqueris. Sicut enim contraria contrariis destituuntur, ita similia similibus augentur. Quare si et aer interior calidus est, et exterior item calore affectus est, necesse est ut cum duo calida coniunguntur maior inde calor progrediatur.

ADELARDUS: Tu te ipsum fallis, dum verba magis quam sententiam opponis. Neque enim in huiusmodi passionibus si minor maiori additur, totum maius efficitur. Illud enim quod inter Aristotelicas asseritur regulas, si quid[213] scilicet cuilibet rei additur, totum maius efficitur, non de effectibus, set de quantitatibus nisi intelligatur prave proceditur. Maius enim et minus modis duobus reperitur, alias secundum quantitatem, alias secundum effectus, ut supradictum est. Esto enim et ut hec duo corpora alba unius sint magnitudinis, totaque eorum superficies alba, una tamen albior alia. Erit igitur ut quantum ad quantitatem quidem albedinis equaliter alba sint, secundum effectum vero magis[214] et minus; est enim una earum magis efficax. Non dissimiliter et in calore vel[215] frigiditate. Verbi gratia: si aque bullienti et calidissime aqua tepida ideoque minus calida addatur, quantitas quidem et substantie et caloris maior efficitur. Effectus vero quantitatem non sequitur. Neque enim inde calidior efficitur. Similiterque

211 teporem G *p.c.*(?)PV, tempore G *a.c.*M, temperiem Q
212 elementum G *p.c.*M, elementatum PQV
213 quid GQ, quis P, quidem MV
214 maius P
215 vel PMV, et in GQ

is less hot not only than the lungs but also than the passageway, it descends from heat into tepidness. Moreover, when you wish to use the breath for cooling, you constrict it between the lips and change it into coolness by making the pursed lips cause a breeze. No element is more changeable into different qualities than is air.

34. Nephew: Since you have mentioned a breeze, there is something about it that I would like to be aired between us. Since it is indubitable that movement is the cause of heat, reason falters and it becomes a matter of doubt why, to produce coldness, a breeze is made in the air by fanning and the air is moved in that breeze.[50]

Adelard: Of course I do not doubt that movement increases heat, but the objection you raise relates to something else. When we use fans, we certainly do that in order to cool our internal heat by mixing it with the outside heat which is not so hot. For in every way our internal air is hotter than the air outside, since that is why we sweat. Hence we harness the outside air by motion, so that, applied in this way and entering through the pores, it may mix with the interior air and moderate it.

Nephew: But you reject all scientific method when you speak in this way. For just as contraries are diminished by contraries, so similars are increased by similars. Therefore, if both the internal air is hot and the air outside is similarly affected by heat, it must happen that, when two hot things are joined, a greater heat arises from them.

Adelard: You deceive yourself when you oppose the words rather than the sense. In affects of this kind, if a lesser affect is added to a greater, the whole is not made greater. For that which is asserted amongst the rules of Aristotle – if anything is added to anything, the whole is made greater[51] – unless it is understood to refer not to effects but to quantities, leads to bad reasoning. For 'greater' and 'less' are found in two ways: one according to quantity, the other according to effect, as was said before. Suppose that these two white objects are of the same size and the whole of their surface is white, but one is whiter than the other. It will be the case that in respect to the quantity of whiteness they are equally white, but in respect to effect, they are more and less; for one of them is more efficacious. It is not unlike this with heat and cold too. For example, if to boiling – that is, the hottest – water, tepid – therefore less hot – water is added, the quantity both of substance and of heat is made greater. But the effect does not follow the quantity; for the water is not made hotter as a result.[52] Let it be

igitur et de aeribus dictum esse concedatur.[216]

NEPOS: Adhuc habeo tibi de eodem aliquid dicere. Sic enim nichil iam quod de incremento caloris conquerar habeo, hac enim ratione minime miror, quod qualitas qualitatem quantum ad effectum non augeat. Cur vero minuat, illud[217] michi ammiratione dignum videtur. Reperio enim in similitudine contrarium. Si quis enim magno cuidam igni ignem minorem addat, et quantitas augetur et effectus acuitur. Totum enim ipsum, si circumstantes consulis, dum altius afficitur,[218] calidius sentitur. Quod quia in aqua usu docente econverso[219] cadit, rationes tuas teque cum eis simul contra rerum ordinem cadere satis patens est.

ADELARDUS: Immo tu dum haut sane similitudinem sectaris, a vera causarum discretione decidis. Dum enim rerum qualitates inesse tantummodo notas, modum inherentie qui in omni tali disceptatione vim habet consequitur ut amittas. Cum enim calor et in igne et in aqua calida insit, igni quidem per se, aque vero per aliud inest. Cum enim ipsa naturaliter frigida sit, pcr ignem sibi vim facientem calefit[220] contrariamque qualitatem assumit. Unde fit ut ipsi igni quodammodo laboriosum sit sua eam qualitate afficere, faciliusque minorem quam maiorem convertit;[221] conversam item si quid ei non adeo affectum additur, quasi indignans relinquere videatur, vires paulatim retrahit, unde ipsi prior tenor decrescit. Que illatio quoniam in additamentum ignis, quod superius opposuisti, non incidit, nec ipsam similitudinem male oppositam in hunc tractatum incidere fas esse debet. Tuque tandem rerum qualitates edoctus satis sane de his dissertum esse cognosce.

NEPOS: At tu impedimentum tuum cognoscere non dubita, dum partem dubitationis extra rationem tuam non incautus relinquis. Sit enim, sit[222] sane ut in aqua esse concedam, ne minor calor maiori additus efficatiam augeat, cum aqua hanc qualitatem sibi contrariam non[223] per se, set per ignem recipiat, cui ut dixisti laboriosius est

216 concedatur PV, credatur GMQ
217 minuat, illud PQV, illud minuat GM
218 afficitur *Dronke*, afficit GMPQV
219 P *add.* 'fiat vel'
220 calefit GMP, calescit QV
221 convertit G *p.c.* P, convertat G *a.c.* (?) MQ
222 sit PQV, GM *om.*
223 ne P

agreed, then, that a similar argument applies to what was said about internal and outside air.

Nephew: I still have something to say to you about this. Put this way, I have nothing now to complain about concerning the increase in heat – for by this line of argument I do not find it at all strange that a quality does not increase a quality in respect to effect – but why it *lessens* the effect, seems strange to me. I find contrariety in the analogy. For if anyone adds a lesser fire to any great fire, both the quantity is increased and the effect is made more acute. If you ask the opinion of those standing round it, the whole fire, when it is affected more deeply, is felt to be hotter. But because, from the lessons of experience, it turns out to be the converse in the case of water, it is clear that your arguments, and you along with them, turn out to be against the order of things.

Adelard: It is rather you who fall away from the true distinction of causes, while you do not follow through an analogy correctly. For in only noticing the fact that qualities of things are in the object, it follows that you fail to notice the *way* they are in it; this is relevant in every dispute of this kind. For although heat is in both fire and hot water, it is in fire through itself, but in water through something else. For since water is naturally cold, only through fire bringing force against it does it become hot and assume a quality contrary to its nature. Hence it happens that it is in some way an effort for fire to affect water with its own quality, and it more easily changes a smaller quantity than a larger one. Similarly, when the water has been changed, if something which has not been affected so much is added to it, the fire would seem to leave the water, as if indignant, and it withdraws its strength a little, so that its earlier hold on the water decreases. Since in adding fire to fire, which you used as an objection before, this application of force does not arise, this analogy, misguidedly introduced as an objection, should not occur in this treatise. You have at last been thoroughly taught the qualities of things; so acknowledge that their discussion has continued for quite long enough!

Nephew: And you, too, do not hesitate to acknowledge your stumbling-block, since you deliberately leave part of the problem outside the realms of your reason! Fine. Granted that I agree that it is the case in water that a lesser heat added to a greater does not increase the efficacy, since water receives this quality contrary to itself not through itself, but through fire, for which, as you have said, it is more of an effort to change a greater than a lesser quantity. But I do

maiorem quam minorem convertere. In aere vero cur idem recipiam non video. Cum enim et interior et exterior utrique naturaliter calidi sint, neque aliqua vi inferente set complexionali qualitate calescant, dubitari non nequit, cur ex alterius additamento alter refrigeretur.

ADELARDUS: Quoniam iam compertum habeo non posse tibi vesicam pro laterna[224] restitui, quid inde animo sentiam sentire te breviter volo. Eam enim causam, ut dicis, que in aqua recipitur, in aere recipi minime convenit. At vero istam recipias,[225] si libet, ut cum uterque aer calidus sit, exterior tamen minus calescit, cum flabellatio adhibetur, ex repercussione quidem interior quodammodo a superficie retunditur,[226] exteriorque recipitur. Neque enim iam de ammixtione vel ammixtionis temperatione loquar, immo de interioris repulsione exteriorisque receptu. Sit itaque flabellantis intentio ut illum reducat, hunc inferat. Unde si quando nos calor vehementius urget, ut eum velocius removeamus, contra faciem nostram recta oppositione ventilamus. Quamobrem ille statim recluditur, hic recipitur.

NEPOS: Quoniam ista satis pro estimatione explicata sunt, illud michi enodari vellem: cur calidiores escas, quas in ore pre calore ferre non possumus, transglutiendo melius feramus.

ADELARDUS: Quia inter Saturnalia Macrobii illud disseritur satisque item ex supradictis patet, quoniam interiora exterioribus calidiora sunt, his nunc supersedendum putamus, cum Terentiano precipue urgeamur[227] proverbio 'ne acta agamus'.

.XXXV. NEPOS: Satis michi occupando ipsa excusatione rem aperuisti. Et quoniam de his que circa vultum sunt non improbabiliter dictum est, de his que ad manuum compositionem attinent breviter tangendum puto. Est enim circa eas quod dubitem. Cum enim ipsa etiam nominis dignitate id innuente inequalitate equalitas optabilior sit, ut quid digiti inequales sint aptati disserendum est.

ADELARDUS: Cum protervo michi quantum intelligo sermo susceptus est. Set ne michi inscitie[228] occasio imputetur, querenti mos agendus est, et quoniam huius[229] rei causam in manus cava inequalitate ponimus, huius ipsius concavitatis[230] causam prius ponemus.

224 laterna P, lanterna GMQV
225 recipies P
226 retunditur PV, reconditur GMQ
227 urgeamur PMQ, urgeamus GV
228 inscitie PQ, inscientie GMV
229 eius P
230 cavitatis P

not see why I should accept the same argument in the case of air. For since the internal and outer air are both naturally hot, and do not become hot by the application of some force, but through a change of complexion, I cannot but doubt why the one air becomes colder by the addition of the other.

Adelard: Because I already know that you won't take a bladder for a lantern,[53] I want you to feel briefly what I feel about this in my mind. For, as you say, the cause which is accepted in the case of water, cannot at all be fittingly accepted in the case of air. But, if you like, you should accept this one: although each of the airs is hot, the outer air is less hot. When a fan is used, as a result of the beating, the internal air in a certain way is driven away from the surface and the outer air is received. I do not now speak of the mixture or the tempering of the mixture, but rather of the driving back of the internal air and the reception of the outer air. So take the intention of the fanner to be to drive the first back, and to lead the second in. Hence, if at any time heat oppresses us more fiercely, to take it away more swiftly, we make a breeze right in front of our face. Therefore the internal air is immediately shut away and the outer air is received.

Nephew: Since these things have been sufficiently explained in my estimation, I should like this problem to be unravelled for me: in the case of rather hot foods which we cannot hold in our mouths because of their heat, why do we manage better by gulping them down?

Adelard: Because that is discussed in the *Saturnalia* of Macrobius,[54] and it is sufficiently obvious from what has been said before that internal things are hotter than outside things, I think that these matters can be left aside now, especially since we are obliged by the Terentian proverb 'not to do what has already been done'.[55]

35. Nephew: By engaging me with your very excuse, you have provided sufficient explanation of the problem, and, since not improbable arguments have been given concerning those things around the face, I think that one should briefly touch upon those things which relate to the composition of the hands. For there is something about them that I am uncertain about. Since the even is more to be chosen than the uneven, as the very worth of the name indicates, we must discuss why it is that uneven fingers have been fitted to the hand.

Adelard: I have become involved, as far as I can see, in a discussion with an impudent man. But lest this give him an opportunity to accuse me of ignorance, I have to humour the questioner. Since I posit the cause of this thing in the uneven hollow of the hand, we shall first posit the cause of this hollowness.

.xxxvi. NEPOS: Sit igitur prius quesitum cur manus interior superficies inequaliter concava sit.

ADELARDUS: Anima nimirum non parum[231] sue opportunitatis provida, ad quedam que digitis attrahuntur opportuna in hac cavatura parari duxit vascula. Set et cum sui corporis non ignara sitibundam intelligeret futuram necessitatem, non incommoda in eadem concavitate sibi[232] paravit pocula. Ut enim quantum ad nature dignitatem nichil ei deesse voluit, ita et nullum penitus ei superfluum adiunxit. Ne igitur extrinsecus pocula petas, habes, si nescis, innata tibi vascula unde nature sufficienter aquam proponas.

NEPOS: Cum concavitatis commoditatem intelligam, de digitorum inequalitate audiam.

ADELARDUS: In retinendis exterioribus que a digitis occupata sunt, claudendos esse eosdem necessarium duco.[233] Atqui hec eadem clausura, cum palma inequalis sit, nisi digiti inequales item forent, hians inutilisque foret. Quod quia quisque[234] in se ipso discere potest, hactenus inde dicendum[235] est.

.xxxvii. NEPOS: Hactenus certe. Neque enim de his amplius ambigo, aliudque proponere desidero. Solet enim vulgariter communis esse questio, cum[236] brutorum animalium pleraque quam cito nata sunt[237] statim ambulant, cur homines, qui eis precellere debent, hac facultate privati debiles reperiantur? Quod quia ad compositionem hominis attinet, in philosophico tractatu tangi non pudebit.

ADELARDUS: Quoniam usque ad vulgarem Minervam tractatus noster suo ordine descendit, ut ad quesita respondeam, in hac ratione respicias volo, tum membrorum teneritatem tum ipsam erecte incessionis difficultatem; ex his enim procedit ambulandi impossibilitas.

231 parvum P
232 sibi *ante* in eadem concavitate P
233 duco PQ, dico GMV
234 quisque PV, GMQ *om.*
235 discendum P
236 cum GQV, cur PM
237 sunt PMQV, sint G

36. Nephew: Yes, let the first question be: why is the inner surface of the hand unevenly hollow?

Adelard: The soul, certainly not grudging in providing for her own convenience, thought up the idea of providing small vessels in this hollow shape for certain things taken in the fingers. But since she was not unaware of the needs of her own body and realised that it would necessarily be thirsty in the future, she provided not unsuitable cups for herself in the same hollowness. As, in accordance with the dignity of nature, she wanted nothing to be lacking for it, so also she added no superfluous part to it at all. Therefore, lest you seek cups outside, you have (in case you do not know) vessels inborn to you, from which you can provide all the water that nature requires.

Nephew: Although I understand the suitability of the hollowness, I still wish to hear about the unevenness of the fingers.

Adelard: For holding the outside things which are seized by the fingers, I think that they must necessarily be closed. But since the palm is uneven, that same closing would be gaping and useless, unless the fingers were uneven too. But because anyone can learn this in himself, we need not discuss it further.

37. Nephew: Certainly there is no need. For I am no longer uncertain about these things and I am keen to put forward another question. The common people are always asking the question: when most brute animals walk as soon as they are born, why are men, who should be superior to brute animals, found to be weak and lacking in this ability? Since this is relevant to the fabric of man, we shall not be ashamed to touch on this subject in a work on philosophy.

Adelard: Since our treatise has descended to the intellect of the common crowd, to answer the question, in this reasoning of yours, I want you to pay attention both to the softness of the limbs and to the difficulty of walking upright; for from these proceeds the impossibility of walking.

.xxxviii. NEPOS: Que autem huius teneritatis crurum necessitas? Ut quid etiam hec erigendi se difficultas?

ADELARDUS: Utrumque ab agentis nature dignitate procedit. Agentem autem naturam animam voco, que quanto in humano corpore digniores exercere potentias proponit, tanto ipsa ad hoc membra habiliora efficit. Minime igitur ossosa vel molesta; hec enim ut brutis conveniunt, ita rationabilis virtutis exercitio incongrua sunt. At vero carnea, ideoque tenera, rationi amica, eiusque exercitiis non incommoda. Que,[238] ne terrene fecis quandoque illecebri ingluvie affecta, ipsam illam a qua reguntur declinent, ab hac pernitie, dum ea erexit, sibi precavit. Superiora igitur, que numquam a sua ratione fatiscunt, intueri suum corpus voluit. Cum igitur non incausalis[239] sit vel crurum teneritas vel incedendi sublimitas, consequens est ut non incongrua videatur primitivorum[240] vagibunda ambulandi difficultas. Sicut itaque brutis membrorum confusa molestia virtuti non attinet, sic et rationalium animalium membrorum elimata nobilitas vitio non accedit.

.xxxix. NEPOS: Vix istis animo accedo, cum iam michi alia surrepit dubitacio. Cur enim primam lactis alimoniam teneris proponi primis parentibus placuerit querere non displicet. Nisi enim usus id mitigaret, quodammodo crudele et inhumanum parentis parte vesci videatur. Oblita itaque paulisper consuetudine, non inmerito videbitur sub questione cadere.

ADELARDUS: Similibus similia nutriri ipso vestibulo libelli nostri edocuimus. Atqui, si nescis, membra humana in ipso utero ex materno menstruali constituuntur sanguine. Idem vero sanguis, cum iam amplius in constitutione tali fetu iam perfecto non recipitur, in lac eadem providentia convertitur. Unde post partum commodius nutritur. Quod sive philosophica perceptione edoctum, sive bestiali exemplo ammonitum, non irrationabiliter genus humanum sequitur.

.xl. NEPOS: Cur igitur eisdem vel iuvenibus vel senescentibus incommodum esse a phisicis perhibetur?

ADELARDUS: Partim negotio, partim etate, partim instrumento prohibetur. Aliud enim membra volunt informanda, aliud informata. Item aliud infantia postulat, aliud iuventus affectat. Postremo aliud

238 Que G *a.c.* PMQ, Qua G *p.c.*V
239 incausalis GV, incasalis P, causalis MQ
240 primitivorum *scripsimus*, primitiorum GPQV, primitiarum M

38. Nephew: What need is there for this softness of the legs? Why too is there this difficulty in standing up?[56]

Adelard: Both proceed from the worthiness of the active nature. I call 'the active nature' the soul. The more worthy the powers that it makes effective in the human body, the more suitable for this it makes the limbs themselves. They are, then, not at all bony or hard; for to be so is appropriate to brute animals, just as it is inappropriate to the exercise of the rational power. But fleshy, and therefore soft, limbs are friendly to reason, and not unsuited for its exercise. Lest they, affected at any time by the enticements of their gluttony towards earthly filth, should drag down that very soul by which they are ruled, she took precautions for herself by lifting them away from this destruction. Thus she wished her body to gaze at the higher realms which never decline from her own rationality. Since, then, both the softness of the legs and walking with the body erect are not without cause, it follows that the beginners' hesitant difficulty in walking does not seem incongruous. Just as for brute animals the confused hardness of the limbs does not attain to virtue, so also the polished nobility of those of rational animals is not in agreement with vice.

39. Nephew: Scarcely do I agree to these opinions in my mind, when another doubt creeps in. It does not displease me to ask why it pleased the first parents that the nourishment of milk should be the first to be offered to children. For unless custom exonerated it, it would seem somehow cruel and inhuman to feed on a part of the parent. So, if we briefly put custom out of our mind, it will quite rightly seem worth asking the question.

Adelard: We thoroughly taught in the very hall-way of our little book that like is nourished by like. But, in case you do not know, human[57] limbs are made in the uterus from the menstrual blood of the mother. The same blood, when it is no longer used for making such a fetus, it being now complete, is converted into milk by the same providence. Hence, after birth, it is nourished more appropriately. Whether taught by philosophical insight or prompted by the example of the beasts, the human race not irrationally follows this pattern.

40. Nephew: Why, then, is it said by natural scientists that milk is not appropriate to the same human beings when they are fully developed or growing old?[58]

Adelard: It is prohibited partly by function, partly by age, and partly by the instrument. For limbs which are to be given form want one thing, those which are already formed, another. Similarly infancy demands one thing, youth desires another. Finally, softness of the

teneritas mandibularum petit, aliud dentium perfectio deposcit. Ad ultimum non ego sum ille qui asserere ausim humanum quidem lac alicui etati esse[241] incongruum; at vero inhumanum.

.xli. NEPOS: Cedo, et quia de humana natura sermo nobis susceptus est, accidit in ea quiddam quod cum taciturnitate pretereundum non est: cur scilicet ad mulierem sanam si vir elephantiosus[242] accedat, non ipsa mulier, set qui eam deinde primus cognoverit, morbum sustinebit.

ADELARDUS: Plangenda quidem non ammiranda est hec translatio. Cuius causam si cognoscere cupis, utriusque sexus proprietatem prius cognoscas. Est igitur virilis quidem natura secundum esse calidior, muliebris vero frigidior, et virilis quidem ad siccitatem pertinet, muliebris vero ad humiditatem. Unde si quando elephantiosi semen suscipiat, frigida et humida proprietas a tali eam affectione defendat. In qua tamen cum pars seminis inefficax remaneat, cum ad eandem vir accedit,[243] cum siccitate tum calore ingruente ex similitudine qualitatum affectionem sibi accidit ut inducat. Sicut enim frigiditas et humiditas expellunt, ita[244] caliditas et siccitas suapte natura attrahunt.

.xlii. NEPOS: At vero si, ut dicis, muliebris natura frigidior est, ut quid libidinis petulantior et dicitur et reperitur?

ADELARDUS: Petulantior certe, set libidinis eiusdem non frigiditas immo humiditas in causa est. Cum enim in ea male digestus sanguis habundet, neque[245] enim propter frigiditatem digerere eum potest, fit inde ut per coitum purgari desideret. Semen enim nichil aliud est quam sanguis in albedinem[246] mutatus. Huc accedit[247] ut quoniam frigida est calori virili associari affectet. Nam et serpentes etiam ob suam frigiditatem ora dormientium intrant. Quod autem frigidus sanguis in mulieribus habundat menstrua earum testatur purgatio, a natura quidem eadem necessitate constituta. Convenit igitur ut eadem purgationis intentione virilem desiderent copulam. Nec tamen conveniet ut ex eadem copula elephantie ducant maculam, nisi aut inde

241 etati esse MQV, etate esse P, inesse etati G
242 elephantiosus G, elefantiosus PV, elephantuosus MQ
243 accedit PQ, accedat GM, accedere V
244 P *add.* et
245 neque PMV, nec GQ
246 albedinem PMQ, -ine GV
247 accedere P

gums seeks one thing, the completion of the teeth demands another. In fine, I am not the kind of man who would dare to assert that human milk is incongruous to any age, but I would say that it is inhuman.

41. Nephew: I agree, and because we are involved in a conversation about human nature, something occurs in it which must not be passed over in silence: namely, why, if a man suffering from leprosy has intercourse with a healthy woman, it is not the woman, but the man who has intercourse with her next, that suffers from the disease.

Adelard: One should grieve over this transference of the disease rather than marvel at it. But if you wish to know its cause, you should first know the property of each sex. The man's nature, following his essence, is hotter, the woman's, colder, and the man's pertains to dryness, the woman's to moisture. Hence if at any time she receives the semen of a leprous man, her cold and moist property protects her from such an illness. However, since some of the semen remains inactive in her, when a man has intercourse with that woman, because of the increase of dryness and heat it happens that he draws the disease into himself from the similarity of the qualities. For just as the coldness and moisture expel, so the heat and dryness by their very nature attract.

42. Nephew: But if, as you say, the woman's nature is colder, why is it both said to be and found to be more lustful?

Adelard: It is more lustful, certainly, but it is not coldness – it is, rather, moisture that is the cause of sexual desire. Since badly digested blood abounds in the woman's nature – for she cannot digest it because of her coldness – it happens that she desires to be purged through intercourse. For semen is nothing other than blood changed into whiteness. Moreover, since she is cold, she desires to mix with male heat. For snakes too, because of their coldness, enter the mouths of sleepers. That cold blood abounds in women is attested by their menstrual purgation, which was established by nature from the same necessity. It is appropriate, therefore, that they should desire to couple with a man as a result of the same urge to purge themselves. It would not be appropriate, however, that they should contract the stain of leprosy from the same coupling, unless they become pregnant as a

liberos suscipiant[248] aut in convictu[249] aerem corruptum frequenter ducant[250] aut, si muliebri[251] excessa natura, calore exuberent.[252]

NEPOS: Iam michi sententiarum tuarum desiderium surrepit. Sunt[253] etenim, licet non necessarie, tamen et auditu dociles et possibilitate nature probabiles.

ADELARDUS: Satis necessarium est quod refelli non potest.

.xliii. NEPOS: Sit sane. Et quoniam de vivis convenienter actum est, quare moriantur homines michi dubitabile est. Nunc vero de his qui vim apertam patiuntur unde enecentur non quero – illorum enim dissolutionis causa patens est. Istorum vero, quia vis illata palam non est, passionis ratio querenda est. Cum enim tanta intentione fabrica huius corporis et facta et aucta[254] ab anima sit,[255] nullaque convenientior copula[256] quam anime et corporis reperiri possit, quod tam studiose paravit, aut volens aut invita deserit. Quod autem volens id non faciat, quantum ad naturam loquor, ipse exeundi a corpore miserandus ostendit modus; preterea etiam communis fere omnium de futura morte metus. Cum itaque invita exeat, mirum est quid[257] tam potenti nature tam intolerabilem vim faciat. Utrumne in Auctorem cuius iussu intravit tam violenta extrusio referri debeat? Atqui si eo educente exiret, sicut ab eo non iniocundus est introitus, ita et[258] ad eum non insuavis foret exitus. Quid igitur? Anne illa cum hospicium suum extricaret, cavaret,[259] extenderet, materiam quidem unde hoc faceret habebat, nunc vero cum iam perfecte absolutum est, unde perfectum servare queat non habeat? Atqui levius est perfectum servare quam infectum[260] facere, cum sibi unde servetur cibalis materia non deficiat. Quid igitur? Cum nec sua voluntas nec Auctoris potentia nec necessitatis carentia hec duo tam convenientia disiungat, ut quid talis compago perpetuo non perdurat?

ADELARDUS: Tantene tuorum verborum inundationi michi paucula interponere licebit?

248 suscipiant Q, suscipiat GPMV
249 convictu GQV, conventu P *a.c.*, convintu P *p.c.*, coniunctu M
250 ducant Q, ducat GPMV
251 muliebri *Müller*, muliebrem GQV, muliebre P, mulierem M
252 exuberent Q, exuberat GPMV
253 Sint P
254 facta et aucta PQV, acta et facta G, facta et acta M
255 ab anima sit PV, sit ab anima GMQ
256 copulaque nulla convenientior P
257 quid *Müller*, quod GMQV, qui P
258 et GQV, PM *om.*
259 cavaret PG al.MQ *p.c.*, caveret GQ *a.c.* V
260 infectum PV, imperfectum GMQ

result, or frequently take in corrupt air through their living together, or, having departed from a woman's true nature, they abound in heat.

Nephew: A lust for your opinions has now crept up on me. For although they are not necessary, nevertheless they are both pleasant to hear and probable in the light of what is possible in nature.

Adelard: That which cannot be refuted is sufficiently necessary.

43. Nephew: Yes indeed. And since living beings have been discussed as they should be, my problem is 'Why do men die?' In this case I am not asking about those who suffer a violent death; for the cause of their destruction is clear. But one must enquire about the reason for the death of those others, because violence is not obvious in their case. Since the fabric of this body was both created and made to grow with such care by the soul, and no joining can be found to be more fitting than that of soul and body, the soul abandons what she so zealously provided either willingly or unwillingly. That she does not do this willingly (I speak only in respect to her nature[59]) is shown by her pitiful manner of leaving the body; and, moreover, by the fear about future death, which is common to almost all people. Thus, since she departs unwillingly, I wonder what can bring such an intolerable violence to bear on such a powerful nature. Should such a violent expulsion be attributed to the Author by whose command she entered? But if she came out because he led her out, just as her introduction by him is not unpleasant, so also her exit to rejoin him should not lack sweetness. What then? When she with difficulty procured a home for herself, hollowed it out, and built it up, did she have the material from which to do this, but now, when it is perfectly complete, does she not have the means to keep it perfect? And yet it is easier to keep something perfect than to make an unmade thing, as long as a nourishing substance to keep it perfect is not in short supply. What then? Since it is not her own will, nor the power of the Author, nor the lack of what is needed, that separates these two things which fit together so well, why does such a combination not endure perpetually?

Adelard: Will you allow me to interpose a few remarks into your veritable flood of words?

NEPOS: Tanta michi vis questionis esse videtur, ut querendo delecter.

ADELARDUS: Delectare igitur inveniendo quod queris. Sic autem invenies, si primum cur hoc corpus de quo hic agitur anima rationalis ingressa fuerit, ratione duce inveneris.[261] Cum enim quedam corpora[262] sint que numquam ab anima vivificantur, quibus scilicet numquam ipsa succedit, alia vero que numquam deserit, nonnulla etiam[263] que aliquando movet aliquando deserit, esse aliquid in ipsis corporibus arbitror, quod ubi ipsa reperit, diligit et quoniam diligit vivificat. In aliis item quiddam[264] aliud quod fugit, ideoque non accedit ut illud moveat. Diligit autem quod sibi amicum est; quod vero natura simile, id amicum esse necesse est. Naturalem igitur similitudinem diligere conveniens est. Est autem ei simile quicquid in corporibus vel equale vel equalitati proximum est. Equalitatem igitur illam[265] diligere consequens est. At vero equalitas ista in elementari compositione attendenda est, ut scilicet nec ignea excrescat qualitas, nec terrea supplodat ponderositas. Cum enim ipsa anima ab Equali equaliter condita sit, inequalitatem sectari, discordiam inhabitare quomodo possit? Equalem igitur elementorum proportionem, vel ad equalitatem accedentem,[266] ex innativa equalitate anima et postulat et cum invenit vivificat. Corpora igitur que inequalitate[267] proprietatum a seipsis dissident, inaccessa relinquit, ut lapides ceteraque que immobilia semper perdurant. Illa vero que numquam ab equalitate sue compositionis fatiscunt, numquam et movere desinit, ut supera sidereaque corpora.[268] Cetera vero que aliquando ad equalitatem accedunt, aliquando recedunt, aliquando animata sunt, alias vero ab anima destituuntur, humana dico, ceteraque mortalia.

.xliv NEPOS. Quoniam, quantum ex verborum sensu concipio, rationis huius tota vis[269] in equalitate proportionum consistit, illud quidem necessario subdendum est: unde hoc accidat ut semel equatis huius compositi qualitatibus et ad motus anime elementari pace

[261] inveneris Q, invenis GPM, invenias P
[262] quedam corpora PQV, corpora quedam GM
[263] etiam P, ratione G, item MQV
[264] quiddam *Müller*, quoddam GPQV, quidem M
[265] igitur illam PMV, illam igitur GQ
[266] accedentem QV, ascendentem GM, accendentem P
[267] inequalitate PQ, inequalitatem GMV
[268] sidereaque corpora G, sideixaque (*sic*) corpora M, sideraque corpora P, sideraque corporea Q, desidereasque corpora V
[269] vis tota P

Nephew: It seems that the force of the question is such that I delight in asking it!

Adelard: Well, delight in finding an answer to what you ask! You will find it in this way: if first you find, under reason's guidance, why the rational soul has entered the body under discussion. For since there are some bodies which are never vivified by a soul – that is, those which she never enters – but others which she never deserts, and some which she sometimes moves, at other times deserts, I think there is something in the bodies themselves to which she is attracted when she finds it, and because she is attracted to it, she vivifies that body. Again, in other bodies there is something else which repels her, and so she does not approach such a body to move it. She is attracted to what is friendly to her; but that which is similar by nature must be friendly. Therefore it is appropriate that she is attracted to similarity of nature. 'Similar to her' is whatever in bodies is equal or close to equality. It follows that she is attracted to that equality. But that equality must be looked for in the elemental composition of the body, so that neither the fiery quality should exceed its boundaries, nor should earthy weight stamp out the other elements. Since the soul herself is created from the Equal in an equal way, how can she pursue inequality and make discord her home? So the soul both demands as a result of her inborn equality an equal ratio of the elements (or one that is approaching equality), and, when she finds it, vivifies it. But bodies which are at war with themselves because of the inequality of their properties the soul leaves unapproached – such as stones and the other things which always remain immobile. Those which never depart from the equality of their composition, she never ceases to move – such as the higher and stellar bodies. But the others, which sometimes approach equality, at other times depart from it, sometimes are animated, but at other times are deprived of the soul – I refer to human and the other mortal bodies.

44. Nephew: Since, as far I understand the sense of your words, the whole force of this argument rests on the equality of the ratios, this must necessarily be added: whence does it happen that when the qualities of this composite body have once been made equal and prepared for the movements of the soul by a peace treaty between the

paratis – unde, inquam, tam necessarii federis disiunctio, unde tam honeste aptationis inutilis separatio?

ADELARDUS: Equum postulas, si attente audis quod interrogas. Ades igitur animo; ego enim[270] quam breve possum[271] facio. In rerum usu universali quicquid aliud convertendo mutat, ipsum etiam ab eodem mutari, et[272] patens et necesse est. Mutat enim securis ligna secando, mutatur et ipsa a lignis hebetando. Mutat cultellus panem dividendo; mutatur idem ab eodem minuendo. Quicquid enim hoc modo in aliud agit, licet modo insensibili, a patiente tamen ibi agens[273] agitur. Agit autem corpus animale in escis – mutat enim eas earumque qualitates convertit. Agit item[274] esca in corpore; eius enim proprietatem mutat, equalitatemque evertit.[275] Licet enim escarum qualitas per corporis qualitatem[276] sensibiliter vincatur, vincentis tamen vis et equalitas a victo paulatim et minoratur et evertitur. Quod in omni continuo usu spectari licet. Fit igitur hoc modo, ut dum velit anima corpus suum cibali sumptione[277] servare, non ex intentione quidem dum reficit, destruit, et dum servat amittit.

.xlv. NEPOS: Ut quid igitur hec ciborum destructiva refectio, semel sane[278] constituto corpore?

ADELARDUS: Necessario certe. Cum enim non in his particularibus animalibus tota suarum partium natura sit, cum item partium exteriorum elementorum, ut dico,[279] maior et quantitas et efficatia sit,[280] interiora nostra ab illis exterioribus dissolvantur, necessario fit. Quo fit ut, quia continua est[281] dissolutio, continua[282] fiat reparatio. In hoc igitur imminenti dissolutionis periculo fit ut anima de suo quidem minus amittat, ut quod maius est, restituat. Verum etenim est: 'qui non dat quod amat, non accipit ille quod optat'. Consequitur

270 enim PV, GMQ *om.*
271 GM *add.* hoc
272 et G *a.c.* MQ, G *p.c.* PV *om.*
273 ibi agens M, in agens GP *p.c.*, agens P *a.c.*, inagens QV
274 item P, enim GM, econverso Q, igitur V
275 evertit PQV, convertit GM
276 qualitatem PV, qualitates GMQ
277 resumptione P
278 sane PQV, sano GM
279 ut dico *Müller*, dico ut G *p.c.*, dico G *a.c.*MQV
280 fit P
281 est PV, fit GQ, sit M
282 continua PMV, -nuo GQ

elements – whence, I say, comes the breaking of such a necessary treaty, whence comes the pointless separation of such a worthy combination?

Adelard: Your question is fair, if you listen attentively to what you ask. So, lend me your attention! I shall be as brief as I can. From the general experience of things, whatever changes something else by converting it is clearly, and must be, changed by that thing. The axe changes the wood by cutting it, but is itself changed by the wood by becoming blunt. The knife changes the bread by slicing it; it is in turn changed by the bread by becoming worn out. For, whatever acts on another thing in this way, in acting there, is acted upon, though imperceptibly, by the patient. The animate body acts on food; for it changes it and converts its qualities. Similarly, food acts on the body; for it changes its property and upsets its equality. Although the quality of the food is perceived by the senses as being conquered by the quality of the body, nevertheless the force and equality of the conqueror is gradually both diminished and overturned by the conquered. This can be observed whenever something is in continuous use. It happens in this way that while the soul wishes her body to be preserved by the consumption of food, she destroys it while she restores it (although not deliberately, of course), and she loses it while she preserves it.

45. Nephew: Why then this destructive restoration coming from food, once the body has been established in a healthy state?

Adelard: There is indeed a necessary reason. For since the whole nature of the elements' parts is not in these individual animals, and since also the quantity and efficacy of the parts of the elements outside, as I say, are greater, our internal elements are necessarily destroyed by those elements outside. Hence it happens that, because the destruction is continuous, continuous restoration should be made. In this imminent danger of destruction, therefore, it happens that the soul loses a little of herself so that she may restore the greater part. For it is true: 'he who does not give away what he loves, does not receive what he wants.'[60] Thus restoration follows destruction, but a

itaque ad dissolutionem refectio, ad refectionem vero equalitatis mutatio, ad hanc mutationem vero anime separatio. Sequitur igitur et tue dubitationis discussio.

.xlvi. NEPOS: Recte procedis, neque enim[283] his amplius hereo. Nunc autem illud communicari vellem, cur defunctorum corpora vivi pertimescamus.[284] Cum enim communis hominum metus ea perhorrescat,[285] scrutandum est, que in hoc quodammodo innativo timore causa lateat. Cum enim leonis vel tygridis cadaver non metuas, ante humanum cur trepidas? Dum vixit, dum nocere potuit, securus amabas, nunc cum[286] inanimatum[287] iacet formidas.

ADELARDUS. Cum immutabilis rerum omnium Causa universa que condidit existere destinaret, quam[288] maximam existendi causam existere velle singulis innativum generavit, quippe nichil eque[289] patrocinaturum eorum essentie vidit, quantum si ipsa[290] esse ament, pernitiem devitent. Quod in sequentibus, cum de causa existentie tractabitur, liquidius explicabitur. Interim vero ea esse quidem amare, non esse vero fugere nativa intentione concede.

NEPOS: Quidni concedam, cum in Consolatione Boetii a Philosophia id per singula demonstratum sit?

ADELARDUS: Demonstratum igitur ad presens sit irrationabilia[291] animalia pre ceteris esse intendere, non esse vitare. Atqui animalis esse nichil aliud est quam anime et corporis coniunctio. Amat igitur animal corporis animeque[292] compagem; odit igitur necessario eorumdem dissolutionem. Quare et eiusdem dissolutionis effectum odire consequens est. Eius autem effectus inanimatum corpus[293] est. Illud itaque animali vivo odiosum atque horrendum est. Neque id quidem nocendi timore, set vivendi naturali intentione. Unde fit ut illum[294] horrorem odium quidem magis quam timorem vocari velim.

283 enim PV, GMQ *om.*
284 pertimescamus PQV, pro- GM
285 perhorrescant P
286 cum GMQ *p.c.*V, PQ *a.c. om.*
287 inanimatum PMQ, inanimum G, in anima V
288 quam MQV, que G *a.c.*, quem P
289 eque MQV, equa GP
290 ipsa PMQV, ipse G *a.c.*, ipsum G *p.c.*
291 irrationabilia PQ, irrationalia GV, rationalia M
292 anime corporisque P
293 corpus inanimatum P
294 illum PQV, illud G, illud scilicet illum M

change from equality follows restoration, and the separation of the soul follows this change. And following this is – the solution of your doubting!

46. Nephew: You proceed in the right way, for I do not get stuck on these points any more. But now I should like to share this problem: why do we, when alive, excessively fear the bodies of the dead? For since it is a common characteristic of men to fear and tremble at them, one should investigate what hidden cause underlies this so-to-speak inborn fear. For since you do not fear the dead body of a lion or tiger, why are you disturbed when faced with a human corpse? While it was alive and could harm you, you loved it and felt at rest; now when it lies without a soul, you fear it.

Adelard: When the unchanging Cause of all things resolved upon putting into existence everything which he created, he generated in each creature, as the greatest possible cause of its existence, an inborn wish to exist, since he saw nothing which would support their essence as much as if they themselves loved to exist and avoided destruction. We shall explain this more clearly later when the discussion is about the cause of existence.[61] Meanwhile, please agree at any rate that they love to exist, and avoid not existing, with a natural desire.

Nephew: Why should I not agree, since in the *Consolation* of Boethius that is demonstrated by Philosophy in each individual case?[62]

Adelard: Let it then be understood as demonstrated for the moment that irrational animals above all other things desire to exist and avoid not existing. But to be an 'animal' is nothing other than to have the conjunction of soul and body. Therefore an animal loves the joining of the body and soul; therefore it necessarily hates their separation. Thus it consequently hates the effect of that separation. But its effect is the inanimate body. Thus the corpse is hateful and horrible to a living animal. This is not, of course, because of a fear that it will do harm, but because of a natural desire for living. Hence it happens that I should prefer that horror to be called hatred rather than fear.

.xlvii. NEPOS: Quoniam hec naturalis intentio a natura non dissidet, ad sequentia transeamus. Et adhuc michi in defunctis dubitabile est quod, dum vivus est, si in amnem vel mare ceciderit, ad fundum gravis tendit; post paucos vero dies ad superficiem aque male natet. In quo nature vis mutata esse videtur.

ADELARDUS: Tua vero magis natura mutata credatur, qui quod cognoscere debes[295] non cognoscis. Cum enim quatuor humores corpus humanum constituant, quatuor elementorum proprietates proferentes,[296] ille quidem qui igneam sortitus est naturam proprietatemque calidam et siccam in felle recipitur, clauditur, et continetur. Quod autem calidum et siccum est, id[297] leve esse necesse est. Quod autem leve naturaliter est, id nisi clausum teneatur superius tendere consequens est. Dum igitur fellea illa cista clausam tenet corporis levitatem, membra levare non potest; cum vero rupto[298] carcere per membra diffunditur, qualitativa necessitate ad superiora ferri suaque membra male levare effectivum est.

NEPOS: Dum[299] omnia que propono vel necessaria vel probabili[300] ratione dissolvas, fit amodo ut sententiis tuis magis animum prebere quam oblatrare velim. Quapropter, quoniam de humano statu que ambigebam enodasti, quid de terra et aqua, quid de aere et igne sentias explicare, et michi non iniocundum et aliis non inutile futurum credo.

ADELARDUS: Quoniam has ferias his gignasiis deputavimus, liga quod potes, ego me pro facultate dissolvam, eo quidem tenore ut in rationibus meis non necesse requiras, set refelli non posse recipias.

.xlviii. NEPOS: De his igitur eorumque statu, motu, cursu, ceterisque circumstantiis agendum est. Primo itaque quod prius occurrit, oppono: quomodo hec terra que pondera omnia sustinet – nunc enim non de simplicibus set de compositis loquimur – quomodo, inquam, ipsa in eodem loco vel a quo sustineatur? Si enim omnia hec pondera, lapis dico, lignum ceteraque talia, sustinentiam[301] querunt neque ab aere pro pondere suo[302] teneri possunt, multo magis terre globus, qui universa gravitate ponderosior est, et sustineri postulat, et ab aere

295 debes GMV, deberes PQ
296 proferentes PM, pre- GQV
297 illud P
298 corrupto G
299 Dum PQV, Cum GM
300 probabili GPV, probabilia MQ
301 sustinentiam GQV, -tia PM
302 suo pondere P

47. Nephew: Since this natural desire is not discordant with nature, let us pass on to what follows. I still have a doubt concerning the dead: namely, the fact that, if a living person falls into a river or into the sea, he sinks to the bottom because of his weight, but after a few days he 'swims' awkwardly to the surface of the water. In this the force of nature seems to have been changed.

Adelard: Your nature rather should be believed to have been changed – you who do not realize what you should realize. Since four humours make up the human body, and these bear the properties of the four elements, that which has obtained the fiery nature and the hot and dry property is received into the gall bladder, and enclosed and kept in it. That which is hot and dry is necessarily light. That which is naturally light consequently tends upwards, unless it is kept enclosed. When that bile-filled bladder keeps the lightness of the body enclosed, it is not able to raise its limbs; but when the prison has been destroyed, the lightness is spread through the limbs, and because of the necessity of its quality, it is effective in rising to the higher regions and in awkwardly 'raising' its own limbs.

Nephew: Since you solve all the questions I put to you by either a necessary or a probable argument, it now happens that I want to apply my mind to your opinions rather than to be cynical about them. Therefore, since you have unravelled what I was uncertain about concerning the human condition, to explain what you feel about earth and water, air and fire, I believe will be both pleasant for me, and not useless for others.

Adelard: Since we have dedicated this holiday to these exercises, tie what knot you can. I shall get myself free to the best of my ability, as long as it is understood that you do not require necessity in my arguments, but accept a position of not being refutable.

48. Nephew: We must then discuss their standing still, motion, course, and the other circumstances. First I bring in opposition what first occurs to me. How does this earth which holds up all weights (for now we are talking not about simple elements but about composite ones) – how, I say, is the earth itself held up in the same place, and by what? For if all these weights – I mean stones, wood and all other such things – need to be held up, and cannot be held by the air because of their weight, much more should the globe of the earth, which is heavier than the weight of all things, both demand to be held

undique verso teneri non potest. Quare ipsum sic stare irrationabile est.

ADELARDUS: Cadere vero inutile, et ne nos quidem simul[303] cum eo cadamus, eum rationabiliter stare monstrabimus.[304] Terram igitur ponderosam esse a primarum affectione qualitatum novimus. Quod autem ponderosum est infimo melius deget loco. Amat autem unumquodque quod vitam suam servat. Atqui[305] ad illud quod amat tendit. Tendat[306] igitur necesse est omne terreum ad locorum omnium infimum locum. Quod autem in rotundo loco medium est, idem et infimum esse patens est. Tendit igitur ad medium locum quicquid terrenum[307] est. Locus autem medius simplex atque indivisibilis localis punctus est. Ad localem igitur simplicem punctum omnia terrena tendere manifestum est. Ille autem localis medius punctus unus, non plures, est. Ab uno igitur terreno occupari, non a pluribus, necesse est. Atqui tamen ad eundem, ut dictum est, cetera tendunt. Premit igitur unumquodque aliud dum ad unum singula festinant. Quo autem singula pondera festinant, illuc et cadunt. Nichil enim ponderum casus est, nisi ad medium locum festinatio. Illud vero ad quod cadunt, loci dico punctus medius, stabilis perseverat. Illa igitur dum ad stabilitatem cadunt, stare non deserunt,[308] nisi si cui eorum vis aliqua[309] inferatur, unde a naturali intentione depellitur. In contrarium igitur res relapsa est, dum quod superius causam casus esse putabas, illud idem stabilitatem coherentiamque ponderibus dare non dubitas. Stant etenim quoniam et[310] illud ad quod tendunt stare[311] reperiunt. A loco igitur ad quem properant quodammodo sustinentur, qui quidem si usquam alio moveatur, omnia que[312] illum amant moveri necesse erit, licet tamen in eo ipso medio loco non sit huius stabilitatis prima causa, set secunda. Erit[313] enim secundum[314] supradictam rationem quietis huius prima causa subiecti proprietas, deinde loci qui appetitur stabilitas.

303 simul P *om.*
304 monstrabimus G, monstramus PQV, videbimus M
305 Atqui GPV, atque MQ
306 Tendit P
307 terrenum GPV, terreum MQ
308 deserunt GQ, deserit PV, desinant M
309 alique P
310 et G *om.*
311 estare (*sic*) tendunt P
312 omnia que M, omniaque ad P, omnia que ad GQV
313 Erit PQV, Erat GM
314 secundum PQV, GM *om.*

up and be unable to be held by the air which surrounds it on all sides. Therefore, for it to stand still – as it does – is irrational.

Adelard: But for it to fall is pointless. And lest we fall with it, we shall show that it stands still rationally. We know that the earth is heavy because of the affects of the first qualities. What is heavy stays best in the lowest position. Each thing loves what preserves its life. But it tends towards that which it loves. Therefore it is necessary that every earthy thing tends towards the lowest of all positions. But in a spherical shape it is obvious that what is in the middle is also the lowest. Thus, whatever is earthy tends towards the middle position. The middle position is a simple point which is indivisible and has a place. It is clear, then, that all earthy things tend towards a simple point which has a place. This simple point which has a place is one, not many. It must therefore be occupied by one earthy object, not many. But all other earthy things tend towards that same point. Each of them presses upon the other, while they all hasten to one point. Where individual weights hurry to, towards that place they also fall. For the falling of weights is nothing other than their hastening to a middle position. But that place towards which they fall – namely, the middle point of the place – remains stable. When they fall into that stable state, they do not cease to keep still, unless some force is brought to bear on any of them, as a result of which they are pushed away from their natural tendency. So the matter has fallen out in the opposite way, since you must not doubt that the same thing that you earlier thought to be the cause of their fall, gives stability and coherence to weights. For they stand still since they find that what they are tending towards is also standing still. They are, then, held up in a certain way by the place to which they hurry, and if this place ever moves elsewhere, it will be necessary for all those things which are attracted to that place to move, although it is not the first cause of this stability that is in this middle place, but rather the second. For according to the previous argument, the first cause of this rest from movement will be the property of the subject, and then comes the stability of the place which is sought.

.xlix. NEPOS: Nec antecedentibus contraire queo, et hec inde consequi ipsa rerum coherentia video, ideoque in quo hereo suppono: si enim ipse terre globus perforatus sit ut ab uno celo in aliud pateat transitus, iniecta mole lapidis quorsum ei futurus sit casus?

ADELARDUS: Quod terrene moli statum prebet, hoc lapideo casui quietem faciet.[315]

NEPOS: Michi quidem satis dictum est. Intelligo enim quod in medio quiescet[316] loco.

ADELARDUS: Ad idem igitur istud intelligas volo, quoniam omnis natura, sicut suum simile diligit, ita suum contrarium fugit. Atqui ignis terre qualitatum efficatia contrarius est. Ipsum igitur a terra vitari consequens est. Ignis autem naturali circumferentia supremo loco undique inhabitans est. Locus igitur ille superior cum suo igne terre fugiendus est. Set quocumque a medio loco fugiet, quod vitat incurret. Ne igitur male fugiens quod fugit incurrat, ipsum locum quidem qui undique equaliter a superioribus distat tenet. Tenet igitur locum talem duplici causarum intentione: una quidem, dum sua ponderositate appetit quod amat; alia vero, dum fugit quod non amat.

NEPOS: Si igitur ita ut superius dissertum est medium imumque in mundana compage idem sunt, quid est quod Statius, non minus quam poeta philosophus, ea opposita posuit his verbis:

Imane tellus
an media et mundo rursus succincta latenti

Dum enim sic disiunxit, opposita iudicavit.

ADELARDUS: Duplici stilo, dum sic loquitur, servit. Nam et vulgaris opinionis sibi mentientis falsitatem tangit, et philosophicam veritatem aperte subdit. Nescit enim, nescit[317] vulgus applanetice continentie circumferentiam. Ideoque lippienti animo, dum corporeis male credit oculis, imperfectionem prave fingit emispericam. Erit igitur hoc inani respectu ima tellus, non media. Hunc autem errorem refellere quia singulis leve est, presentis tractatus non est. Patet enim vel doctrina initiatis huius mundane molis terram et imam esse et mediam.

[315] faciet GMV, faciat P, faceret Q
[316] quiescet PV, quiescit GM, quiesceret Q
[317] nescit PQV, GM *om.*

49. Nephew: I am not able to contradict the premisses, and I see that these consequences follow from them by the very coherence of things; therefore I put forward a topic which causes me to hesitate: if the very globe of the earth is bored through, so that a passage passes from one heaven to the other, when a piece of rock is thrown into the hole, where would it end up?[63]

Adelard: What provides a stopping place for the earth's bulk, will give rest to the fall of the rock.

Nephew: For me at any rate enough has been said: I understand that it will come to rest in the middle position.

Adelard: In addition, then, I want you to understand this: that every nature, as it loves its like, so it flees its contrary. But fire is contrary to earth through the efficacy of its qualities. Therefore it follows that it is avoided by earth. Fire, by a natural quality of surrounding the perimeter, inhabits the highest place. That higher place together with its fire, then, should be avoided by earth. But in whichever direction it flees from the middle position, it will meet what it avoids. Lest, therefore, fleeing inadvisedly, it meet what it flees, it holds the very place which is at the same distance on all sides from the higher places. It holds such a place because of the effect of two causes: one, while it seeks by its weight what it loves; the other, while it flees what it does not love.

Nephew: If then, as was discussed before, the middle and lowest in the fabric of the world are the same, why is it that Statius – as much a philosopher as a poet – posited them as opposite, in these words:

> Is the earth in the lowest position,
> or is it in the middle and girded again by a hidden universe?[64]

For, in making this distinction, he has judged them to be opposite.

Adelard: He serves two styles of writing when he speaks in this way. For he both touches upon the falsity of the opinion of the common people who lie to themselves, and he openly introduces philosophical truth. For the common people are ignorant of the surrounding envelope of the unwandering container. Therefore, for the bleary-eyed mind, while it puts its trust, wrongly, in the eyes of the body, Statius erroneously invents an incomplete hemisphere. From this foolish point of view the earth will be lowest, not in the middle. Because it is easy for everybody to refute this error, it is not the task of the present treatise to do so. For it is clear even to those who have only just been initiated into teaching, that the earth is both lowest and middle in this bulk of the world.

.l.[318] NEPOS: Manifeste quidem. Nam nec litus aras nec semina credis harene. Intelligenti etenim loqueris. Si vero ut superius disseruisti duplici de causa mediam quietem occupaverunt pondera, quid est quod ea ab hac causali necessitate nutare facit, instabilitatem inducit? Nec enim illud abdices – te ipsum testantem notavi – quia cum semel in partibus Antiochenis pontem civitatis Mamistre[319] transires, ipsum pontem simul etiam totam illam[320] regionem terre motu contremuisse, unde parum securior in terra quam in aqua esse potuisti. Que igitur hec est tam imperiosa potentia que rerum loca mutat, terramque ponderum matrem gravissimam ab occupato statu turbat?[321] Num hic iterum commenticiam aliquam invenies qualitativam causam?[322] Atqui si erit aliqua, erit forsitan[323] supradicte contraria,[324] sicque, dum erit falsa, fiet nulla.

ADELARDUS: Dum vera predicas, falsa tamen prophetizas.[325] Ego etenim, sicut pontem trementem illesus transivi, ita questionem tuam inoffensus explicabo. Movetur etenim terra particulariter quidem, non universaliter. Huius autem motus causam non suam qualitatem dico, set continentis effectum. Continentem vero aerem voco qui, cum terram undique versus claudat, ei se etiam intrinsecus visceratim informat. Hic igitur, exteriore eum evocante, cum nisu innativo exire properet[326] adunatusque cavernas terre intentione hac impleat, dum obstaculum reperit, ipsum vi magna concutit nec quiescit donec exitum inveniat.[327] Patet enim quia omne tale compositum licet aliqua opportunitate ad tempus obruatur, tandem tamen ad suam originem erumpere nititur.

.li. NEPOS: Patet certe, ideoque ad aquarum naturas traducamur. In quibus id michi primo dubitabile est, cur marina aqua affecta sit amaritudine salsuginea.

318 V *omits the whole text of Q50.*
319 Mamistre PQ, Manistre GM
320 illam GQ, ipsam M, P *om.*
321 turbat P, mutat GMQ
322 causam qualitativam P
323 G *om.*
324 contraria supradicte P
325 prophetizas Q, proph'as GP, prophetas M
326 properet *scripsimus*, -at GPMQ
327 inveniat GM, invenit PQ

50. Nephew: It is clearly so. For you do not plough a beach, or commit seeds to sand.[65] Rather, you are speaking to an intelligent man. But if, as you discussed before, weights have occupied the middle position of rest for two reasons, what is it that makes them bend away from this causal necessity and induces instability? For you will not deny this (I took note of you attesting it): that once, when you were crossing the bridge of the city of Mamistra in the region of Antioch, that bridge together with the whole of that region trembled with an earthquake, so that you could hardly be more secure on earth than on water. What, then, is this so commanding power, that changes the positions of things, and unsettles the earth – the most heavy mother of weights – from its stationary seat? Will you find some fabricated qualitative cause here too? But if there is one, it will perhaps be opposite to the previous cause, and thus, being false, it will be no cause at all.

Adelard: While your preaching is true, your prophecy is false. For just as I passed unharmed over the trembling bridge, so I shall solve your question without meeting an obstacle. The earth moves in part, not all together. I call the cause of this movement not its own quality, but the effect of the container. I name the container 'air', which, while it encloses the earth on all sides, also shapes itself into it on the inside through its bowels. When the outer air summons it forth, since this air hurries with an inborn striving to get out, and fills the caverns of the earth in a body with this intention in mind, if it finds an obstacle, it shakes it with great force and does not become quiet until it finds a way out. For it is clear that every composite body of this kind, although it is buried for a time because of some circumstance, at length strives to break out to its origin.

51. Nephew: It is clear, certainly, and therefore let us pass over to the natures of waters. In this I am first unsure why sea water is affected with salty bitterness.

ADELARDUS: Salsuginis causam in calore Solis planetarumque pono. Cum enim per torridam mediamque zonam verus feratur Occeanus, perque eandem licet indirectus versetur planetarum cursus, a tanto stellarum calore ipsum mare calefieri necesse est, ex quo et eiusdem caloris effectivam salsuginem accipere consequens est. Quod autem hec ita se habeant, illud asserit, quod in maritimis[328] illis que illi[329] Occeano propinquiora sunt sine omni artificio aqua marina ad Solem super rupes siccata in sal vertitur. In longinquioribus vero maribus, ut sal habeas, ipsam marinam utpote iam a vi caloris remotam ideoque minus coctam, igni adhibere et recoquere necesse est. Set et dulces quasdam aquas in sal verti caloris decoctione artificiosa sepe visum est. Huc[330] etiam illud accedit, quod estate quidem omnis marina salsior est quam hieme, quod si quis operam dederit, re ipsa experiri poterit.

.lii. NEPOS: His quidem non ab re decursis aliud non minus famosum subiungi velim: unde scilicet accessus recessusque marini procedant – utrum videlicet quorundam philosophorum super his probanda sit ratio. Aiunt enim verum Occeanum per torridam fluentem brachia inmense quantitatis fluentia ab orientali et occidentali plaga in Articam et Antarticam refundere regionem. Illis igitur vi magna confluentibus redundationem hanc fieri dicunt, ut ictus nobis accessum pariat,[331] cessatio vero recessum.

ADELARDUS: Philosophorum dictis invidere non ausim. Illud tamen audacter affirmem, si ita ut aiunt marina conveniunt brachia, semel commixtis undis secundo ictum non fieri. Neque enim iterum convenit eas[332] separari, vel certe si iterum[333] collidantur, minor erit secundus ictus quam primus, tercius quam secundus, itaque et quandoque minimus, deinde nullus. Videant igitur illi quid dixerint; ego pro me ipso breviter respondebo. Recursus itaque brachiorum colligo, eorumdem etiam obviationibus non contradico, non tamen ea conflui vel collidi concedo. Impotentie autem huius causam in ipsius terre situ facio. Cum enim ipsa brachia sibi obviare atque confluere impetuose festinent, fit tandem tum montium interpositione, tum ipsius terre situ quodam elatiore, ut ab eodem cursu dum deficiunt, referantur. Itaque

[328] marinis G
[329] illic G
[330] Huc PQV, Huic G, Hinc M
[331] G *add.* quidem
[332] ea P
[333] iterum *Müller*, item GPMQV

Adelard: I place the cause of saltiness in the heat of the Sun and the planets. For since the true Ocean flows through the torrid, middle zone, and the course of the planets is through the same zone, though indirectly, it is necessary that the sea itself is made hot by such great heat of the planets, from which it follows that it receives saltiness as the effect of the same heat. That this is so, is affirmed by the fact that in those sea-side places which are closer to that Ocean, sea water is turned into salt without the use of any artifice, but simply by drying on the rocks in the Sun. But in seas further away, to have salt it is necessary to submit the sea-water (being now remote from the force of the heat and therefore less cooked) to fire and to boil it again. But even some fresh waters are often seen to be turned into salt by the artifice of being boiled down. Added to this is the fact that in summer all sea-water is saltier than in winter, which anyone can experience for himself, if he makes an effort to do so.

52. Nephew: Now that these relevant questions have been run through, I should like to add another no less well-known one: namely, from where do the ebbs and flows of the tide proceed? Is the argument about this given by certain philosophers[66] to be approved? For they say that the true Ocean which flows through the torrid zone pours out flowing branches of enormous quantity from the eastern and western tracts to the Arctic and Antarctic regions. Thus they say that this fluctuation of the seas is caused by these branches flowing together with great force, so that the collision causes the flow for us, and the drawing back, the ebb.

Adelard: I would not dare to begrudge the philosophers their words. I would however affirm boldly that if branches of the sea meet in the way they say, once the waves are mingled, a collision will not occur a second time. For it is not fitting that they should be separated again. But even if they did collide again, the second collision would be less than the first, the third less than the second, and so eventually it would be minimal and then non-existent. Therefore, let the philosophers look to what they have said; I for my part will give my own answer briefly. I accept the flowing back of the branches, and do not speak against their meeting. I do not, however, agree that they flow against each other or collide. I suggest that they cannot do this, because of the situation of the earth's body itself. For when the branches hurry impetuously to meet each other and flow against each other, it happens that eventually, now because of the obstacle of mountains, now because of the higher elevation of the earth itself, they fail to complete their course and are carried back. Thus it

fit ut quo ea paternus ac naturalis impellit motus, ab eodem loci ipsius reducat situs. Licet non ignorem quosdam esse qui hunc motum vi lunari[334] dicant estuare. Quod si verum esset, in maribus illis que torride zone viciniora sunt vis talis non minus valeret. Nunc vero illa omni fere tali carent agitatione, eo videlicet quod ab illa causa quam supra scripsimus procul remota sint, Lune tamen viciniora.

.liii. NEPOS: Quoniam convenit ut possibilia non repudiemus[335] quod sequitur spectandum est. Si enim, ut vulgaris clamat tumultus, omnia volvuntur in mare flumina, mirum est, cum tantus sit eorum fluctus, tam perpetuus cursus, tam fere infinitus numerus, quomodo inde non augeatur Occeanus.

ADELARDUS: Si vulgus sequeris, precipicium offendis. Hic enim rerum naturas dum male intelligit, somniat; dum easdem prave disserit, stertit. Neque enim omnia in mare decurrunt flumina, nec item nulla. Atque ut in illud quedam defluunt, ita et[336] ab illo quedam nascuntur. Itaque si dum recipit reddit, sensibile augmentum non incurrit. Similiter enim a quibusdam dubitatum est, cum multa ex eo subterranea procedant flumina, multam etiam partem stellarum auferat haustus, detrimentum quomodo non incurrat, quodve supplementum recipiat. Unde et apud Statium Phebus cantat: 'Quo fonte inmensum vivat mare.' Vivit autem dum quantum reddit recipit.

.liv. NEPOS: Ecce iterum caligine operior. Cur enim, si a mari flumina exeunt, ad nos aliqua salsa non perveniunt?

ADELARDUS: Quod a patre ceperunt in itinere amittunt. Licet enim a mari salsa prodissent, tamen dum per viscera terre ea quodammodo cribrantis multipliciter transeunt, in eadem salsuginem deponunt.

.lv. NEPOS: Quoniam de fluminibus mentionem incurrimus, est in eis quiddam quod, si perspicaciter intueamur, dubitationem incurremus: unde scilicet quedam eorum tam perpetuum habeant cursum. Que enim est vel esse potest tante aquositatis copia, que iam pridem non foret excursa? Aut si post aquam aqua, et post illam alia, sicque semper precedentis sequens est amministrativa, tota erit hec amministratio infinita.

[334] vi lunari GP, nilimari id est caribdi Q, illuminari V
[335] repudiemus QV, repudiaremus PG
[336] etiam G

happens that the situation of the place itself brings them back from the place to which the natural movement of their father, Ocean, impels them. However, I am quite aware that there are some people who say that this movement ebbs and flows because of the force of the Moon. If this were true, such a force would be no less effective in those seas which are closer to the torrid zone. But now those seas lack that kind of fluctuation almost completely, obviously for the reason that they are far removed from that cause which we have written about above;[67] yet they are closer to the Moon.

53. Nephew: Since it is appropriate for us not to reject what is possible, we must consider what follows. If, as the common crowd clamours, all rivers roll into the sea, it is remarkable, when their flow is so great, their course is so perpetual, and their number is almost infinite, how the Ocean is not increased as a result.

Adelard: If you follow the crowd, you stumble upon a precipice. For while they understand the natures of things wrongly, they dream; while they discuss the same natures crookedly, they snore. For neither do all rivers flow down into the sea, nor do none of them. But as some flow down into it, so also others are born from it. Thus if, while it receives, it gives back, a perceptible increase in its volume does not occur. In fact, since many underground rivers arise from the sea, and the quaffing of the planets takes away a large part of the water, some people have been puzzled about how the sea does not suffer a loss, and how it receives sufficient water in compensation. Hence in Statius Phoebus sings: 'From what source does the sea sustain its life?'[68] It lives as long as it receives as much as it gives back.

54. Nephew: Now I am swathed in darkness again! Why, if the rivers come out from the sea, are some not salty when they reach us?

Adelard: What they have taken from their father they lose on their journey. For although they set out salty from the sea, while they pass in many streams through the bowels of the earth which sieves them in a certain way, they deposit their saltiness in it.

55. Nephew: Since we have come to mention rivers, there is something about them which, if we look at it carefully, will cause us doubt: namely, how it is that some of them flow so perpetually. For what is, or can be, the supply of so much wateriness, which has not already run out long before now? Or, if behind this water there is more water, and behind that, other water, and thus the next water is always administered by the preceding, this whole administration will be infinite.

ADELARDUS: Quicumque principium ignorat, male de fine loquitur. Tu igitur in re circulari nec principium nec finem esse intellige. Quicquid enim id esse potest, in se ipsum reciprocabile est. Flumina vero, si nescis, que perpetuum habent cursum,[337] circularem naturaliter nacta sunt motum. In se ipsa igitur redeunt, et quod eundo fluxerunt redeundo restituunt. Unde Satiricus vulgarem irridens stoliditatem ait:

Rusticus expectat dum defluat amnis. At ille
labitur et labetur in omne volubilis evum.

Tu igitur qui paulo ante finem amniflui cursus rusticus expectabas, rerum edoctus ad navem recte sententie ut amnem transeas redi.

.lvi. NEPOS: De illis quidem que vel in se orbiculariter currunt, vel in mare fluentia ab eodem item fluxum sumunt, unde indeficientia sint, satis patet. At vero de illis que in excelsis nascuntur montibus quid rationabiliter teneam, penitus ignoro. Cum enim circumiacens undique planicies pateat, nec amnem illuc ascendere posse video, nec unde ibi nascatur[338] intelligo.

ADELARDUS: Attonitus audi et intelliges. Cum enim in visceribus terre multiformem flumina cursum dividant, fieri potest ut tale quandoque incurrant territorium quod undique lapideis obstrusum[339] molibus ea sursum quidem, si ita patet exitus, cogat effluere. Dum itaque semper ascendit, semper[340] excurrit. Quod et in humanis etiam operibus quibus aquas cohibent perspicuum est.

.lvii. NEPOS. Si igitur in aquarum cursu hunc quem dicis reciprocationis reditum recipimus, ut ea in se[341] currere credamus, nullosne esse[342] fontes dicemus? Eritque sine re inane nomen, aut si eos esse veteribus etiam placuit, unde tam perhennem eis[343] ortum assignabimus?[344] Num terra ipsa in qua oriuntur in aquam[345] vertitur? Absit!

ADELARDUS: Negare non possum quin solutionis ratione dubitatio

337 perpetuum habent cursum PV, cursum habent perpetuum MQ, cursum habeant perpetuum G
338 nascatur PQV, nascantur GM
339 lapideis obstrusum PV, obstrusum lapideis GM, obstrusum Q
340 super ascendit super G
341 ea in se PQV, in se ea GM
342 nullosne esse GV, nullos necesse PMQ
343 eis GMQ, eius PV
344 assignabimus GMV, assignabunt P, signabimus Q
345 aquam GP, aqua MQV

Adelard: Whoever is ignorant of the beginning, speaks wrongly about the end. Therefore, understand that in a circular process there is neither a beginning nor an end. For anything to which this can apply can be returned into itself. Rivers which flow perpetually, in case you are unaware, have naturally acquired a circular movement. They therefore return into themselves, and what has flowed away in their going, they give back by returning. Hence the Satirist, in making fun of the stupidity of the common people, says:

> The country bumpkin waits for the river to flow away, but it flows and will flow, rolling on for ever.[69]

You, then, who a little while ago were waiting for the end of the flowing stream of water like a country bumpkin, having learnt the facts, return now to the boat of correct opinion, so that you may cross the river!

56. Nephew: It is sufficiently clear why those rivers never fail which either flow into themselves in a circular fashion, or, flowing into the sea, take their flow likewise from it. But as to what to think rationally about those rivers which arise on the tops of mountains, I am completely at a loss. For since a plain lies on all sides, I neither see that a river can ascend there, nor understand how it can arise from there.

Adelard: Listen, in your astonishment, and you will understand! Since the rivers divide into many different courses in the bowels of the earth, it can happen that they sometimes meet a terrain which is obstructed on all sides by rocky outcrops and forces them to flow upwards, if the only exit is in that direction. So when they are always ascending, they always flow out. This is easily seen also in man-made mechanisms, by which the flow of water is regulated.[70]

57. Nephew: If then, in the flow of water, we accept this reciprocal return that you describe, so that we believe that they run into themselves, shall we say that there are no springs? And will 'spring' be an empty word without a real object? Or, if it pleased the Ancients also that they should exist, from where shall we assign their eternal springing?[71] Is the earth itself in which they have their source turned into water? God forbid!

Adelard: I cannot deny that your uncertainty deserves a rational

tua digna sit. Neque enim illa que in se redeunt fontes dicendi sunt, et cum fontes veros esse[346] autumatum sit, unde perhennis eis[347] sit ortus discutiendum est. Ut autem planiori tractemus Minerva, exemplis agatur res. Erit enim, ut mea fert opinio, quod in sudanti corpore gutta, hoc in humectanti terra fons. Cum enim multimodo afficiatur humore terra, dum eundem naturali reddens exalat qualitati, cum ille a diversis partibus in unum convenit, fons esse incipit. Quoniam igitur locus terre exudantem humorem generare non deficit,[348] magna semper inde gutta nascens fons esse non desinit. Unde etiam fit ut locis quibus parva adest humiditas, eadem estivali calore desiccata, fontiva scaturigo estate cesset, hieme vero redundet.

.lviii. NEPOS: Adhuc michi de aquarum naturis quiddam dubitabile restat. Cum enim tempore, ut scis, iam preterito anum prestigiosam studio incantationis discende adissemus, ibique anilibus imbuti sententiis, nescio an sentibus, aliquot diebus moraremur, eadem in domo vas quoddam mirabilis efficatie ad horas pransiles efferebatur.[349] Quippe cum idem et superius et inferius perforatum multipliciter foret, aqua etiam ad manus abluendas infusa, dum minister aquarius superiora foramina pollice obturabat, nichil aque ab inferioribus emanabat; ablato a superioribus pollice, statim nobis circumstantibus per inferiora foramina aqua redundabat. Quod ego prestigium ipsum[350] esse putans, 'Quid mirum', inquam, 'si anus domina incantatrix est, cum aquarius servulus monstra pretendat?' Tu vero more tuo quoniam incantationibus studiosus eras, minime illi rei vacare dignatus es. Nunc igitur quid de aqua illa sentis? Aperta semper erant subteriora foramina, et nichil tamen nisi ad arbitrium aquarii fluebat.

ADELARDUS: Si prestigium erat, nature potius quam aquarii violentia id incantatum est. Cum enim huius mundi sensilis[351] corpus quatuor elementa componant, ita ipsa naturali amore conserta sunt, ut cum nullum illorum sine alio existere velit, nullus locus ab eis vel vacuus sit vel esse possit. Unde fit ut quam cito unum illorum a loco suo cedat, aliud absque intervallo eidem succedat. Neque[352] potest a loco cedere, nisi aliud quod substantiali quodam affectat amore, possit ei succedere. Clauso igitur introitu succedentis, frustra patebit exitus

346 esse PQV, GM *om.*
347 ei P
348 defecit P
349 efferebatur PV, aff- GMQ
350 prestigium ipsum PV, GMQ *transp.*
351 sensilis PQ, sensibilis GMV
352 G *add.* enim

solution. For it is the case both that those rivers which return into themselves should not be called 'springs', and that, since it has been averred that there are true springs, one should discuss the source of their eternal springing. So that we treat it by plainer teaching, let the matter be dealt with through examples. In my opinion, a spring in moist earth will be what a drop of sweat is in the case of a perspiring body. The earth is affected in many ways by moisture, and, when it exhales it, restoring it to its natural quality, and the moisture comes from different places into one spot, then it begins to be a spring. Since the place does not cease to generate the sweated moisture of the earth, the large drops always arising there do not cease to be a spring. Hence, too, it happens that in some places in which there is little moisture, the same outpouring of the spring ceases in summer when dried out by the summer's heat, but is abundant in winter.

58. Nephew: There still remains for me one uncertainty about the natures of waters. In the days now past,[72] as you know, we had approached an old sorceress to learn spells and lingered there for a few days, immersed in the thoughts (not to say thorns) of the old woman. In the same house some kind of vessel with a wonderful power was brought out at meal times. This vessel had many holes in the top and bottom, and, after water had been poured into it for washing our hands, when the water steward stopped the upper openings with his thumb, no water flowed out of the lower openings. But when the thumb was taken away from the upper openings, immediately the water flowed out through the lower openings over us as we surrounded it. I, thinking this to be a trick, said: 'If the old lady of the house is a caster of spells, what is surprising about her water steward presenting us with marvels?' But you, in your manner, although you were interested in spells, hardly deigned to pay attention to the matter. Now, then, what do you think about that water? The lower openings were always open, but nothing flowed out, except on the decision of the water steward.

Adelard: If it was a trick, the spell was cast by the force[73] of nature rather than that of the water steward! Four elements compose the body of this world that can be perceived by the senses, but these are so woven together by natural love that, since none of them wishes to exist without another, no place is, or can be, empty of them. Hence it happens that, as soon as one of them leaves its place, another takes its place without a break. Nor is one able to leave its place unless another which it desires with a kind of substantial love can succeed it. If, then, the entrance for the successor is closed, the exit for the element

succurrentis. Hoc itaque amore, hac expectatione, in cassum aque exitum aperies, nisi aeri introitum prestes. Hec[353] enim, ut supradictum est, cum non pura sint, ita coniuncta sunt, ut sine se, vel esse non possint vel nolint. Unde fit ut si in vase superius penitus integro inferius fiat apertio, non nisi cum intervallo quodam et quasi[354] murmure liquoris fiat effusio. Tantus enim aer intercedit, quantum inde liquoris exit. Qui quidem cum ipsam aquam porosam inveniat, innativa sibi et tenuitate et levitate penetrando, superiorem vasis locum qui vacuus videtur occupat.

.lix. NEPOS: De aquis igitur pro dubitatione quidem satis dictum sit. Illud enim cur earum quedam nitrum sapiant, quedam etiam per se calescant, quoniam ab aliis quocumque modo dictum est, nobis tractandum non est. Nunc vero quoniam huic ordini insistimus, ut ad altiora ascendamus, post aquarum naturam de aere agendum est. Quero itaque unde ventus nascatur, cum Statio dubitans,[355] qui in rerum naturis querit: 'Que pabula ventis?'

ADELARDUS:[356] Ventum quidem partim a terrarum superficie, partim ab aquarum planitie surgere estimo. Nam et multam densi aeris[357] partem naturali intentione ad superiora surgentis paludes vallesque reddunt, multam etiam aquarum humiditates, quam clausam modo tenebant,[358] cum dissolvuntur ad locum naturalem remittunt, licet etiam ipsum aerem terram continentem ab hac descriptione non removeam. Quod ut intelligi facilius sit, id ipsum de quo hic queritur, comprehensive diffiniam. Est igitur ventus aer densus usque ad offensionem quidem motus. Esse enim venti genus aerem estimo; ut autem illum qui a Luna superior est discernam, 'densum' addo.[359] Quoniam vero, dum quietus manet, ventus esse non potest, movendum esse non dubito, ideoque 'motus' nomen subdo. Postremo, ut illos aeres densos qui levissime[360] moventur separem, offensivam motionem in hac diffinitione posco. Hic igitur et illos aeres qui ab inferioribus multipliciter dissolvuntur, et continentem etiam qui nos ambit, si prefinitas habeat differentiarum notas, colligo.

353 Hec GQ, Hic PV, Hoc M
354 quasi et P
355 cum Statio dubitans MQ, Adelardus: Statio condubitas G, Adelardus: Statio cum dubitas PV
356 ADELARDUS MQ, GPV *om.*
357 densi aeris PV, GMQ *transp.*
358 modo tenebant GMQ, tenebant modo P, modo tenebant modo V
359 addes P
360 levissime PMV, leni- GQ

running out below will be open in vain. Because of this desire, this expectation, you will uncover the exit for the water fruitlessly, unless you provide an entrance for the air. For these elements, as was said before, since they are not pure, are so joined that they either cannot, or refuse to, be without each other. Hence it happens that, if, in a vessel which has no opening whatever in the top, an opening is made at the bottom, the liquid does not flow out except intermittently and with a kind of gurgling sound. For however much liquid goes out from the bottle, so much air enters in the mean time. But because the air finds the water itself full of empty spaces, penetrating the water by the fineness and lightness inborn to it, it occupies the higher part of the vessel which seems to be empty.

59. Nephew: Enough has been said about the uncertainties surrounding waters. The question about why some of them taste of soda and others are hot in themselves, since it has been discussed in some way or other by others, need not be dealt with by us. Now, since we have kept to this order – that we should ascend to ever higher things – , after the nature of waters, one must deal with air. So I ask what is the origin of wind, sharing an uncertainty with Statius, who seeks among the natures of things 'What is the nourishment for winds?'[74]

Adelard: I think that wind arises partly from the surface of the earth, partly from the expanse of the seas. For, marshes and valleys produce a great part of the thick air which rises by natural desire to the higher regions, but also the moistures of waters, when they are dissolved, return to its natural place a great part of that air, which they had previously held enclosed. However, I do not exclude the air surrounding the earth from this account. So that this should be more easily understood, I shall define fully the topic of this question. Wind is 'thick air, moved to such an extent that it can hit things'. I think that the genus of wind is air. But to distinguish it from the air above the Moon, I add 'thick'. But since, as long as air remains quiet, it cannot be wind, I do not doubt that it must be moved and therefore I add the word 'moved'. Finally, to separate it from those thick airs which move very gently, I require 'hitting movement' in this definition. In this way, if they have the clarifications of the definitions beforehand, I bring together both those airs which are dissolved from the lower elements in many different ways, and the all-containing air which surrounds us.

NEPOS: Ego igitur ad id quo[361] tendebam propero.
ADELARDUS: Sic propera ne cadas.

.lx. NEPOS: Si itaque,[362] ut superius diffinivisti, ventus aer motus est, ut ventus esse possit, aerem moveri necesse est. Ponamus itaque quantum ad talem motum quietum esse aerem, ut videamus utrum aliqua eius pars predictum possit habere motum. Quicquid enim movetur, ab alio movetur. Quod autem ab alio motum habet, aut a quiescente movebitur, aut ab aliquo in quo etiam erit motus. Atqui quod penitus in quiete est, rerum pacem abrumpere et ad motum impellere non potest. Movebitur ergo ab eo cui inest motus. Atqui hoc idem cui motus inest, ab alio item moveri necesse est. Set infinita sunt alia. Infinitus igitur erit[363] aerum motus. Aut igitur nullus talis ventus esse potest, aut si semel esse constiterit, et universalem motum faciet, et metam ipse non incurret. Amoveatur itaque hec dampnosa diffinitio, que rebus quietem tollit, intolerantiam inducit.

ADELARDUS: Qui prius prosilit quam videat, prius cadit quam debeat. Prius igitur vide non importunitates sophisticas, set rerum exiles naturas. Ut enim communiter agam atque universaliter, omnium que moventur quietem[364] esse causam mea sententia est; ut enim existentiam variationis, ut eternitatem temporis, ut unitatem pluralitatis, ita quietem causam pono motionis. Quicquid tamen movetur, si proprie ac passive accipitur, ab alio certe movetur. Aliud autem, inquies, aut in quiete aut in motu esse dicetur: in quiete quidem a passione, in motu vero ad actionem. Movebit enim et non movebitur; erit enim forma non patiens, set agens – erit causa motus, non eiusdem effectus. Que quidem forma vel causa[365] in igne cum sursum fertur, vocatur levitas; in lapide cum deorsum trahitur, gravitas. In aere igitur vocetur agilitas. Hec etenim, cum quiescens ipsa sit, suum fundamentum ad motum impellit. Itaque nec impossibilis erit diffinita motio, nec infinitatem incurret cause requisitio. Videant itaque quidam quid de motu dixerint. Ego vero formas rerum passivorum effectuum causas esse duco. Hee enim sua quandoque habundante

361 quo P, quod GMQV
362 itaque PV, GMQ *om.*
363 igitur erit PQV, erit igitur G, erit M
364 quietam P
365 forma vel causa GV, formalis causa P, causa vel forma MQ

Nephew: I, then, am speeding towards my destination.

Adelard: Make sure you don't fall over in your haste!

60. Nephew: If wind is moved air (as you defined it before), in order that it can be wind, the air must be moved. Let us posit, then, for such a movement a certain amount of air at rest, so that we can see whether any part of it can receive the said movement. Whatever is moved, is moved by something else. Whatever has movement from another, will be moved either by something which is at rest, or by something in which there will also be movement. But whatever is completely at rest cannot break the peace of things and force anything into movement. It will, therefore, be moved by something in which there is movement. But this same thing in which there is movement must be moved by yet another thing. But the 'other things' are infinite. The movement of airs, therefore, will be infinite. Thus either there can be no wind of this kind, or, if it is once established in existence, it will both cause everything else to move, and, itself, never reach an end of its movement. This faulty definition, then, which removes rest from things and brings on intolerance, should itself be removed.

Adelard: He who leaps before he looks, lands in the wrong place. First, then, consider not unsuitable sophistries, but the unadorned natures of things. To deal with the matter in a general and universal way, my opinion is that rest is the cause of all things which move; for as I postulate existence as the cause of variation, eternity as the cause of time, and unity of plurality, so I postulate rest as the cause of movement. However, whatever is moved (if 'is moved' is taken in a proper way and as a passive[75]), is, of course, moved by something else. You will assert that that 'something else' will be said either to be at rest or in movement: 'at rest' from the point of view of its passivity; in movement in respect to its action. For it will move and not be moved; it will be an active form, not a passive one: it will be the cause of movement, not its effect. This form or cause is called 'lightness' in fire when it is carried upwards, 'heaviness' in a stone when it is dragged downwards. In air, then, it will be called 'agility'. For although agility is itself at rest, it impels its subject into motion. Thus the movement as defined will not be impossible, nor will seeking the cause lead to infinity. Let some people look to what they have said about movement; I personally think that the forms of things are the causes of passive effects. When their own effectiveness sometimes

efficatia[366] ceterorumque[367] compositorum evocante eas simili natura, ad diversos passionum effectus impellunt subiecta, cum ipse tamen eiusdem passionis ignare permaneant. Unde etiam et ipsa universorum prima causa, cum omnia aliquo modo moveat, ipsa tamen nulli varietati est subdita.[368] Non igitur sequitur,[369] si quid movet, ut[370] moveatur.

.lxi. NEPOS: Ego vero, licet supradicta non moveantur, tamen a dubitatione non removeor. Illud enim abnegare non poteris quin quicquid moveatur moveat aliud. Moveri enim localiter, dico, nichil aliud est, quam suum dimittere aliumque locum occupare. Quicquid autem locum mutat, movet illud cuius locum occupat. Itaque quicquid movetur movet aliud. Quare, uno moto corporis athomo, in infinitum procedet motio. Sicque ventorum motus qui nunc quocumque modo[371] principium invenit, finem invenire non poterit.

ADELARDUS: Sic autem inveniet. Cedo equidem ut quicquid movetur moveat aliud. Procedet itaque hac necessitate motus ille non extensive nec in infinitum, set orbiculariter, et ad primum motionis reciprocabitur locum. Eritque ventorum motus orbicularis motui aquarum non dissimilis. Nec tamen unus atque idem erit perhennis: erit enim quiddam quod simili exigente forma ex transverso illum impellet. Unde universum illum athomorum coherentem cetum a proposito motu necesse erit declinare. Eundem item qui modo efficatiori inpulsu predictum divertens vincebat, si paulo post aliunde venientis aeris efficatior offendat inpulsus, procul dubio ab eodem vincetur. Itaque si in diversis huius sensilis partibus diversi frequenter ratione predicta nascuntur motus, eosdem se invicem offendere necesse est; sic igitur divertuntur. Eorum igitur quantum ad illum ordinem finitur inpulsus. Cum itaque reciprocus possit esse athomorum impulsus, cum item ex aliorum offensione incepti finiantur motus, nec igitur ad motum unius sequetur universorum impulsus, nec eorum semel inceptus, aliquis erit perpetuus. Quod et[372] in Solis radio perspicuum[373] esse poterit.

366 efficatia GP *a.c.*MQV, efficaria P *p.c.*
367 ceterorumque GP, ceterorum MQV
368 subdita est P
369 G *add.* quod, M *add.* ut
370 non P
371 modo GMQ, PV *om.*
372 et PMV, GQ *om.*
373 perpetuum G

overflows and the similar nature of the other composite bodies calls them out, these forms impel subjects into the effects of different passions, while the forms themselves nevertheless remain unaffected by that passion. Hence, too, that First Cause of all things, although it moves everything in a certain way, nevertheless is itself subject to no variation. It does not, then, follow that whatever moves, is moved.

61. Nephew: Although the things you mentioned are not moved, I personally am not removed from doubt. You will not be able to deny that what is moved, moves another. For I say that 'to be moved in place' is nothing other than to leave one's own place and occupy another. Whatever changes places, moves that whose place it occupies. Thus whatever is moved, moves another. Therefore, when a single atom of matter is moved, movement will go on for ever. And so the movement of winds, which now has found in some sort of way a beginning, will not be able to find an end.

Adelard: It will find it in this way. I agree that whatever is moved, moves another, but, under this necessary condition: that movement will go on neither in an extended way nor infinitely, but in a circle, and will return to the first place of the movement. The movement of winds will be a circular movement not dissimilar to the movement of water. However, it will not be one and the same for ever; for there will be something which impels that movement obliquely, because of a similar form drawing it aside. Hence that universal crowd of atoms sticking together will of necessity turn away from that movement we have proposed. Similarly, in the case of that movement which just now overcame the said movement by deflecting it with a stronger impulse, if a little later a stronger impulse of air coming from another direction hits it, the first impulse will without doubt be defeated by it. Thus, if in the different parts of the sensible world different movements frequently arise, for the previous reason, it is necessary for them to hit each other; and so they are deflected. Therefore, according to this disposition, their impulse comes to an end. Since the impulse of the atoms can be reciprocal, and since, too, movements which have begun, are terminated by hitting others, neither will the setting-into-motion of all matter follow the movement of one atom, nor will any movement among them, once begun, be perpetual. This can also be observed in the rays of the Sun.

.lxii. NEPOS: Licet ea que reddis intellectu difficilia sint, tamen quoniam ab absurditate[374] tuentur obiecta,[375] non censeo esse refutanda, magisque cetera de quibus subdubitatur subdere affector. Si enim, ut asseris, exteriorum passionum similitudine evocante propriaque qualitate impellente, talis aerum cetus in motum cogitur, cur non magis ad superiora movetur quod eiusdem natura postulare videtur, set terram verrens eique adherens, naturalem deserere mavult sedem?

ADELARDUS: Immo, quod minime putas, in hoc maxime rerum ordinem natureque rationem servat. Sicut enim ille qui ignea habundante natura levissimus, ether dico, igni proximum locum obtinet,[376] ita iste humoris densitate superante propinquam aque servat sedem. Illum raritas suffert, istum densitas subruit. Illum calor alleviat, istum increta humori frigiditas gravat. Itaque quoniam moveri necesse est, nec superius ferri potest, circa nostram regionem cursans, pacem nostram inquietat.

.lxiii. NEPOS: Quam bene dixisti! Nam nec verum tacuisti, et locum questioni subdende dedisti. Nichil enim iam de regione locove conqueror. Est aliud quod magis miror: unde videlicet ei tante vires, tantus incumbat impetus, ut domicilia deicere, silvas etiam eradicare, pacem rerum abrumpere possit.

ADELARDUS: Huius quidem tam damnose motionis et celeritas inpetuosa et ipsius corpulentia in causa sunt. Cum enim non modicus aer in visceribus terre complosus multo temporum spatio exitum reperire nequeat, tandemque vel propriis viribus materia quidem magis habundante, vel ipsa terrarum habitudine ei iter prestante eruptionem inveniat, quanto diutius clausus extitit, tanto violentius exit. Qui si etiam humore increto densitate corpulentus sit, cum celeritate[377] corpulentia mixta fuerit, offensionem parit. Quocumque igitur tendat, quantum eius vires possunt, lesionem facit. Non ergo mirum videri debet, quod dum erumpere nititur, terram nutare cogit, id ipsum post eruptionem si obstantia obruit.

374 ab absurditate GQ, absurditatem P, absurditate MV
375 obiecta GQV, abiecta PM
376 locum obtinet PV, GMQ *transp.*
377 celeritati Q

62. Nephew: Although the answer you give is difficult to understand, nevertheless, it protects its objectives from nonsense, I do not believe it should be rejected. Rather I desire to slip in the other matters concerning which doubt creeps in. If, as you assert, because the similarity of external passions calls them out, and their own quality pushes them out, such a crowd of airs is forced into movement, why does it not move rather to the higher realms, which its nature seems to demand, but, sweeping the earth and clinging to it, prefers to desert its natural position?

Adelard: On the contrary, it is above all in what you think completely unnatural, that it keeps to the order of things and the law of nature. For just as that air which is lightest because of its abundant fiery nature – I mean the ether – obtains the place closest to fire, so this one, because of the excessive thickness of its moisture, keeps a position close to water. Thinness carries that air aloft; thickness thrusts this one downwards. Heat lightens that air, coldness increased by moisture weighs this one down. Thus, since it must move, and cannot be carried upwards, rushing round our region it disturbs our peace.

63. Nephew: How well you have spoken! For you have not kept silent concerning the truth, and you have presented an opportunity for adding a question. I have no complaint now about the region or position. There is something else which I am more surprised about: namely, how so much strength, so much force is laid up in wind, that it can throw down houses, pull even forests up by their roots and break off the peace between things.

Adelard: The cause of such a harmful movement is both the rushing swiftness and the thickness of the wind. For imagine a fair amount of wind compressed in the bowels of the earth and for a long space of time unable to find an exit. When at length – whether by its own strength because its matter is more abundant, or by the very condition of the earth which provides a way for it – it finds a means of breaking out, the longer it has been enclosed, the more violently it rushes out. If the air is also corpulent and thick because of increased moisture, when its corpulence is mixed with swiftness, this gives rise to a hitting power. Thus, wherever it blows, it causes harm in relation to its strength. It should not, then, seem surprising that, in attempting to break out of the earth, it forces it to rock, if after breaking out it destroys things in its way.

.lxiv. NEPOS: At illud certe omnibus populis mirandum videtur, quod in aere fieri manifestum est, tumultum dico timoriferum[378] quem tonitruum vocant, unde non solum omnes terrentur populi, verum etiam et irrationabilia hec universalis formido premit animalia. Huius etenim sonitu et aves aerem metuentes ad sua foliata redeunt, et bruta que vocamus domicilia petunt, et conche marine imperfectas linquentes gemmas, dum ad fundum properant, conchilia relinquunt. An igitur huius tonitrui causam originemque dare tua philosophia confidat, an potius quia cum ceteris philosophis etiam trepidat, tante turbationis causam difficillimam ignorat?

ADELARDUS: Nichil est difficile, nisi cum desperes. Spera igitur et facultatem invenies. Ego quantum potero rem dilucidabo. A constanti itaque et indubitato principium sumendum est. Constat autem et indubitatumque est, nebulas ab inferioribus ad superiora surgere. Id enim et frequenter accidit, et visus testatur. Earum vero sicut quedam tenues sunt, ita alie dense, alie densiores, alie densissime, prout unaqueque humiditate conferta est. Densiores itaque harum cum ascendunt, si in superioribus frigoris intensionem offendunt, in nubes primo concrescunt. Quod si eis frigiditatis coherentia ventorumque articorum constrictio validiuscula incubuerit, bifaria coacte causa coguntur in glatiem. Hec itaque glaties dum sua sibi constat latitudinis coherentia, ab aere subiecto sustentatur illesa. Quod si violentus ventorum oppositorum concursus incubuerit, item si estate adolescente calor disiunctivus[379] accesserit, per illum necessario frangitur, per hunc solvitur. Privata itaque continenti sua latitudine, cum ab aere frustatim[380] sustineri nequeat, subruitur. Sicque et in hieme fractiva audiuntur tonitrua, in estate vero dissolutoria. Cum igitur singula huius eventus accidentia non miraris[381] – nam et nubium ascensum visus affirmat, et earumdem glatialem concretionem immoderata superiorum ventorum frigiditas probat, et eas vel frangi vel dissolvi collisionem fragor sequens auditumque nostrum offendens indicat – quid est quod ipsum totum usque adeo ammireris? Quid stupes? Quid dubitas? Quid nunc hac nunc illac nutans inconstantie vultum prestas? Atqui scio qua tenebra teneris, que universos qui de rerum ordine dubitant involvit, errorem inducit. Ammiratione enim

[378] timoriferum GP, mortiferum MQV
[379] disiunctivus G *p.c.*QV, -tius G *a.c.*P, -tus M
[380] frustatim P, frustratim GMQV
[381] mirari P

64. Nephew: But what certainly seems surprising to everybody is what clearly happens in the air – I mean the frightening commotion which they call 'thunder'. Not only are all peoples terrified by this, but also even irrational animals are affected by this universal dread. At the sound of thunder birds, afraid of the air, return to their nests, what we call brute animals[76] seek their dwelling places, and shellfish abandon their shells, leaving their pearls unfinished as they hurry to the bottom of the sea. Is your philosophy brave enough to give the cause and origin of this thunder, or rather, because it trembles along with the other philosophers, is it ignorant of such a difficult-to-find reason for such a disturbance?

Adelard: Nothing is difficult except when you give up hope. Hope, then, and you will find a way! I shall clarify the matter as much as I can. A beginning must be taken from something established and irrefutable. It is established and irrefutable that mists rise from the lower realms to the higher. For this both happens frequently and is attested by sight. As some mists are thin, so others are thick, others thicker and others very thick, in relation to the amount of moisture each one is stuffed with. The thicker of these, when they ascend, if they meet with an increase of coldness in the upper realms, first grow into clouds. But if a cohesion of coldness and a slightly stronger constriction of Arctic winds bears down upon them, they are forced by this two-pronged attack to become ice. This ice, while its own horizontal coherence is intact,[77] is held up unharmed by the air below it. But if a violent collision of opposing winds bears down upon it, or alternatively if, when summer is advancing, heat with its separating quality approaches it, it is necessarily shattered by the one, or dissolved by the other. Being therefore deprived of its horizontal extension, which kept it in place, since it cannot be sustained in a fragmentary state by the air, it rushes downwards. Thus in winter 'shattering thunder' is heard, but in summer, 'dissolving thunder'. Since you, then, are not surprised at the individual accidents of this event – for our eyesight confirms the rising of the clouds, the intemperate coldness of the upper winds proves their freezing into ice, and the noise following their collision and meeting our sense of hearing indicates that they are either shattered or dissolved – why is it that you find the whole business so strange? Why are you stunned? Why are you in doubt? Why now swaying this way, now that, do you give the appearance of a lack of constancy? But I know by what darkness you are held captive – it is that which envelops all those who doubt about the order of things and brings in error. For, veiled by the

insolentiaque indutus, animus, dum rerum effectus sine causis abhorrens a longe aspicit, numquam se[382] dubitatione exuit. Propius intuere, circumstantias adde, causas prepone, et effectum non mirabere. Ne sis ille qui mavult nescire quam accedere. Tales nempe in hac sententia fere omnes repperi. Unde cum eis aliquid tale promitto, nec promittentem recipiunt nec explicantem audiunt, meque digito ostendunt, et insaniam imponunt.

NEPOS: Ego etiam certe vix ad causas animum adducere queo, ita libens ad effectus ammirationem recurrit.

ADELARDUS: Et tu ex illis es.

NEPOS: Non adeo tamen ut putas, quin michi omnia que in supradicta confirmatione posita sunt, firma constent,[383] si de glatierum essentia aliquid quod ad probabilitatem accedat adhuc addas.

ADELARDUS: Equum postulas. Considera itaque et ab effectu causam perpende. Nonne post talia sepissime tonitrua grandinum videmus subsecutum esse strepitum?

NEPOS: Immo sine illis raro accidunt.

ADELARDUS: Tales etiam grandines nonnumquam vidimus, quarum ictibus et bruta animalia et hominum quidam interempti sunt. Quid igitur? Nonne illa magis glatierum frusta[384] quam grandinum dicenda erant? Elisis quippe illis non parvis superiorum glatierum corporibus, cum frustatim[385] ad inferiora decidant, in ipso casu magis magisque minuuntur, eorsumque[386] tandem cum perveniunt, si aerem inferiorem frigidum reperiunt, in sua constrictione grandinisque specie descendunt; si vero calidum, dissolvuntur in pluviam. Unde fit ut omnem talem collisionem perfectam[387] aut grando sequatur aut pluvia.

.lxv. NEPOS: Iam satis michi sententia insinuata est; illud quod inde sequitur subtractari vellem: unde videlicet procedant fulmina. Licet enim hoc a nonnullis quesitum sit – unde Statius inter cetera: ‘Tunc aperit quis fulmen agat’ – tamen adhuc minime memini an sit ab aliquo patenter explicatum.

[382] PQV, GM *add.* a
[383] constent G *a.c.*PMQV, consistent G *p.c.*
[384] frusta G *p.c.*QV, frustra G *a.c.*P, M *om.*
[385] cum frustatim G *p.c.*V, cum frustratim PMQ, confrustratim G *a.c.*
[386] eorsumque GQ, deorsumque PMV
[387] collisionem perfectam PMV, GQ *transp.*

wonderfulness and strangeness of the matter, the mind, while it recoils from effects in things which do not have causes and gazes from afar, never casts off its doubt. Look more closely! Add the circumstances! Lay down the causes first, and you will not be surprised by the effect! Do not be the kind of man who prefers not knowing, rather than coming nearer. Unfortunately, I have found almost all men to be in this state of mind. That is why, when I promise them anything like this, they do not accept my promise, nor listen to me when I explain it to them, and they point their fingers at me, accusing me of madness.

Nephew: It is true, I am hardly able to put my mind to examining the causes; so ready is it to rush into marvelling at the effect.

Adelard: 'You too are one of them'![78]

Nephew: Not, however, as much as you think, because everything which has been posited in the previous argument would be convincing to me, if you went on to add concerning the essence of ice something which approaches probability.

Adelard: Your demand is fair. So consider and weigh up the cause from its effect. Do we not see that the crashing noise of hailstones very often follows such thunderings?

Nephew: Indeed, thunder rarely happens without them.

Adelard: We have sometimes seen, too, such hailstones that their blows have killed brute animals and some people. What then? Surely they should have been called pieces of ice rather than hailstones? After the large bodies of the higher sheets of ice have been broken up, when they fall in pieces to the lower realms, in the very act of falling they become smaller and smaller, and when they finally reach that point, if they find the lower air cold, they descend in their constricted state and in the form of hail; but if they find it hot, they are dissolved into rain. Hence it happens that at the end of such a collision, either hail or rain follows.

65. Nephew: Now that the point has been sufficiently brought home to me, I should like the question which follows as a corollary to be dealt with: namely, where thunderbolts come from. For although this question has been asked by some people – hence Statius says, among other things: 'Then he reveals who brandishes the thunderbolt'[79] – nevertheless, up to now, I cannot recall whether it has been clearly explained by anyone.

ADELARDUS: Tractatu quidem digna est[388] questio, et ut tota tibi huius rationis vis breviter innotescat, hoc tibi propone, ex omni violenta corporum collisione quod in eis levissimum est primum dissolvi. Est autem omnium sensibilium ignis levissimus; excuditur itaque ex eorum impetuosa collisione. Atqui nulla maior in hac elementari aula fit[389] collatio, quam supradictarum glatierum fragosa collisio. Quare et inde velocitate mira ignem emicare necesse est.

.lxvi. NEPOS: O necessitatem pravam quam cessare videmus. Quicquid enim necesse est, semper est. Nos vero multa audimus tonitrua ex quibus non semper secuntur fulmina. Sic nec tu semper solvis quodcumque[390] minaris.

ADELARDUS: Tu igitur adhuc audi ut errore solvaris. Ego etenim dum necessitatem nominavi, non universalitatem intendi, set causalem huius eventus efficatiam expressi. Licet tamen pauca aut nulla rear fieri tonitrua ex quibus non elidantur fulmina, non tamen semper ignea. Eorum enim, si nescis, alia findunt,[391] alia scindunt, alia urunt. Quandoque etenim ex tali enisu parturitur lapis qui a Grecis 'ceraunium' dicitur, qui a Latinis fulmen findens nominatur. Quandoque aer tantum inde eliditur, qui a quibusdam fulmen scindens[392] appellatur. Frequenter etiam ignis excuditur.[393] Hoc autem secundum complodentium naturam ictuumque differentiam. Si enim soli conferantur venti, aerem pernitiosum elidunt. Cum vero glaties vel compelluntur vel dissolvuntur, si ictus maior fuerit, lapis enititur; si maximus, ignis elimatur.

.lxvii. NEPOS: Unde autem huic nequam generationis nefando partui, sive aer, sive lapis, sive ignis dicendus est, cum tantillus sit, tanta violentia innascitur, ut quandoque spissitudinem nubium subditarum usque ad terram scindat, aliquando etiam arbores ligneasque moles radicitus findat, nonnumquam lapideas eneasque structuras penetrans perurat?[394]

ADELARDUS: Quicquid obstantia penetrat, magis id celeritate sumpta quam substantia propria efficit. Unde et in bellicis negociis ligneis missilibus admodum gracilibus sine omni ferro densissima

388 digna est P *transp.*
389 fit *ante* in hac PV
390 quicumque P
391 alia findunt PQV, GM *om.*
392 fulmen scindens MV, fulmen findens P, scindens fulmen GQ
393 Frequenter . . . excuditur G *om.*
394 perdurat P

Adelard: The question is worth dealing with. So that the whole force of this argument should become known to you briefly, propose this to yourself: that from every violent collision of bodies, what is lightest in them is released first. Of all sensible objects fire is the lightest; thus it is struck out from their forceful collision. But no greater example of this can be given in this arena of the elements than that of the crashing collision of the ice mentioned before. Therefore it is necessary for fire to flash from it at an astonishing speed.

66. Nephew: Oh false necessity, which we see to fail! For whatever is necessary, always happens. But we often hear thunder which is not always followed by thunderbolts. In the same way, you do not always hit every target[80] that you threaten.

Adelard: Then continue to listen, so that you are not hit by error. When I referred to 'necessity', I did not mean something applying universally, but I described the efficiency of the cause of this event. Although I think that there is rarely or never thunder from which thunderbolts are not struck off, these are not always fiery. For, in case you are unaware, some of them split, others tear, and others burn. Sometimes from their throes a stone is born, which is called 'ceraunium' by the Greeks,[81] and named 'a splitting thunderbolt' by the Latins. Sometimes only the air is struck by it; this is called by some a 'tearing thunderbolt'. Frequently, too, fire is struck out. All this happens according to the nature of the colliding bodies and the differences of the blows. For if only winds meet, they strike out harmful air. But when ice is either broken up or dissolved, if the blow is greater, a stone is brought forth, if it is very great, fire is ignited.

67. Nephew: How does this bastard child of a vile coupling, whether it should be called air or a stone or fire, although so small, generate such violence that it sometimes tears through the thickness of the clouds below and reaches the earth, sometimes splits even trees and thick trunks from the very roots up, and sometimes penetrates stone and metal structures and burns them up?

Adelard: Whatever penetrates what is in the way, does this rather by the speed that it has gained than by its own substance. Hence also in warfare the thickest shields can be pierced by wooden missiles which are quite slender and have no iron added to them – so much

scuta perforantur, adeo in huiusmodi magis velocitas quam molestia valet. Ad mittentis autem vires velocitas refertur. Ad easdem igitur et penetrabilitas referatur. Itaque tenetur quod queritur, ut, quia tonitruorum ictus universorum violentissimus sit, tantam et velocitatem et penetrabilitatem in suo ut ita dicam missili parere possit.

.lxviii. NEPOS: Si itaque, ut supra dictum est, ex tali fragore ignis ille procedit, quid est quod sepenumero illos vidimus ignes, et quandoque non statim cum illis, quandoque etiam nec consequenter, aliquos inde audimus[395] fragores?

ADELARDUS: Quod non statim audimus causam cum videmus effectum, non ad ipsas res, set ad auditus et visus naturam refertur. Hec enim non eiusdem comprehensive celeritatis sunt ut supradicta satis ostendunt, adiecta similitudinis dulcedine que minus provectis semper blanditur, ut si quis ab altissima montis specula in ima valle percussorem[396] notet, prius actum rei visu quam auditu arguet, quod item quandoque nec consequenter aliquos audimus fragores, id ex intervenientibus conficitur. Intercurrunt[397] enim frequentes nubium spissitudines, que[398] facilius sonitum intercludunt fragoris quam claritatem obturent ignis.

.lxix. NEPOS: Nisi vera dissimulavero, veris obsistere nequeo. Et quia in hac sententia spinas dubitationum ab agro mentis mee eradicasti, lege tractatus nostri ad ulteriora procedamus. Elactasti etenim ignorantiam meam uberibus rationis, ideoque[399] capistrali opinione abiecta, quocumque voluerit precedat illa, ego sequar. Ut enim, exhortante Socrate, virtus et amicitia res pene impossibiles ad facultatem redigunt, ita ratio rerum desperabilem intelligentiam ad deliberationem reducit perspicuam. Huius itaque induti pennis, ascendamus ad astra, et de Luna que nobis propinquior est, prima questio habenda est: ea scilicet cur sola de stellarum tanta chorea luminis sui venustate curtata sit usque adeo ut nec[400] integra esse possit nec detrimenti sui certa saltem qualiscumque[401] forma permanserit. Sive enim universus[402] hic

395 audimus G *p.c.*, audiri G *a.c.*PMQV
396 percussorum G
397 Incurrunt P
398 qua P
399 ideoque PV, ideo GMQ
400 ne P
401 qualicumque P
402 universalis G

more effective is speed rather than bulk in such a situation. Speed is related to the strength of the one who throws the weapon. To this same strength is also related the penetrating force. Thus one has what is sought: that since the blow of thunder is swifter than any other blows, it can produce so much both of speed and of penetrating force in what we might call its weapon.

68. Nephew: If, as was said before, that fire arises from that kind of crashing, why is it that we have often seen those flashes, and sometimes we hear some crashes not immediately with the flashes, sometimes not even afterwards?

Adelard: That we do not hear the cause immediately on seeing the effect, has no bearing on the things themselves, but rather relates to the nature of hearing and sight. For these are not of the same swiftness in comprehending things, as has been shown sufficiently before. One can add the sweetening of an example, which always pleases less advanced students. If anyone from the viewpoint of a very high mountain notices a man murdering somebody in the bottom of a valley, he proves that the deed is being done by his sight before his hearing. Similarly, that sometimes we do not hear any crashes afterwards, happens because of what is in between us and the collisions. For many thick clouds roll in between, which more easily dampen the sound of the crashing than dull the brightness of the flash.

69. Nephew: Unless I shall pretend that what is false is true, I cannot object to the truth, and because you have pulled out the tares of doubt in this subject from the field of my mind, following the rule of our treatise, let us proceed to the next matters. For you have nursed my ignorance at the breasts of reason, and since halter-like opinion has been thrown away, let reason lead the way, wherever she wishes! I shall follow.[82] For, in the encouraging words of Socrates, as virtue and friendship bring things which are almost impossible into the realms of possibility,[83] so reason leads the one who has despaired of understanding things, into thinking about them clearly. Having put on the wings of reason, let us ascend to the stars; and the first question should be about the Moon, which is nearer to us: namely, why, out of such a great band of stars, the Moon alone is left short of the beauty of its own light, to such an extent that it cannot be whole or (having come to terms with its lack) at least retain some sort of fixed shape? For I have not yet discovered whether this sensible universe is formed

sensilis[403] de aliquo sive de nichilo formatus sit, nondum invenio. Nam si de nichilo, ut et cetere stelle perfecte sunt, ita et hec perfici potuisset. Neque enim nichilum deficere novit vel potest. Quicquid enim esse caret, nichilum est. Ex eo igitur quod non erat, similiter[404] ut cetera, facta esse poterat. At vero si de aliquo hec aula facta est, aut ex materia sufficiente, aut ex ea que posset ad perfectionem deficere. Si ex habundante, non posset in Luna deesse. Si vero ex avara et insufficiente, ut quid non potius ceteris defecit[405] stellis quam Lune? Cum enim certum sit ea scrutantibus quia hominum causa hec signa elata sint, numquid utilius nobis Saturni beneficium quam Lune? Nonne ea, ut nobis propinquior, ita etiam[406] effectu validior, et visu iocundior? Nam et humoribus imperat, et tenebras nostras[407] illuminat.

ADELARDUS: Utrobique conclusionem satis facete complecteris. Nec tamen adeo cornutus procedis quin possint cornua ista, ne dicam frangi, dico ligari. Sit enim ut de nichilo, sit ut de aliquo. Utrumvis dono. At deinde istud dabo, ut in universo opere magis efficientem spectes causam quam materiam. Unde Philosophus: 'Operi porro fortunam dat opifex suus'. Est autem huius mundi, quicumque sit, auctor optimus. Qui si talis est, et omnia secundum se optima esse necesse est. Nichil igitur se ipso melius esse potest. Si itaque mutetur, convertitur in peius. Nunc igitur, ut familiarius agamus, ad causam redeo. Lunam conquereris non esse perfectam. Perfectum autem vocas esse rotundum et idem. Vide quid inde sequatur. Si rotunda et semper eiusdem forme foret, propriam lucem haberet. Set[408] quoniam non habet propriam, nec habet eandem et rotundam. Quod si propriam haberet lucem, proprium haberet et ignem. Set[409] quia non habet ignem, non[410] habet et lucem. Quod si proprium haberet ignem, nocivum haberet ardorem. Nam et fructus exureret, et fontes siccaret, et tellurem a vicino flagrans in pulverem redigeret. Si enim Sole in Cancro locato non parva fit desiccatio, cum in eodem signo futura[411] foret Luna singulis mensibus, sequeretur rerum destructio. Hoc igitur cavit optimi Auctoris consideratio.

403 sensilis PQV, sensibilis G, sensibilis mundi M
404 similiter G *p.c.*Q, similis G *a.c.*P, simul MV
405 defecit G *p.c.*PV, deficit G *a.c.*MQ
406 G *add.* in
407 P *add.* et
408 Set G *p.c.*MQV, Si G *a.c.*P
409 Set G *p.c.*QV, Si G *a.c.*P, Sicque M
410 nec P
411 futuro P

from something or from nothing. If from nothing, as the other stars are perfect, so the Moon too could have been perfect. For 'nothing' does not know how to, or cannot, fall short, since whatever lacks being, is nothing. From that which did not exist, then, the Moon could have been made in the same way as all other celestial bodies. But if this court of the universe was made from something, it was made either from enough matter or from matter which could fall short of its full quantity. If from abundant matter, then that matter could not be lacking in the Moon. But if from a miserly and insufficient amount of matter, why did it not cause a lack to the other stars rather than to the Moon? Since it is certain to anyone investigating this, that these signs have been raised up in the heavens for the sake of man, is the benefit to us of Saturn greater than that of the Moon? As the Moon is closer to us, should it not also be stronger in its effect and more pleasing in its appearance? For it both rules over moistures and lightens our darkness.

Adelard: You grapple with the conclusion on both sides in an elegant enough way. However, you do not advance so protected by your horns that those horns cannot be bound (not to say broken off). Let the world be made from nothing, let it be made from something. I shall allow you either possibility. But then I shall present you with this point: that in every process you should look rather at the efficient cause than the material. For the Philosopher says: 'The craftsman then gives to his work its destiny'.[84] Whoever is the author of this world is supremely good. If he is such, then it is necessary for all things following him to be supremely good. Nothing then can be better than itself. If it changes, it changes for the worse. Now then, to deal with the matter more closely, I return to the case. You complain that the Moon is not perfect. You call 'perfect' being round and the same. See what follows from this. If it were round and always of the same form, it would have its own light. But since it does not have its own light, neither does it have a form which is the same and round. If it had its own light, it would have its own fire also. But because it does not have fire, neither does it have light. If it had its own fire, it would have a harmful burning. For it would burn up the fruit and dry up the springs, and, igniting the earth from its proximity, it would reduce it to ashes. For if when the Sun is in Cancer there is a great amount of drying out, since the Moon would be in that same sign every month,[85] the destruction of things would follow. The consideration of the supremely good Author thus guarded against this.

.lxx. NEPOS: Esto, quoniam ita esse possibile est. At vero quoniam de Luna tractatus incidit, quod me pre ceteris urget in communi subponam: illud videlicet, unde ei illa umbra universis notabilis umbreque tam obscura nota incubuerit. Neque illud nunc arguo, cur eclipticam patiatur obscuritatem. Tractatum etenim hoc a multis est. Hoc autem inveho quod sive plena, sive diminuta sit, obscurum quiddam[412] secum gerit, idemque eodem modo semper formatum. Quodque magis miror, Sole in eodem existente emisperio, tamen illa tenebras istas non discutit, velut[413] aut nolit aut non possit.

ADELARDUS: Immo et nolit et non possit. Quod sic conicio: in locorum enim occupatione unumquodque animans illum elegit,[414] qui sue habundantie conveniens est. Atqui lunare corpus pre ceteris stellis in ima delapsum est. Magis igitur terrestre, aquaticum, aerium, quam igneum esse arbitrandum est. Set et terrestris et aquatica et aeria essentia umbre generativa[415] est. Lunare igitur corpus umbratile esse necesse est. Umbra autem illa cum spissitudine subiecti sui, si per totum corpus equaliter diffusa esset, claritatem non usque adeo reciperet. Cum itaque et materiarum densitas ad loci obtentum necessaria, et claritas item eam expectantibus sit commoda, ne sibi obessent alterutra, in unum regesta sunt et densitas et umbra. Nam et philosophi etiam Lunam aeriam vocaverunt terram. Non itaque illis consentio qui illam umbrositatem ab inferioribus in Luna[416] apparere confirmant, magisque essentiali necessitati quam extrinseco accidenti id imputo. Cumque ipsi eam cum ceteris sex[417] stellis erraticam vocent, magis eos quam stellas errare puto. Quicquid enim in ipsis vel ab ipsis fit, ratione provida fieri estimo.

.lxxi. NEPOS: Quoniam de errore stellarum mentionem fecisti, quas Greci planetas vocant, inde ammonitus habeo quod de omnibus conquerar, precipue tamen de Sole mirum videri potest:[418] que huius obliquitatis necessitas, que erroris opportunitas[419] (ita enim illi vocant). Neque id quero cur contra applanon[420] cursus figant – id enim post dicendum est – set quocumque currant, cur oblique incedant? Neque tu quidem depellere potes planetarum anfractus et reditus. Id

412 quoddam P
413 velut G *a.c.*P *p.c.*Q, velit G *p.c.*P *a.c.*MV
414 elegit G *a.c.*PMQV, eligit G *p.c.*
415 generativa GPQV, genitiva G *i.m.* M
416 Luna PV, Lunam GMQ
417 sex G *p.c.*P, septem G *a.c.*MQV
418 potest videri P
419 opportunitas G *a.c.*PMQV, importunitas G *p.c.*
420 applanon GMQ, aplanon P, aplanum V

70. Nephew: Let that be so, since it is possible for it to be like this. But since the discussion happens to be about the Moon, I shall bring into the open what worries me more than anything else: namely, why it has been enshrouded with that shadow known to everyone and with such a dark blot of a shadow. I do not now make this complaint: why does it suffer the darkness of eclipse? For that has been dealt with by many. Rather, I bring in this charge: that whether it is full or diminished, it bears a kind of duskiness in itself which is always formed in the same way. What I am more surprised about is that when the Sun is in the same hemisphere, the Moon still does not shake off that darkness, as if it is either unwilling or unable to do so.

Adelard: In fact, it should be both unwilling and unable to do so. I look at it in this way. When occupying their positions, each animate being chose the place which is appropriate to what is abundant in it. But the lunar body sank to the lowest position compared with all the other stars. Therefore it should be thought to be more earthy, watery and airy, than fiery. But the earthy, watery and airy essence is productive of shadow. Therefore the lunar body must be shadowy. But if that shadow were spread equally throughout the whole body, considering the thickness of the subject, it would not receive brightness so much. Since then, both density of materials is necessary for spatial extension, and brightness is appropriate for those expecting it, lest one should get in the way of the other, the density and shadow have been concentrated in one place. For the philosophers too have called the Moon 'the airy earth'.[86] So I don't agree with those who maintain that that shadowiness appears in the Moon as a result of lower things, and I attribute it more to the necessity of its essence than to an accident coming from outside. Since those men call it, together with the other six planets, 'wandering', I think that *they* wander rather than the planets. For whatever happens in the planets or by them, I think happens through provident reason.

71. Nephew: Since you have mentioned the wandering of the stars which the Greeks call 'planetes', prompted by this I have something to complain about concerning all of them, but something that can seem especially strange about the Sun: what necessity is there for its obliquity, what's the use of its wandering (for this is what they call it)? I do not ask why it fixes its course in the opposite direction to that of the *aplanos*. For that should be spoken about afterwards. But in whatever direction it runs, why, I ask, does it progress in an oblique way? Even you certainly cannot get rid of the windings and backtrackings of the planets. This is shown now by summer, now by

enim tum estas, tum hiemps arguunt. Quod si circuli sui obliquam circumferentiam in causa facis, dicisque quia zodiacus a Cancro ad Capricornum obliquatus sit, nichil in eo pugno. Id enim ipsum ad nodum incidit. Ut quid enim et[421] ipse zodiacus ita distractus sit, queritur. Potuit nempe per medium duci recta linea emisperium. Nam et inde etiam sequeretur non parvum humano generi commodum. Obolita[422] etenim foret estatis et hiemis iniocunda distemperantia. Qua excessione caloris ac frigoris eliminata vigeret mortalis[423] res publica mansueta mediocritate exhilarata.

ADELARDUS: Exhilarata certe foret; neque enim hilaris esse posset. Tu vero quod sequitur[424] specta. Hec enim mea est in causis inquirendis doctrina. Si ab antecedenti non repperis, a consequenti consideres. Considera igitur quid ex hac tua prava[425] commoditate secuturum sit. Si enim recta linea, et zodiacus et stelle sue per etherem diducta forent, is[426] semper[427] esset rerum status, quicumque esse potest Sole in equinoctiali degente linea. Ver itaque[428] haberemus perpetuum, sicque numquam estatem vel hiemem. Ablata vero hieme, auferretur[429] et seminum plurima putrefactio. Ea[430] absente, periret eorum vivificatio. Nullum enim semen ad vitam nascitur, nisi prius corrumpatur. At vero estate abrasa, superflue humiditatis rerum periret desiccatio. Ea sublata, nulla earumdem sequeretur maturatio. Abeat igitur hinc in malam rem vernantis incommodi tui prava insinuatio!

.lxxii. NEPOS: Recordor quoniam absque ratione vernam introduxi delectationem. Nunc vero quedam que superius tacite preterita sunt, ad tractatum libet revocare. Que scilicet necessitas, que commoditas a recto itinere et naturali circuitu planetas revocavit? Nonne rectius, nonne naturalius, nonne ad cursum eorum notandum, atque ad disciplinam perceptibilius foret, si eodem circuitu cum applano velut cum suo principio rotarentur? Hoc enim quod quidam inde[431] scripserunt intelligo, non probo. Aiunt enim ut temperent rapidum spere

421 et G *a.c.*PV, G *p.c.*MQ *om.*
422 Obolita GPV, Abolita Q, Oblata M
423 mortalis G *p.c.*PV, G *a.c.*MQ *om.*
424 sequetur G
425 tua prava PMQ, GV *transp.*
426 is GQV, his PM
427 semper MG al.Q, super GPV
428 itaque G *a.c.*PMV, ita G *p.c.*Q
429 auferetur P
430 Ea GPV, Et ea G al.MQ
431 inde G *p.c.*PV, G *a.c.*MQ *om.*

winter. If you make the oblique circumference of their circles the cause, and say that the zodiac is oblique from Cancer to Capricorn, I find nothing in that to attack. But that very thing cuts at the knot: why is the zodiac itself so skewed? – that is the question. Surely it could have been led in a straight line round the middle of the hemisphere? From this, too, would follow a considerable convenience for the human race. For the unpleasant intemperateness of summer and winter would have been destroyed, and if the excess of heat and cold had been eliminated, the human commonwealth would thrive out of joy, having a moderate climate as the norm.

Adelard: 'Out of joy' it would certainly have been; for it could not have been joyful at all! You must look well at what would follow. For this is my method when enquiring after causes. If you do not find the cause from the antecedent, you should look for it from the consequent. Consider then what would follow from this misguided 'convenience' of yours. If both the zodiac and its planets were led through the ether in a straight line, the condition of things would always be whatever condition can occur when the Sun is dwelling on the Equator. We would have a perpetual spring, and thus never summer or winter. Do away with winter, and the putrefaction of seeds, for the most part, would be removed. When that is missing, their vivification would perish. For no seed is born into life unless it is first corrupted. But wipe out summer, and the drying out of the excess moisture in things would perish. Take that away, and no maturing of them would follow. So to hell with the misguided suggestion of your inconvenient spring!

72. Nephew: I recall that I introduced the delights of spring without reason. But now I should like to summon back to the treatise certain things which we passed over in silence before: namely, what necessity, what convenience summoned the planets from their straight courses and natural orbits? Surely it would have been more right, more natural, more easy to perceive for noting down their courses and for teaching, if they turned in the same circuit as the *aplanos* as if with their principle? I understand what certain people have written on this – I do not approve of it. For they say that the planets temper the rapid

extime cursum. Que vero huius temperantie expetibilitas? Quid nostra refert? Poteratne illius cursus timeri precipitium?

ADELARDUS: Sane. Atqui nostrum intercurrere poterat exitium? Ut enim illa superiora animalia – applanos dico et sidera – corporis animeque[432] perfectione naturaliter mobilia sunt, ita nos terrena animantia ingruente rerum fece turbida stabilia esse quodammodo necesse est. Licet enim frequenter ad motum alienemur, ad quietem statim naturali desiderio revocamur. Si itaque extime spere subditarumque stellarum idem esset cursus, dum totus aer subiectus moveretur, nos etiam secum aut impediret aut implicaret impetus. Nunc vero interiecto in media regione planetarum occursu, et ab illis motus innativus observatur et nobis quietis refectio non aufertur.

.lxxiii. NEPOS: Rationem recipio. Nunc vero recipere vellem quod pridie michi accidit. Nam cum celestia signa nocturnus et notares et divideres, Yadasque et Pliadas loco numeroque distingueres, subito ex earum cetu quidam fulgor emicuit, qui cum stellarem[433] haberet forme similitudinem, longeque a loco tractu vestigante se transtulisset, quandam de Pliadum numero rebar esse semotam. Cum vero ad earum cetum rimabundos oculos retulissem, easque numero non pauperatas esse percepissem, ammiratione perculsus sum.

ADELARDUS: Non ita currunt cursus illi ut hoc modo cadant. Quippe iam pridem exacto tempore viduatum foret celum stellarum lumine.

NEPOS: Quid igitur est quod tanta similitudine per inane ductum stelle formam oculis vigilantibus immo notantibus exprimere potest?

ADELARDUS: Quod leviter capis, non firmiter tenes. Cum enim de tonitruis superius tractaremus, id inter nos ratione duce convenit, ignem scilicet quandoque ex glatierum fracturis, quandoque etiam ex ipsorum ventorum nisu enasci. Scintille itaque ille que per aerem trahi videntur, non stellarum corpora, set inferioris aeris et ventorum progenies. Unde et si nautas consulas, ipsi quidem ex tali effectu communiter ventorum motus concipiunt. Quod autem quasi a corpore ipsius stelle talis scintilla videtur procedere, inde est, quod ex tam vicina stellarum positione que undique celum obsident, quocumque in

[432] animeque GQV, anime P, et anime M
[433] stellarem G al.PQV, stellarum G, stellaris M

course of the outermost sphere. What is the need for this tempering? What has it to do with us? Could the collapse of its course have been feared?

Adelard: Certainly – but could not our destruction have occurred in the meantime? For as those superior animate beings – I refer to the *aplanos* and the stars – by the perfection of their body and soul, are naturally in movement, so we earthy animals ought in some way to be stable as the turbulent filth of things assails us. For although frequently we are enticed into alienating motion, nevertheless we are immediately summoned to a state of rest by our natural desire. If, then, the course of the outermost sphere and the stars beneath it were the same, then the whole air below them would be moved, and the force would either get in our way or sweep us along with it. But now, because the planets thrown into the region in between bar the way, they keep to an inborn movement and at the same time we are not deprived of the refreshment of rest.

73. Nephew: I accept the reason. But now I should like to accept what happened to me the other day. For at night-time you were noting and separating the celestial constellations, and distinguishing the Hyades and the Pleiades by place and by number. Suddenly from the company of those stars a certain spark flashed out. Since it had the likeness of the form of the stars and carried itself far from the place, leaving a trail behind it, I thought that one star had been removed from the number of the Pleiades. But when I brought my searching eyes back to their company and found that they had not been made any poorer in their number, I was struck with wonder.

Adelard: They do not run their courses in such a way that they fall like this. Otherwise in the march of time the heavens would have been widowed of the light of the stars long ago.

Nephew: What is it, then, which, when carried through the void, is able to imitate the form of a star so closely for wakeful, or rather for observant, eyes?

Adelard: What you lightly take in, you do not firmly retain. For when we were dealing before with thunder, we agreed on this, with reason as our guide, that fire is born sometimes from the shattering of ice, sometimes even from the throes of the winds themselves. Those sparks then which are seen to be carried through the air are not the bodies of stars, but the progeny of the lower air and of winds. Hence if you consult sailors, they generally deduce the movements of the winds from this kind of effect. That such a spark seems as it were to come from the body of the star itself, is because, in whatever place the lower

loco inferior ignescat aer,[434] a stella exire falso putatur, quemadmodum et planetas in signifero ferri putare solemus. Nec tamen id ipsum penitus urgeo, quin aliquando etiam vi[435] ipsarum stellarum huiusmodi flammas fieri concedam. Cum enim tanta corpora tali ferantur velocitate, consequens est ut tam violenti volutione exalescat ether in flammam.

.lxxiv. NEPOS: Teneo quod querebam. Nunc itaque ad disciplinam revocato animo, id quod superius tetigisti audire desidero; de Luna enim cum diceres, animantem vocavisti. Ego itaque licet stolidus, non tamen oblitus, ad rem verbum reduco: videlicet utrum stellas animatas an inanimatas esse intelligas.

ADELARDUS: Qui eas inanimatas esse opinatur, eum potius sine anima esse existimo. Assertiones vero quibus eas animatas[436] opinari inducor, partim a loco, partim a compositione, partim etiam ab earumdem actione concipio. Si enim hec regio grandine turbida, nubibus horrida, feda tenebris, rationem providentiamque sustinere debet, quoniam potest, quanto magis etherea planities omni immunditia purgata, menti ac rationi obediens est? Amplius: si in hac ipsa tenebrosa regione nostra ille locus philosophorum fertilior est, qui serenitate quadam a densa fece elimatus purgatior est, qua fronte illam plagam celestem que quanto a nobis remotissima, tanto omni tali sorde intactissima est, animali motu expertem esse dicere sine summa insania quisquam potest? Item, si rerum creatarum[437] animo nichil melius esse potest, isne in tenebris infodiendus est, locusque ei aptissimus ab eo quem[438] nature similitudine vehementer expetit viduandus? Atqui si ita est, rerum ordo transpositus est. Nam et contraria coniunguntur, et similia disparantur.[439] Qui vero rerum ordinem tollit, insipiens est. Ab insipiente igitur hanc confusionem constitui necesse est. Rerum vero Dispositor sapientissimus est. Minime igitur rerum ordinem tollere vel vult vel potest. Quare nec hanc transpositionem stare possibile est. In philosophantis itaque animum id incidere minime conveniens est.

[434] inferior ignescat aer PV, ignescat aer inferior GMQ
[435] vi GMQ, in PV
[436] GM *add.* esse, PQV *om.*
[437] creaturarum P
[438] quem PQV, quam G *p.c.*, quod G *a.c.*(?)M
[439] disparantur G al.PQV, dissipantur G, M *om.*

air is ignited, it is so close to one of the stars which throng the heavens on all sides, that one thinks falsely that the spark comes from a star. In the same way we are accustomed to think that the planets too are borne on the zodiac. I do not press this point too far, because I agree that sometimes flames of this kind occur because of a force in the stars themselves. For, since such great bodies are carried along with such speed, it follows that the ether bursts into flame because of such a violent sweeping round.

74. Nephew: I have what I was seeking. So now, having called my mind back to instruction, I want to hear what you touched upon before; for when you were speaking about the Moon, you called it an animate being. Although I am dull-witted, I haven't forgotten, and I want the conversation to return to that point: namely, whether you understand the stars to be animate or inanimate.

Adelard: Whoever thinks they are inanimate, is himself rather without a soul in my opinion. The premisses by which I am led to believe them to be animate I take partly from their position, partly from their composition, and partly, too, from their action.[87] For if this region which is churned about with hail, bristling with clouds, and murky with darkness, ought to sustain reason and providence – since it can – how much more is the ethereal plane, purged of all uncleanliness, obedient to mind and reason? Another point: if in this shadowy region of ours that place is more productive of philosophers which, in a certain serenity, is the more purified, cleansed of dense pollution, with what impudence can anyone say that that celestial realm, which is as much untouched by all such soils as it is remote from us, does not participate in the movement of the soul?[88] That would be the height of folly! Again, if of created things nothing can be better than the mind, should this be buried in darkness, and should the place which is most suitable for it be deprived of that which it strenuously seeks because of the similarity of its nature? But if it is so, the order of things is reversed. For contraries are joined, and like is separated from like. But whoever takes away the order of things is a fool. So it is necessary for this confusion to be established by a fool. But the Disposer of things is supremely wise. He, therefore, neither wishes to, nor is able to, take away the order of things. Therefore this reversal cannot stand. For this, then, to occur to the mind of a philosopher is not at all appropriate.

Nunc vero de compositione considerandum est. Id enim in omni disceptatione non parvam vim habet, ut consideretur cuius nature sit id de quo agitur. Nichil enim est[440] quod ad consequentes effectus magis annotescat, quam proprie essentie compositio, preter efficientis cause inexplicabilem illationem. Compositionem itaque in materia et forma accipio. Circa materiam igitur prima[441] sit intentio. Hec autem stellarum corpora ita proportionaliter compacta sunt, ut cum ex[442] quatuor elementis componantur, eo magis habundent[443] quod ad vitam rationemque magis aptum est. Magis enim ignea sunt quam terrestria. Ignis autem illarum, ut Cleantes arguit, aut peremptorius dicendus est, ut ignis exterior, aut mulcebris et innoxius, ut ignis corporis nostri interior. Atqui, si peremptorius esset, iam pridem suum subpositum disparasset.[444] Ad formam etiam ignis exterioris, si talis esset, se accommodaret, atque in conum surgeret. At neque suum subiectum rodit, neque in conum surgit. Non est igitur peremptorius. Est itaque mulcebris et innoxius. Sequitur ergo ut et sensui et rationi sit obediens et obnoxius. De forma vcro quid attinet dicere, cum Latinis etiam pateat omnium formarum eam anime[445] esse aptissimam, quam videant plenam esse et rotundam.

De actione itaque earum notandum est. In quo non meam set Aristotilis accipe sententiam – immo, quia ipsius, ideo meam. Quicquid movetur, ait, aut natura aut vi aut voluntate moveri convenit. Quod autem natura movetur, aut sursum movetur ut ignis, aut deorsum ut terra. Non autem ita moventur stelle. Non igitur moventur natura. Vi autem non moventur. Que enim maior vis esse potest? Non igitur vi moventur. Sponte igitur voluntarieque moveri eas necesse est. Quod si spontaneo, et animali sequitur ut moveantur motu. Viderit ille quid dixerit. Ego vero ita nectam: si quandoque moventur, aliquando vero stant, cumque moventur, aliquando ad anteriora, quandoque retrograde convertuntur, nulla natura moventur. Quod enim natura fit, nec cessat, nec convertitur. Restat itaque ut aut vi aut voluntate moveantur. Vis autem in rerum natura nulla molestior est quam applanetici corporis conversio. Atqui[446] ab illo non moventur; contra illud enim citantur. Voluntate itaque moventur. Quid inde sequitur, vides.

440 enim est PMV, GQ *transp.*
441 primo P
442 cum ex M, e G *p.c.*, cum G *a.c.*PQ
443 habundent G *a.c.*, habundant G *p.c.*PMQV
444 disparasset G *a.c.*PQ, dissipasset G *p.c.*M
445 anime G *p.c.*PMV, animam G *a.c.*Q
446 Atque P

But now one must consider the stars' composition. In every discussion it is most important to consider the nature of that which is discussed. For nothing reveals the consequent effects of anything better than the composition of that thing's own essence (leaving aside the inexplicable contribution of the efficient cause). I understand composition to consist in matter and form. Let our first consideration be about the matter. These bodies of the stars are put together in such a proportional way, that, although they are composed of the four elements, they have more of that which is more suitable to life and reason. For they are more fiery than earthy. Their fire, as Cleanthes proves, should either be called destructive, like the fire around us, or soothing and harmless, like the fire inside our bodies.[89] But if it were destructive, it would have demolished its subject long ago. Also if it were like this, it would fit itself to the form of the fire around us, and rise up in a cone. But it neither consumes its subject nor rises up in a cone. So it is not destructive fire. It is, therefore, soothing and harmless. It follows, then, that it is obedient and subject to both sense-perception and reason. As for their form, what is the point of speaking about it, when it is clear even to the Latins that, of all forms, that of the stars, which they see is full and round, is most appropriate to the soul?

One must, then, turn one's attention to their action. In this accept not my own opinion, but that of Aristotle – or rather, because it is his, it's my own.[90] Whatever moves, he says, must move by nature, force or will. What moves by nature either moves upwards, like fire, or downwards, like earth. The stars do not move in this way; so they do not move by nature. They do not move by force. For what greater force can there be? So they do not move by force. They must move spontaneously, then, and voluntarily. But if they move spontaneously, then it follows that they also move by the soul's movement. Let him look to what he has said; I shall weave the following into the argument. If sometimes they move but at other times stand still, and when they move, they sometimes move forwards and at other times turn back, then they are moved by no nature. For what is done by nature does not stop or turn back. It remains, then, that they move either by force or by will. In the nature of things no force is more irresistible than the revolution of the body of the *aplanos*. But they are not moved by that; for they are roused into a motion contrary to that. So they are moved by will. You see what follows from this.

Amplius: si eorum actio inferiorum animalium mortis viteque causa est, quid de his[447] censendum est? Atqui recedente a Cancro et a nostra regione Sole, primo herbida emarcescit viriditas; deinde caput inclinat floralis iocunditas; postmodum nec etiam foliorum truncis suis perdurat honestas; postremo non pauca animalia esse tam terrestria quam volatilia, que hieme naturaliter commoriuntur, estate reviviscunt, satis notum est. Quid igitur dicendum est? Id quod aliis vitalem prestat effectum, ipsum vite expers esse putandum, non nisi scurre iocantis est.

Item: quicquid in motu suo numerum determinatum atque ordinem constantem servat, id ratione uti certum est. Atqui stellarum cursu nichil est numero[448] determinatius, nichil ordine stabilius. Quando enim planetarum licentia zodiacum excessit? Cum Sol[449] Capricorni extrema tetigit, cursum cohibet, ad nos recurrit. Set nec extra Cancrum nos[450] ipsos respicit. Idemque cetere observant stelle. Nichil igitur eis inter creaturas rationabilius.

Item – per deos immortales oro, obsecro, animum adhibe! – cum ille que modo summa celerrimaque concitatione ad Capricornum obliquate properaverant, ab eodem absque vi illata revertuntur, aut sciunt revertendum inde esse, aut nesciunt. Si revertendum esse nesciunt, cur ultra non procedunt? Cur tam propere[451] redeunt? At si sciunt, scientia ditate sunt. Hanc autem sine anima nature nulle habere possunt. Animate igitur scientieque et rationis compotes sunt.

.lxxv. NEPOS: Dum crebras usque adeo coacervasti rationes, nec solum michi verbum interponere licuit. Prius enim defecit tibi anhelitus quam argumentum. Nunc itaque, dum respiras, michi fistula accommodanda est. Si enim ita[452] esse potest ut a te disputatum est, quod inde sequitur recipiendum est. Cum enim omne animal cibo esuque degat, si stellas animalia esse autumas, eisdem oportet cibos ut invenias.

ADELARDUS: Preclare. Ut igitur illa animalia loco, compositione, forma, ratione, his inferioribus diviniora sunt, ita esu quodam cibis

447 his PV, ipsis GMQ
448 nichil est numero GMQ, et numero nichil est PV
449 Sol Q, GPMV *om.*
450 nos G *p.c.*PQV, nosmet G *a.c.*M
451 tam propere M, tantopere GPQV
452 ita G *p.c.*PQ, *om.* G *a.c.*MV

Another point: if their action is the cause of the death and life of lower animals, what should be thought about these stars? When the Sun withdraws from Cancer and from our region, first the greenness of the grass fades; then the delightful blooms of the flowers droop; soon not even the distinction of leaves remains on their branches; finally it is well known that there are not a few animals – both of the earth and of the air – who naturally die in winter and come to life again in the summer. What then should be said? To believe that what provides the effect of life for others is itself without life, can only be the belief of a frivolous jester.

Again, whatever in its movement keeps to a fixed measurement and constant order certainly uses reason. But in its measurement nothing is more fixed than the course of the stars, or more stable in its order. For when had the planets a licence to go outside the zodiac? When the Sun touches the edge of Capricorn, it restrains its course, and runs back to us. But it does not look towards us after reaching Cancer. The other planets observe the same law. Nothing, then, amongst creatures is more rational than they.

And again – I beg you in the name of the immortal gods, pay attention! – when those planets which had just hurried on an oblique course to Capricorn, with the greatest and swiftest movement, return from there without any force being brought to bear, either they know they have to return from there, or they don't know. If they don't know they have to return, why do they not continue further? Why do they return so readily? But if they know, they are endowed with knowledge. No natures can have this without a soul. They are, therefore, animate, and in possession of knowledge and reason.

75. Nephew: While you have gone on heaping one reason on top of another, I have not been allowed to insert a single word! You have run out of breath before running out of arguments. So now, whilst you recover your breath, the pipe should be passed to me. If the situation as you have described it can be like this, what follows from it must be accepted. Since every animal lives from food and nourishment, if you claim the stars are animals, you must find food for them.

Adelard: Obviously. As those animals are more divine than these lower ones in place, composition, form and reason, so they both ought to and are able to take a kind of nourishment which is considerably

nostris admodum puriore uti et debent et possunt. Habent itaque humiditates terrarum et aquarum longinquissimo intervallo, dum ad superiora trahuntur, extenuatas. Quibus dum utuntur, nec pondere gravantur nec ratione prudentiave hebetantur.

.lxxvi. NEPOS: Quoniam de his que circa stellas ambigua michi parebant dubitantem firmavisti animum, quid de ipso applano qui universa formata sinu continet tenendum sit, insinuari opere pretium duco. Cum enim philosophorum multorum multa scripta legerim, varias dissidentesque[453] de hac continenti universitatem spera colligo sententias. De quibus cum adeo diverse sint, utrum plures vere esse possint, id quidem dubito, falsas autem omnes esse, id non iuro. Alii enim inanimatum esse hunc ambitum dicere presumpserunt; alii non animatum[454] modo, set et animal rationale esse disseruerunt; quidam[455] altius vel intelligentes vel furentes deum vocare ausi sunt. Quare que harum diffinitionum tibi sedeat, sedens expecto.

ADELARDUS: Intelligendo magis[456] quam sedendo expecta. Et extimam quidem speram, de qua sermo susceptus est, sicut ad unam divisionis naturam descendere indecens est, ita ad aliam ascendere impossibile. Nam quod inanimata sit dicere, non vulgarem set disciplinarem pudeat hominem. Eam enim animatam esse, tum[457] forme figureque exterioris exornatio, tum subtiliorum elementorum materialis compositio, tum loci supremi ideoque dignissimi, quo nichil inanimum repit,[458] naturalis occupatio, tandem vero orbicularis motus indefesse vite, ineffabilis velocitatis arguit conversio. Deum vero esse eandem continentiam quodam modo philosophantis, alio modo insanientis[459] est. De deo enim animali, rationali, immortali, si queritur, hoc modo applanon deum esse concedendum est. De Deo vero a quo universalis rerum causa incomposito, informi, immutabili, infinito, si investigatur, hoc modo extimam speram deum dici abhominandum est.

453 dissidentes P
454 animatum G *a.c.*MQ, inanimatum G *p.c.*P, V *om.*
455 GMQ *add.* autem
456 magis M, GPQ *om.* (*V gives a different text*)
457 tum GMQ, tunc P, tam V
458 repit G, $^{r}_{c}$epit (*sic*) P, repperit MQ, recepit V
459 insaniantis P

purer than our foods. Thus they have the moistures of the earth and the waters, thinned out by the very long distance they travel when they are drawn up to the higher regions.[91] When they take these foods, they are not weighed down by their heaviness, nor are they blunted in their reason or wisdom.

76. Nephew: Since you have strengthened my doubting mind concerning these things about the stars which appeared uncertain to me, I think it worth while for you to suggest what should be thought about the *aplanos* itself, which holds all created things in its embrace. For, since I have read many writings of many philosophers, I get different and incompatible opinions about this sphere which contains everything. Since they are so diverse, can several of them be true? This I doubt, but I do not swear that all of them are false. For some philosophers have been bold enough to say that this surrounding being is inanimate; others have argued that it is not only animate, but even a rational animal. Some, of deeper intelligence or deeper madness, have dared to call it a god.[92] Therefore, which of these definitions is settled in your view, I shall settle down and wait to hear.[93]

Adelard: Wait for the answer through understanding rather than through settling yourself. For, just as it is unfitting for the outermost sphere (about which we have taken up the conversation) to descend to one kind of category, so is it impossible for it to ascend to another. To say that it is inanimate should shame not the common man, but the educated person. That it is animate is proved by the adornment of its form and exterior shape, by its material composition consisting of the more subtle elements, by its natural occupation of the highest and therefore worthiest place (where nothing inanimate can creep in), and finally by its turning, consisting of a circular movement, not tiring in its liveliness, and of indescribable speed. That this same container is a god is in one sense the statement of a philosopher, in another, that of a madman. If one is looking for an animate, rational, immortal god, in this sense it must be agreed that the *aplanos* is a god. But if one is seeking the incomposite, formless, unchangeable, infinite God from whom is the cause of all things, in this sense, to speak of the outermost sphere as God should be abominated.

NEPOS: Abhominandum certe ipsa videtur appellatione, set quia in principio disputationis nostre eo tenore convenimus, ut rationem magis quam auctoritatem sequeremur, quia etiam hactenus dissertum est ut a rebus infimis usque ad summitatem ascensum sit, quia demum ita sermo processit, ut de Deo efficiente omnium causa mentio facta sit, hic precipue mentis intentio acuenda est. Scientia etenim mea nulla est, nisi de illo sciam. Nunc igitur quia deum quidam esse perhibent, alii vero id ipsum negant, qui vero esse asserunt, alii materiam, alii formam, alii mundum, quidam applanon, quidam Solem deum esse dissentiunt, a te quidem utrum sit vel non sit, et quid etiam sit, quidve agat, ratione sola auctoritatis adulatione exclusa audire desidero.

ADELARDUS: De his michi questiones inducis de quibus magis quid non sit inprobem, quam quid sit probem. Facilius enim in rebus talibus falsa divertere quam vera reperire soleo. In hac enim difficultate tractandi de Deo, de NOY, de hyle, de simplicibus formis, de puris elementis disserendum est, que, sicut simplicitate propria naturam compositorum excesserunt, ita et de eis disputatio alias omnes dissertiones et intellectus subtilitate et sermonis difficultate precellit. Nos igitur quoniam quedam de compositis diximus, vespere iam somnos suadente, quiete naturali mentes reficiamus. Mane autem, si tibi idem sedet, conveniamus de initio vel initiis ut disputemus.

NEPOS: Michi vero nichil magis sedet. De deo etenim mentem instruere, quoniam Patrem omnium fatemur honestissimum; de eodem etiam argute dicere, quoniam auctoritatem non recipio difficillimum; de his vero que id ipsum comitantur discutere, quoniam multi multa inde turbaverunt, utilissimum est. Quietis igitur refectionem libens accipio, ut ad tractatum novum novi veniamus.

Nephew: It certainly seems to be an abomination to use this term, but because at the beginning of our disputation we agreed on this course, that we should follow reason rather than authority, because, too, up to now the discussion has taken the form of an ascent from the lowest things to the heights, because, finally, the conversation has advanced to the point of mentioning God, the efficient cause of all things, here especially the mind's attention should be sharpened. For my knowledge is nothing unless I know about him. Now, therefore, because some people say that God exists, but others deny that very fact, and those who assert his existence differ amongst each other, some saying God is matter, others, form, some, the world, others, the *aplanos*, yet others the Sun – from you, then, I want to hear, using reason alone and keeping away the flattery of authority, whether he exists or not, and what he is, and what he does.[94]

Adelard: You bring me questions concerning matters in which I would rather disprove what is not, than prove what is. For I am more easily accustomed in such cases to turn away what is false, than to find what is true. In this difficult situation of dealing with God, one must also discuss *noys*, *hyle*, the simple forms and the pure elements; as they exceed the nature of composite things because of their own simplicity, so a disputation concerning them excels all other discussions both by the subtlety of its intellectual content and the difficulty of its expression. But now that we have said something about composite things, and since the evening is encouraging us to sleep, let us refresh our minds with natural rest. Tomorrow, if you are agreed, let us meet to dispute about the beginning or beginnings.[95]

Nephew: I cannot agree to anything else more readily. For to instruct the mind about God, since we confess him to be the most honourable Father of all, and to talk in a persuasive way about him (since I do not accept authority), is most difficult, but to discuss those things which accompany him, since many people have muddied the waters in this, is most useful. Therefore, I gladly accept the refreshment of rest, so that we may approach a new discussion as new men.

Notes

1 Southern (*Robert Grosseteste*, p. xlvii, n. 26) wrongly interprets this as 'at the command of king Henry'.
2 Only in MS L.
3 The Renaissance editions and Müller add the question: 'Why do we tolerate certain very hot foods more easily by gulping them down than by holding them in the hand?'
4 The numbers in brackets are those of Müller when they diverge from our numbering.
5 These two questions are dealt with together; Müller omits both.
6 'This is how the causes of things work' ('Sic faciunt cause rerum') is a kind of motto at the head of Adelard's text; see pp. xxx–xxxi above.
7 Plato, *Timaeus*, trans. Calcidius, 49D.
8 'Inferiora' should be in distinction to the supralunary elements. Either 'interiora' is expected, in contrast to 'exteriora', or 'superiora' in contrast to 'inferiora'. The first alternative suits the previous sentence better.
9 Plato, *Timaeus*, trans. Calcidius, 33C.
10 For the choice of the term 'sensilis' (which is Calcidian) rather than 'sensibilis' (which is Boethian), see R. D. Crouse, '*Hic sensilis mundus*: Calcidius and Eriugena in Honorius Augustudunensis', in *From Athens to Chartres*, ed. Westra, pp. 283–8.
11 A proverbial expression meaning 'to all the world'; see Horace, *Satires*, 1, 7, 3.
12 Cf. Cicero, *De natura deorum*, II, 15 (42); the 'fiery animals' (i.e., stars) are discussed by Adelard in Q74.
13 The scribe of MS G originally wrote 'araborum' (for 'arborum'): he obviously had Arabs on his mind.
14 In what follows, although claiming the Arabs as his models, Adelard echoes an argument at the beginning of Cicero's *De natura deorum*, in which Cicero agrees to use rational arguments, rather than tradition, to discuss the nature of the Gods; e.g., I.v (10), p. 13: 'non enim tam auctoritatis in disputando quam rationis momenta quaerenda sunt. Quin etiam obest plerumque iis qui discere volunt auctoritas eorum qui se docere profitentur. Desinunt enim suum judicium adhibere: id habent ratum quod ab eo quem probant judicatum vident.' ('In discussion it is not so much the weight of authority as force of reason that should be demanded. Indeed the authority of those who profess to teach is often a positive hindrance to those who desire to learn; they cease to employ their own judgement, and

take what they perceive to be the verdict of their chosen master as settling the question.') Similar, too, are Beringerius of Tours' arguments in his *Rescriptum contra Lanfrannum*, pp. 85 and 90, where he places reason before authority and states that if man does not take refuge in reason he abandons his own dignity ('honor'), since it was in respect to his reason that man was made in the image of God: see Speer, *Die entdeckte Natur*, pp. 40–41 and references there.

15 'Scriptorum' could equally be translated 'writers' (from 'scriptor').

16 An echo of Juvenal, *Satire* 1.6–7: 'et in tergo necdum / finitus Orestes' ('The *Orestes* written also on the back [of the roll] and not yet finished').

17 There is a double metaphor here: that of the prostitute who attracts the client by adorning her face, and that of the written word 'painted' on the skin of an animal.

18 'Iecur', of course, means 'liver' in Classical Latin, but Adelard elsewhere uses the word '(h)epar' for liver, and seems uncertain whether some animals have an '(h)epar' (Q9). A late Medieval Latin-German glossary claims that 'iecur' properly refers to the liver in brute animals, while '(h)epar' refers to the human liver (see *LLNMA*, s.v. iecur). However, the description of the organ is clearly that of the gizzard, 'the thick-walled part of a bird's stomach in which hard food is broken up by muscular action and contact with grit and small stones' (Collins English Dictionary, s.v.). It is difficult to see that the liver and gizzard of birds could actually have been confused in the Middle Ages, and the confusion is probably only in terminology. For the word 'gizzard' comes from French *guisier*, which meant either liver or the 'third stomach of birds' (i.e., gizzard; see v. Wartburg, IV, p. 135, s.v. gigerium). In Old French glosses this word ('gisier, jusier') is glossed as 'iecur' (Tobler-Lommatzsch, s.v. gisier). In the two Hebrew translations of the *QN* 'iecur' is translated as 'qorq^{e}ḇân', which is the Classical Hebrew word for the gizzard (Gollancz, *Dodi ve-Nechdi*, Hebrew pp. 16 and 50 – Gollancz erroneously translates the word 'crop' on p. 24); and, in the abbreviated Hebrew version, 'qorq^{e}ḇân' is glossed as 'gusier in the Romance language' (ibid., p. 50). Note also Adelard's (or an early copyist's) confusion over the gender of 'iecur'; in Classical Latin it is always neuter.

19 The 'albugineous humour' is mentioned by William of Conches in the same context: *Dragmaticon*, ed. Ronca, VI.20, p. 251.

20 A reference to the questions on sight, Q23–30 below.

21 This view is criticized by William of Conches in *Dragmaticon*, ed. Ronca, VI.24, p. 261: 'inde est quod bruta animalia dominos suos videntur agnoscere, unum fugere et aliud appetere, quod non ex discretione, ut quidam autumant, faciunt, sed ex imaginatione' ('hence it is that brute animals seem to recognize their masters, to flee one thing and make for another; this they do not from discernment, as some people assert, but from their imaginative faculty').

22 The same definition is found in William of Conches' *Dragmaticon*, VI.11, ed.

Ronca, p. 217: 'Aristoteles igitur in suis Metaphysicis sic sensum diffinit: sensus est animati corporis, applicatione exteriorum non levis mutatio'. The source has not been found.

23 I.e., human bodies should be able to use opinion without the soul.

24 For the definition of soul as 'the active nature' ('natura agens') see Q38 below.

25 This advocacy of the immortality of animals' souls recalls Johannes Scotus Eriugena's discussion of the same topic in *Periphyseon*, Book III, ed. I. P. Sheldon-Williams, pp. 296–8; see P. Dronke, 'La creazione degli animali', in id. *Intellectuals and Poets in Medieval Europe*, Rome, 1992, pp. 193–217 (see pp. 199–202). Scotus' statements are considerably toned down in the abbreviation of this passage in Honorius Augustodunensis, *Clavis physicae*, c. 226, pp. 177–8.

26 The Nephew is referring to Mt Etna, which Adelard must have seen during the period when he dedicated his *DED* to William, bishop of Syracuse.

27 A similar argument is found in Cicero's *De re publica*, III, 2, *apud* Lactantius, *De opificio dei*, III, 16–19: 'Ita fit, ut plus homini conferat ratio quam natura mutis, quoniam in illis nec magnitudo virium neque firmitas corporis efficere potest quominus aut opprimantur a nobis aut nostrae subiecta sint potestati.' ('So it happens that reason confers more to man than nature does to mute animals, since in these neither their greatness of strength nor their sturdiness of body can prevent them from being overcome by us or subject to our power.')

28 The terms are those of Calcidius' translation of the *Timaeus*, 17C–D: 'bellicis negotiis . . . in pacis officiis'.

29 The translation follows the emendation of Müller ('nobis a phisicis'). The reading of GPM – 'a nobis' – is clearly wrong, but that of QV – 'a vobis', identifying Adelard with the natural scientists – is plausible.

30 For this paragraph compare *DED*, pp. 70–1.

31 The reference to Aristotle's *Physics* is wrong; Adelard is rather indebted to medical sources, and above all to Nemesius, *Premnon phisicon*, c. 13, p. 89, which included the proofs of the locations of the faculties from the effects of injuries. Adelard was probably confused by Nemesius' own reference to 'Aristoteles in Physicis' three pages later (c. 14); see T. Ricklin, *Die* Physica *und der* Liber de causis *im 12. Jahrhundert*, Freiburg/Schweiz, 1995, pp. 39–41 and Lawn, *Salernitan Questions*, p. 22. The whole of this paragraph recalls the last paragraph of the speech of Philosophia in *DED*, pp. 70–1 above.

32 I.e., Aristotle, as mentioned (erroneously) in the Nephew's question. In William of Conches' version of the same answer, the authority is the 'physici' ('natural scientists').

33 Plato, *Timaeus*, trans. Calcidius, 29D–E. The reference is to p. 121 above.

34 The answer to this question reproduces similar arguments to those in Macrobius, *Saturnalia*, 7.10; see Lawn, *Salernitan Questions*, p. 24.

35 Here Adelard seems to be explaining a Classical Latin term ('craneum'), by the non-Classical 'testa' which had given rise to the Romance words for the head.
36 Boethius, *De musica*, I.1, p. 179.10–11.
37 Adelard is not so much quoting Boethius as summarising the arguments in the *Glosule ad Priscianum*, which are a development of Boethius, *De musica*, I.14, p. 200. The passage in the *Glosule* reads: 'Aer in ore loquentis naturalibus instrumentis formatus vicinos impellit aeres et in sua afficit forma. Ita fit ut vox eadem secundum formam sit in auribus diversorum, sed quantum ad materiam diversorum, ut ita dicam, aerum diversa.' ('The air formed by the natural organs in the mouth of the speaker impels the neighbouring airs and affects them with its form. Thus it happens that the sound in the ears of different people is the same in respect to its form, but different in respect to the matter of the different airs, so to speak.') The addition of 'parcel (of air)' in the translation has been made for the sake of clarity, but can in fact be found already – as 'particula (aeris)' – in William of Conches' *Dragmaticon*, VI.21, ed. Ronca, p. 253 for which Adelard's discussion is the source. For further discussion of the whole of this Question see Burnett, 'Adelard, Music and the Quadrivium', pp. 175–80.
38 Virgil, *Eclogue*, 3.104.
39 Adelard may be recalling inaccurately Boethius, *De musica*, I.1, p. 179: 'Adest enim cunctis mortalibus visus.'
40 Boethius, *Consolation of Philosophy*, V, m. 4.
41 Boethius uses the phrase 'argumentandi locus' in two places in his Commentary on the *Topics* of Cicero – see PL, 64, cols 1113 and1169 – but not 'argumentandi gratia'.
42 Plato, *Timaeus*, trans. Calcidius, 45D–E.
43 MS Q adds the missing phrases from the *Timaeus*.
44 There is a pun here which depends on the double meaning of 'decidere' = 'to fall away from' and 'to die'.
45 In MS P only.
46 In MS P only.
47 There is an intricate set of overlapping inferences here. The Stoics take their name from the Stoa – a covered colonnade which Adelard hints at in his 'dark doorway'; the square is the forum, which is the place where the advocate put into practice his rhetorical skills in Classical times; the old Roman forum was still a conspicuous feature of medieval Bath. In addition, the 'porta nigra' recalls Boethius' words just quoted by Adelard 'quondam *Porticus* attulit /*obscuros* nimium senes'.
48 An imprecise reference to Calcidius, c. 242, ed. Waszink, p. 255.
49 In MS P only. The question is answered in the Hebrew versions: see Gollancz, *Dodi ve-Nechdi*, pp.36 (Q33) and 77 (Q19).
50 This question is posed in Macrobius, *Saturnalia*, 7.8.14; the answer there is similar to that of Adelard, but much briefer.
51 This reference has not been found.

52 In MS J, fol. 50r, the hand that writes the text adds another solution in the margin: 'Solutio alia: tepida et ideo minus calida infusa ebullienti aqua, partem sibi superflui caloris attrahit sicque calor per plura dispersus sentitur in singulis partibus merito minor et minus.' ('Another solution: tepid and therefore less warm water poured into boiling water attracts part of the superfluous heat to itself and thus the heat, dispersed more widely, is felt in single parts rightly to be less and <is felt> less strongly.')

53 'To take a bladder for a lantern' is proverbial, and occurs in Martial (*Epigrams*, XIV, no. 62, ed. Shackleton Bailey, III, pp. 250–1: 'Lanterna de vesica'). Müller (p. 40) cites a French proverb: 'prendre des vessies pour des lanternes'.

54 Macrobius, *Saturnalia*, 7.12.22; here the contrast is between holding in the hand (as in the *capitula*, p. 228, n. 3 above) and swallowing.

55 Terence, *Phormio*, 418–9. This question is answered in the two Hebrew versions, which give the fact that the throat is nearer to the heart and therefore warmer than the mouth as the solution (Gollancz, *Dodi ve-Nechdi*, pp. 41 and 78); the Italian version gives three answers, of which the first is similar to that in the Hebrew versions (Geymonat, *Questioni filosofiche* IV.3.21). None of these texts mentions Macrobius and they refer, as does Adelard, to holding food in the mouth rather than in the hand.

56 These are listed as two separate questions in the *capitula* at the beginning of the book, and are numbered as such at this point in MS G.

57 Although the manuscripts are in agreement in including 'humana', the argument more naturally refers to brute animals, to which human beings are compared in the last phrase of the paragraph.

58 'Iuvenes' and 'senes' are two of the ages of men, defined by Isidore as being from the years 28–50 and 50–70 respectively: see *Etymologies*, 11.2.5–6.

59 Adelard may be guarding himself from objections that a Christian should rejoice to leave his body.

60 This is a saying found widely in Medieval proverb literature; see Walther, no. 24391.

61 This seems to be a reference to the sequel to the *QN* mentioned at the end of the text.

62 Boethius, *Consolation of Philosophy*, III, pr. 11.

63 This is the question that the Greek philosopher in Magna Graecia posed to Adelard at the end of the *DED*.

64 This is the first of several references to the same passage of Statius (hence, here, Statius is introduced: as a philosopher and poet). This quotation is from his *Thebaid*, VI, v. 364.

65 These two activities are mentioned as typical of futile endeavour in Ovid, *Heroides* 5, v. 115.

66 Adelard is referring to Macrobius' account in *Commentary on the Dream of Scipio*, 2.9.3, p. 123: 'ex ipsa aquarum collisione nascitur illa famosa Oceani accessio pariter et recessio' ('from the very collision of the waters there arises that famous flowing and ebbing of the Ocean').

67 This inference must derive from Adelard's experience of the lack of tides in the Mediterranean.

68 Statius, *Thebaid*, VI, v. 362.

69 Horace, *Epistles*, I, 2, vv. 42–3.

70 Adelard is referring to irrigation mechanisms, such as the shadoof and *sāniya*, which were particularly characteristic of Arabic irrigation systems, but not exclusively so.

71 I owe this pun to Ron Thomson.

72 Adelard seems to be distancing himself here from an interest in magic: it was an aberration of his younger days. He did, however, translate from Arabic a work on the topic which may be described here. The work is called the *Liber prestigiorum Thebidis secundum Hermetem et Ptholomeum* ('The Book of Magical Tricks of Thābit, following Hermes and Ptolemy'). It tells one how to make talismans for achieving beneficial and harmful results and what prayers to address to the spirits of the planets to make the talismans effective; see Burnett, 'Talismans: Magic as Science?'

73 The manuscripts are agreed in giving 'violentia' ('force'), although 'volentia' ('will') may seem a more natural reading. 'Violentia', the *lectio difficilior*, would refer to any movement that is caused contrary to the (normal) course of nature. Thus 'violentia aquarii' is unsurprising, but 'violentia nature' would be a (perhaps deliberate) contradiction of terms.

74 Statius, *Thebaid*, VI, v. 361.

75 Here Adelard is referring to the two ways of interpreting 'movetur': (1) as a middle form, meaning 'it moves' (intransitive); (2) as a passive form, meaning 'it is moved'.

76 It is unclear why Adelard should qualify the term 'bruta animalia' which he uses frequently (as do other writers). Is it because he thinks that 'bruta' is not an appropriate adjective to describe animals which have the same passions (such as fear) that human beings have?

77 I.e., while it forms an unbroken sheet.

78 This is a quotation from Matthew 26, 73 (the servant-girl's accusation that Peter is a disciple of Christ).

79 Statius, *Thebaid*, VI, v. 360.

80 Adelard plays on two meanings of 'solvere': 'solve (a problem)' and 'release (a thunderbolt)'.

81 Isidore, *Etymologies*, 1.21.21.

82 These words cleverly recall the discussion of reason and authority in Q6, and, by using parallel constructions, replace the earlier phrase – that beasts are led in whichever direction the halter is pulled – by the statement that the Nephew wishes to be led in whichever direction reason leads.

83 Rather, Calcidius' preface to his translation of Plato's *Timaeus*, ed. Waszink, p. 5: 'vis amicitiae . . . impossibilium paene rerum extricatio'.

84 Plato, *Timaeus*, trans. Calcidius, 28A.

85 The Sun is in the zodiac-sign Cancer once a year, in midsummer, and this combination causes maximum heat; the Moon passes through Cancer once

a month, so if the Moon were hot like the Sun, there would be an excess of heat every month.

86 Macrobius, *Commentary on the Dream of Scipio*, I, 11.8.

87 Several of Adelard's arguments for the animation of the stars are similar to those of Cicero for the divinity, rationality and animation of the stars in *De natura deorum*, II, xv–xvi (39–44).

88 Cf. *De natura deorum*, II, xv–xvi (42); Cicero refers to those who have 'keener wits and greater powers of understanding' ('acutiora ingenia et ad intellegendum aptiora') living in a purer climate; Adelard calls these men 'philosophers' and hence his language recalls a passage from his description of the 'domus philosophica' in his *De opere astrolapsus*: 'In the first clime (i.e., near the Equator), they say, the home of philosophers has its natural position. For there, all seeds spring up spontaneously and the inhabitants always do the right thing and speak the truth, obeying only God first, and the spirits of the planets second' (translated from the text edited in Dickey, pp. 169–70).

89 The terminology comes from Plato, *Timaeus*, trans. Calcidius, 45B (see Q23 above). The argument, however, comes from Cicero's citation of the opinions of the Stoic philosopher, Cleanthes, in *De natura deorum*, II.xv (40–41).

90 Adelard's citation of Aristotle's opinion also comes from *De natura deorum*, II.xvi (44).

91 In this answer Adelard reproduces the argument and several of the phrases of *De natura deorum*, II.xvi (44).

92 The Latin 'deus' could mean either 'a god' or 'God'.

93 MS V, in which several passages towards the end of *QN* have been rephrased, has dropped the pun here, substituting: '(*Nephew*) Quare quid inde tibi placeat audire volo. Adelardus: Igitur accipe quod cupis . . . ' ('So I would like to hear what your opinion is on this. Adelard: Therefore receive what you are seeking').

94 The Nephew's question is similar to that posed at the beginning of Cicero's *De natura deorum*; cf. I.i (2), pp. 3–5, with which it shares its phrasing: 'Velut in hac quaestione plerique quod maxime verisimile est et quo omnes duce natura vehimur, deos esse dixerunt, dubitare se Protagoras, nullos esse omnino Diagoras Melius et Theodorus Cyrenaicus putaverunt. Qui vero deos esse dixerunt tanta sunt in varietate ac dissensione ut eorum molestum sit dinumerare sententias' ('As regards the present subject, for example, most thinkers have affirmed that the gods exist, and this is the most probable view and the one to which we are all led by nature's guidance; but Pythagoras declared himself uncertain, and Diagoras of Melos and Theodorus of Cyrene held that there are no gods at all. Moreover, the upholders of the divine existence differ and disagree so widely, that it would be a troublesome task to recount their opinions').

95 The subject-matter 'beginning or beginnings' ('initio vel initiis') is that mentioned in Plato, *Timaeus*, trans. Calcidius, 48C: 'Nostra igitur haec est

professio nihil ad praesens de universitatis vel *initio vel initiis*, ut quidam putant, esse dicendum.' ('Our intention, then, is that nothing should be said at present about the beginning or beginnings – as some think – of universal being.') This disputation, which appears also to be referred to in the *DED* (p. 23) is not extant. In three manuscripts of the *QN* (MSS MCJ) a sequel follows (described in J as 'secundus liber'), which does deal with these subjects, drawing material from Macrobius, Helperic and the Pseudo-Clement *Recognitiones*. However, this short piece, which is not in dialogue form, does not have the hallmarks of a work of Adelard; see Burnett, '*Omnibus convenit Platonicis*'.

De avibus tractatus

HIC EST DE AVIBUS TRACTATUS[1]

(MSS AW) <1 NEPOS> Quoniam in rerum causis disserendis animus noster admodum est fastiditus,[2] ad eiusdem dilatationem[3] et fastidii relevationem aliquid magis delectabile quam grave interponendum est.[4] Intellectus enim non pausans sicut et[5] arcus: si numquam cessas[6] tendere, lentus erit.[7] Quare meo iudicio tale aliquid quod et[8] iocundum et utile sit, eligendum est.[9] Id autem recte fieri puto[10] si de accipitrum natura et usu quicquam elegantius[11] sentis aperias,[12] precipue cum nos Angli[13] simus genere et eorum sententia[14] pre ceteris gentibus sit probata, et eiusdem sententie constet qualitas ut[15] quanto pluribus dividitur, tanto magis efflorescat.[16]

ADELARDUS: Sit sane, ne aut inscitia aut invidia[17] arguamur. Ea igitur disseremus[18] que modernorum magistrorum usu didicimus, et non

(MS R) <1> Nepos carissime, hic disseremus de natura anciptris et usu. Quicquid elegancius a modernorum usu et que Aroldi regis in libris scripta reperimus, tui causa et aliorum utilitate disseremus. Tuum igitur sit querere que volueris, nostrum autem explicare.

1 Hic est de avibus tractatus W, Adelardus de acciptrum natura et usu (*late hand*) A, *no title in* R
2 fastiditus A, fatigatitus (*sic*) W
3 dilatationem *scripsimus*, dictationem A, W *om.*
4 est W, et A
5 sicut et A, similiter ut W
6 cessas W, cesses A
7 lentus erit *scripsimus*, lentus eris A, mollis erit W
8 et W, est A
9 sit eligendum est W, sic est eligendum A
10 puto A, spero W
11 elegantius W, eligentius A
12 aperias W, apetias A
13 Angli W, Anglici A
14 sententia A, scientia W
15 eiusdem sententie constet qualitas ut *scripsimus*, eadem sentencie constet qualitas et A, ea de inde qualitas scientie et stet ut (*or* et) W
16 efflorescat W, afforescat A
17 inscitia aut invidia *scripsimus*, sciencie aut invidi A, inscitia aut individua W
18 disseremus W, sic disserimus A

<1> (MSS AW) Nephew: Since our mind has become quite sated with discussing the causes of things,[1] something of delight rather than gravity should be interposed, to give joy to the mind and relieve its weariness. For the intellect which does not pause is just like a bow: if you never cease to stretch it, it will become slack.[2] Therefore, in my opinion, something must be chosen that is agreeable as well as profitable. I think the right action is if you reveal whatever you feel is more finely expressed concerning the nature and handling of hawks, especially since we belong to the race of Englishmen and their judgement[3] is approved beyond that of all other peoples,[4] and it is an established quality of that judgement, that the more widely it is shared, the more it flourishes.

Adelard: Certainly, let us do that, so that we are not accused either of ignorance or of spite.[5] We shall, then, discuss what we have learned from the practice of the present-day masters and no less what we have found written

<1> (MS R) Dearest nephew, we shall discuss here the nature and handling of hawks. Whatever we have found more finely expressed by the practice of the moderns and what we find written in the books of King Harold, we shall discuss for your sake and for the usefulness of others. So let your part be to ask the questions you wish, ours, to explain.

minus que in[19] Haroldi[20] regis libris
scripta reperimus, ut quicunque his
intentus hanc disputationem habeat,[21]
si negotium exercuerit, peritus esse
possit. Tuum itaque sit inquirere,
meum vero sit[22] explicare.

(MSS AWR)

<2> NEPOS: In primis[23] audire cupio[24] quales eos esse velis qui huic studio conveniant.

ADELARDUS: Sobrios, patientes, castos, bene anhelantes,[25] necessitatibus expeditos.

NEPOS: Quare?

ADELARDUS: Ebrietas enim est mater oblivionis, ira lesiones generat, meretricum frequentatio tineosos ex contactu accipitres facit, fetidus vero anhelitus eos osores hominum reddit[26] et egro aere implet unde reumatici efficiuntur. Necessitas demum, quando legem non habet, id efficit ut vel per pluvias vel per turbines ferantur, vel nimis vexentur vel parum teneantur. Ideoque huiusmodi artifices a necessitatibus expeditos esse convenit.

<3> NEPOS: Convenit certe, et quoniam ista patent, amodo de eorum[27] natura et usu dicendum est. Quod ut perfecte fiat, a nido incipiendum esse censeo[28] et usque ad senectutem vitam eorum[29] tractandam.

ADELARDUS: Cedo. Transactis igitur septem diebus, absentibus parentibus, nudis manibus, hora matutina, suaviter a nidis excipiendi sunt.

NEPOS: Quare?

ADELARDUS: Post septem enim dies et sensus eorum perfecti sunt et membra perfecta[30] sine plicatione tractari possunt. 'Absentibus

19 in *scripsimus*, in *ante* modernorum A, W *om.*
20 Haroldi W *p.c.*, Haraldi W *a.c.*, Aroldo A
21 disputationem habeat W, dispositionem intellexerit A
22 vero sit A, W *om.*
23 RW *add.* illud
24 cupio AR, desidero W
25 anhelantes *scripsimus*, hanelantes R, redolentes A, hanhelantes W
26 reddit AR, facit W
27 eorum W, earum A, ipsorum R
28 censeo AR, sentio W
29 eorum W, esse A, R *om.*
30 membra perfecta A, membra R, W *om.*

down in the books of King Harold,[6] so that whoever, being interested in this subject, has this disputation at hand, if he practises this activity, can become an expert. So let your part be to ask questions, but mine to explain.

<2> Nephew: In the first place I would like to hear how you wish men should be, in order to be fit for this activity.

Adelard: Sober, patient, chaste, pleasant smelling, free from preoccupations.

Nephew: Why?

Adelard: Drunkenness is the mother of forgetfulness, anger causes injuries, visiting prostitutes transmits parasites to birds when they are touched, a bad breath makes them haters of men and fills them with bad air, so that they suffer rheum. Moreover, a preoccupation that is not under control will result in the birds being carried through the midst of rain or gales, or treated with excessive violence, or not being carried enough. Therefore it is appropriate that this kind of practitioner should be free from preoccupations.

<3> Nephew: It is certainly appropriate and, since this is clear, now we should talk about their nature and handling. In order to do it in a complete way, I think we should begin from the nest and deal with their life until their old age.

Adelard: I agree. The birds must be taken from the nest after seven days, whilst their parents are absent, with bare hands, in the morning, and gently.[7]

Nephew: Why?

Adelard: Because, after seven days, their senses have fully developed[8] and their limbs,[9] being fully formed, can be handled without risk of bending them. 'Whilst their parents are absent', I say, because

parentibus' dico, quoniam presentes exceptores ledunt et nidos non adeo[31] diligunt. 'Nudis manibus' dixi, quia sic expeditius tractantur. 'Hora matutina', quia tunc ieiuni sunt nec magno fervore diei[32] leduntur. Sedibus etiam eorum aut herba recens aut iuncus supponendus est. Quod si ferri de loco ad locum convenit, etiam cum ramusculis eis umbram facere expedit. Idem enim ab ipsis patribus in nidis fieri sepe videmus. Set etiam passim ire portantes convenit, ne ex nimio[33] motu teneritas eorum[34] concutiatur. Cum autem ad locum quietis pervenerint, dum infirmi sunt et nondum super perticam sedere possunt, casam eis sine trabe fieri oportet, quam vulgo firmam dicunt,[35] eo quod in ea penne firmantur.[36]

<4> [De victu accipitrum][37]

NEPOS: His ita dictis, amodo de eorum victu tractandum esse censeo.[38]

ADELARDUS: Victus eorum sint volucres parve, et pre ceteris passeres, deinde gallinarum caro, vel[39] etiam vervecum corda. Cavendum est autem dum infirmi sunt ne vervecina caro frequenter eis detur;[40] nervi enim inde contrahuntur. Oportet[41] ipsas carnes quas eis dederis[42] primo minutatim[43] incidere et incisas super asserem porrigere, frequenter etiam eis dare escam – fames etenim mirabiliter[44] eos debilitat et pennas venturas corrumpit. Patellatas[45] etiam eis facere convenit;[46] nam et infirmi inde bene nutriuntur et firmi cum fuerint commode eis utuntur. Fiunt autem sic: ova in primis dura coquantur, deinde pellis exterior auferatur – illa enim ab accipitre non bene inducitur[47] – postea in scutella[48] sine omni odore, melius tamen in

31 nidos non adeo W, nidum tantum non A, nidos non adeo deinde R
32 nec magno fervore diei A, et magno diei calore non RW
33 ex nimio *scripsimus*, eximio A, nimio RW
34 eorum R, AW *om.*
35 vulgo firmam dicunt R (*cf.* vulgo firmam vocant *SdH*), vulgus dicunt firmam A, vulgus universaliter firmam vocat W
36 firmantur RW, firmentur A
37 De victu accipitrum *rubric in* A, RW *om.*
38 esse censeo A, est RW
39 vel A, et R, set W
40 dum infirmi sunt ne vervecina caro frequenter eis detur W, ne vervecine carnes cum infirmi sunt dentur A, ne detur eis dum sunt infirmi vervecina R
41 W *add.* etiam
42 eis dederis A, comederint R, dabis W
43 primo minutatim W, minutim AR
44 etenim mirabiliter A, autem R, etiam W
45 Patellatas R*SdH*, Patellas AW
46 convenit A, expedit RW
47 inducitur AW, digeritur R
48 W *add.* aut in vase

if they are present they injure the men who are taking the birds and come to lose their attachment to the nest.[10] 'With bare hands', I said, because in this way they can be handled better. 'In the morning', because then their stomachs are empty and they are not damaged by the great heat of the day. Under them there should be a bed of fresh grass or rushes.[11] If they have to be carried from place to place, it also helps to use small twigs to make a shade for them. For we often see the parents themselves doing this in the nests.[12] But those who carry them should also go with slow step,[13] so as not to endanger their fragile state by excessive motion. Once they arrive at the resting place, as long as they are weak and cannot sit on a perch, one should provide for them a small shed without a beam, which they call a 'firma' in the vulgar language, because it is in this that they get firm feathers.

<4> Nephew: This being said, I think we should now deal with their food.

Adelard: Their food should be little birds, especially sparrows, then hens' meat or also wethers' hearts. One should beware of giving them wethers' meat frequently when they are still weak, because this tightens their sinews.[14] When giving them meat, you should first cut it into small pieces and present the cut meat on a board. You should give them food frequently, because hunger makes them amazingly weak and harms their growing feathers. One should also make them cooked preparations, because these both give them good nourishment when they are still weak, and are appropriate for them to have when they have become strong. One prepares them in this way. First one cooks hard-boiled eggs. Then one takes away the outer skin,[15] because this is not easily digested by the hawk. Afterwards, the remainder is mashed in a cup which does not have any smell – a vessel made of

argenteo quam eneo vase,[49] conteritur et cum lacte dulci parumper[50] colligatur et accipitri[51] porrigatur. Quod si in primis abhorruerit, de nocte ei vel[52] sanguine superasperso[53] offeratur. Cum itaque pennas eius[54] maturare cognoveris, ad solem eum frequenter habere facias, et non nimis eum in sole teneas, ne ex calore sincopizet.[55] Dimidia igitur casa solem recipiat, altera[56] vero pars umbram habeat. Deinde aqua ei ad balneandum frequenter adhibenda[57] est. Postea vero pertica apponenda[58] est, in primis demissa,[59] postea altior, prout pennarum vigor postulat.[60] Illud autem cavendum est:[61] ne antequam perfecte firmus fuerit,[62] abstrahere ipsum[63] properes. Nullatenus enim in pertica[64] tam bene firmatur ut in casa.

<5> NEPOS: Quomodo igitur firmum esse cognoscam?

ADELARDUS: Dumo abiecto et caude pennis ordine perfectis. Cum autem firmum esse cognosces, octo tamen diebus in firma ad pennas indurandas eum permitte esse.[65]

<6> NEPOS: Cum firmus[66] fuerit, quomodo abstrahendus est?[67]

ADELARDUS: Iacti primo faciendi sunt, qui sic[68] fiunt ut nec nimis sint stricti nec nimis laxi, inferius tamen ex parte pedis largiores.[69] Set etiam de tali corio fiant quod post humectationem non indurescat; deinde corrigiam longam[70] adhibeas, quod melius videndo quam legendo disces. His igitur peractis,[71] sic[72] abstrahes. Post illius diei

49 vase A, R *post* argenteo, W *om.*
50 A *add.* conteratur vel
51 R *add.* totum factum
52 de nocte ei RW, dentur eis de nocte vel A
53 superasperso A, superaddito RW
54 eius A, ei RW
55 sincopizet *scripsimus*, sincopiret A, languescat R, tumescat W, evanescat *SdH*
56 altera W, alia A, reliqua R
57 adhibenda A, apponenda RW
58 apponenda A, aptanda RW
59 demissa A, remissa RW
60 postulat RW, exiit (*for* exigit ?) A
61 cavendum est A, cave RW
62 fuerit RW, sit A
63 ipsum A, RW *om.*
64 in pertica A, super perticam RW
65 octo...esse A, .7. diebus eum in firma ad pennas indurandas permitte R, octo dimitte W
66 firmus A *p.c.*R, infirmus A *a.c.*W
67 abstrahendus est W, sit abstrahendus A, abstrahendus R
68 sic A, hoco modo RW
69 largiores AW, laxiores (*supra*: vel largiores) R
70 corrigiam longam A, et corium et linguam R, et cornum et longam W
71 igitur peractis AR, itaque paratis W
72 sic A, ita RW

silver rather than bronze; this should be mixed with some sweet milk and offered to the hawk.[16] But if at first he has an aversion to it, it should be given to him at night or with blood sprinkled on top.[17] When you see that his feathers are full-grown, let him sit more frequently in the sun, and do not keep him too long in the sun, lest he should faint from the heat. Half the shed should receive the sun, but the other half should be in the shade. Then one should frequently provide him with water to bathe in. Afterwards, a perch should be set up, at first a low one, then a higher one, as the strength of the feathers demands. One should be particularly careful not to be in a hurry to take the bird out[18] before he is perfectly strong. He can by no means become as strong on the perch as he would in the shed.

<5> Nephew: How shall I recognize that he is strong?

Adelard: When his down has been cast off and the feathers of his tail are in perfect order.[19] When you recognize he is strong, let him nevertheless stay eight more days in the *firma*, so that his feathers can be hardened.

<6> Nephew: Once he is strong, how should he been taken out?

Adelard: One should in the first place make jesses,[20] which are fashioned in such a way that they are neither too tight nor too loose, but are broader underneath at the place where they meet the feet. Let them be made of such a kind of leather that it does not become hard after being wet. Then you should provide a long leash[21] – but you will learn this better by seeing than by reading.[22] Once these tasks have been completed, you will take him out in this way. After the third

terciam horam[73] cibi nichil ei porriges ne capiendo gorgia[74] ledatur. De nocte capiatur – quevis autem[75] avis de nocte mansuetior est.[76] Accede itaque ad casam cum lumine. Primum quomodo et ubi sedeat[77] notato.[78] Deinde lumen absconde et in firmam tacitus intra, set nec album pannum habeas nec anhelitu terreas ipsum.[79] In manibus chirotecas habeas, et sic eum capias.

De capiendo autem apud diversos diversi sunt mores. Alii enim per gambas eum capiunt, digitis ex anteriori parte interpositis.[80] Angli vero a parte dorsi eum capiunt manibus alis superiectis,[81] dicentes quoniam magis penne eorum quam manus hominum in illo negotio servande sunt.[82] Cum autem abstractus fuerit, iacti ei[83] apponendi[84] sunt. Deinde usque in mediam[85] noctem tenendus. Deinde super perticam non firmam set mobilem ponendus. Itemque ad[86] lucanar et in ipso lucanari usque ad claram lucem[87] est tenendus, frequenterque manus sinistra movenda est, ut sedere discat, dextra vero ante illum huc et illuc frequenter feratur;[88] cum virga semipedali et alas et caudam tangat et aptet. Illud autem cave, ne antequam privatus fuerit[89] ad manum aptes nec clames. Prius enim oportet privatum esse quam aptatum. Si enim properaveris,[90] malum aliquem motum[91] addiscet. Id quoque[92] cave, ut firmiter ante aptatus sit quam[93] eum volare facias.[94] Cum autem primum volaverit, si volucrem ceperit, ad libitum suum eum plumare permitte[95] et deinde paulatim accede et

73 horam A, RW *om.*
74 gorgia *scripsimus (following SdH)*, gorgiam R, corrigiam A, W *om.*
75 quevis autem A, omnis enim RW
76 R *add.* quam de die
77 sedeat A, sederit RW
78 notato *scripsimus*, vocato A, nota RW
79 terreas ipsum A, R *om.*, eum terreas W
80 ex anteriori parte interpositis W, ex anterioribus impositis A, R *om.*
81 a parte dorsi eum capiunt manibus alis superiectis R, manibus a parte dorsi vel super eiectis A, a parte dorsi manibus alis subiectis eos apprehendunt W
82 in illo negotio servande sunt RW, sunt conservande A
83 ei A, RW *om.*
84 apponendi A, aptati R, aptandi W
85 mediam *scripsimus (following the β-version)*, multam ARW
86 ad A, ante R, apud W
87 lucem A, diem RW
88 feratur RW, ferat A
89 fuerit A, sit RW
90 properaveris W, probaveris AR
91 motum AR, morem W
92 Id quoque A, Illud etiam RW
93 ante...quam W, antequam aptatus sit A, ante aptatus quam R
94 facias RW, A *om.*
95 eum plumare permitte A, dimitte plumare R, eum plumare dimitte W

hour of that day you will not give him any food, so that the throat does not get injured when he is taken. He must be taken at night, because every bird is calmer at night. Go to the shed with a light. Notice first how and where he sits. Then hide the light and enter the *firma* in silence, but do not wear anything white and do not frighten him by your breathing. On your hands you should wear gloves, and you will then take him in this way.

As to the manner of taking him, different practices are followed amongst different people. Some take him by the legs, placing the fingers between them from the front.[23] But the English take him from the back, putting their hands over the wings, saying that in this task their feathers should be protected rather than men's hands. Once he has been taken out, one must put jesses on him and then hold him <on the fist> late into the night. After this he must be put on a perch that is not fixed, but mobile.[24] Again, let him be put next to a louvre (?)[25] and kept at this louvre until broad daylight. The left hand must be moved about often, so that he learns to sit, but the right hand should often be waved this way and that in front of him and with a stick half a foot long one should train and touch the wings and the tail.[26] You should beware of training him and calling him to the hand before he is manned. He must indeed be manned before being trained. If you hurry him, he will learn some wrong movements.[27] You must also make sure that he is firmly trained before you let him fly. When he has made his first flight, if he has taken a bird, let him plume it as

leniter tangendo, ne rapax fiat, eum comedere[96] permitte, et cum comederit, apte et securiter tracta, et non amplius illa[97] die volare permitte.[98]

<7> NEPOS: His ita dictis, ad egritudines eorum et medicinas tractandas accedamus, ita ut a capite incipientes usque ad pedes que singulis membris accidere possunt disseramus.

ADELARDUS: Quoniam quidem ordine procedere vis, de reumate[99] et fungis que capiti accidunt dicamus.

<8> [Si patiatur reuma][100]

Si accipiter reuma patiatur,[101] †equum† pinguem domesticum vel silvestrem[102] interfice, et de lumbis adhuc calidis tres morsellos accipe, singulisque morsellis pulverem duorum granorum piperis impone et sic ei prebebis preparatum.[103] Deinde tegulam rubeam calidam[104] lineo panno involve et accipitrem inde[105] calefac circa caput et gorgiam[106] et cor et sub alis. Postea lineum pannum in vino calefactum[107] accipe et accipitrem inde[108] calefac in[109] ore et in naribus et circa caput et pectus.[110] Et cum sic calefeceris, tres[111] pillulas ei fac ad modum avellane (id est nucis corili[112]) de pipere, butiro et sale; et post pillulam unam[113] in ore impone[114] in superiori maxilla et os firmiter claude; deinde calamum uni[115] nari appone et cum ore tuo sugendo reuma extrahe[116] quamdiu exierit.[117] Postea in alia nare idem fac et tunc ipsum aliquantulum quiescere[118] permitte, et eodem modo alias

96 comedere RW, quiescere A
97 amplius illa W, plus illo A, amplius in illa R
98 permitte A, dimitte RW. R *inserts chapter 29 here.*
99 raumate W
100 Si patiatur reuma *rubric in* A, de reumate *rubric in* R, A(delardus) W (*no rubric*)
101 patiatur A, patitur RW
102 silvestrem A, silvaticum RW
103 ei prebebis preparatum A, morsellos ei prebe RW
104 calidam RW, A *om.*
105 inde RW, A *om.*
106 et gorgiam RW, A *om.*
107 calefactum AR, tepefactum W
108 inde RW, A *om.*
109 in R, etiam A, et in W
110 et circa caput et pectus RW, A *om.*
111 tres RW, quatuor A
112 id est nucis corili AW, R *om.*
113 post pillulam unam A, primam pillulam RW
114 impone A, pone RW
115 uni R, una W, A *om.*
116 extrahe AW, adtrahe vel extrahe R
117 quamdiu exierit RW, A *om.*
118 tunc ipsum aliquantulum quiescere A, iterum (R *add.* vel interim) eum quiescere aliquantulum RW

he wishes; then approach slowly and let him eat, touching him gently, so as not to make him rapacious. When he has finished eating, handle him suitably and firmly and do not let him fly any more that day.

<7> Nephew: Having talked about these things in this way, let us turn to dealing with their illnesses and medications, in such a way that, beginning with the head and proceeding to the feet,[28] we explain what can occur to every single member.

Adelard: Since you want to proceed in an orderly manner, let us talk about rheum and the fungi which affect the head.

<8> If the hawk suffers rheum, kill a fat domestic or wild †horse†,[29] and offer him three morsels of the loins when they are still warm and on each of them put the powder of two grains of pepper, and you will give it to him prepared in this way. Then wrap a hot red tile in a linen cloth and warm the hawk with this around his head, throat and heart and under his wings. Then take a linen cloth warmed in wine and warm the hawk with this in his mouth and nostrils and around his head and breast. And when you have warmed him in this way, make three pills for him in the shape of a hazelnut[30] (i.e. the nut of a hazel-tree) with pepper, butter and salt; next put one pill in his beak, in the upper mandible and firmly close the beak. Then put a reed at one nostril and extract the rheum by sucking with your mouth until it stops coming out. Afterwards, do the same with the other nostril; then let him relax a while; and do the same operation with the other two

duas pillulas adde.[119] Postmodum eum surgere fac et aliquam escam confortativam[120] ei tribue.[121] Hoc fac donec sit sanus.

Est et alia eiusdem rei medicina, que, licet immunda sit, tamen commodam esse constat, ut accipitris caro pueri masculi lactentis[122] egestionibus[123] intingatur, et sic accipitri porrigatur.

<9> [De fungis][124]

NEPOS: His dictis, de fungis dicamus.

ADELARDUS: Sunt igitur[125] fungi alii sicci, alii humidi.[126] Si siccos fungos patitur, accipe eum et in agnina pelle colloca et os aperi et siccum fungum abrade aut[127] extrinsecus abscide[128] et postea vulneri tale medicamentum adde: accipe limaturam cupri minutissimam[129] et piperis pulverem et viscum arboris et per pannum cribra et salem adde et butirum[130] et vulneri applica. Si vero humidos fungos patitur, accipe mellicratum et vinum,[131] utriusque cifulum unum, et simul misce. Postea accipe tegulam rubeam calidam et partem supradicti liquoris impone et accipitri[132] aperto ore[133] per pannum lineum in os instilla et deinde os claude. Postea calamum naribus applica et liquorem simul cum fungis extrahe et hoc frequenter fac. Deinde ipsum surgere dimitte[134] et hoc de tercio in tercium diem prosequere.

Item,[135] sunt due fossule iuxta extremas oris partes. Illas fossulas incide et liquorem predictum[136] infunde et deinde per vulnera sugendo egritudinem extrahe. Quod si tibi videtur vulnera velle recludi,[137] accipe sericum[138] pannum vel filum et rade et rasuram in superiori

119 adde AR, ei prebe W
120 escam confortativam A, R *transp.*, confortivam salvam W
121 tribue A, prebe RW
122 lactentis AW, lactantis R
123 egestionibus AR, liquide egestioni W
124 De fungis *rubric in* AR, W *om.*
125 igitur A, autem RW
126 alii sicci, alii humidi R*SdH*, alii humidi, alii sicci AW
127 RW *add.* etiam
128 abscide W*SdH*, abscinde AR
129 minutissimam W, minutissimi A, subtilissimam R
130 RW *add.* appone
131 vinum RW, vitrum A
132 acciptri AW, accipitris R
133 RW *add.* liquorum tepidum
134 ipsum surgere dimitte A, permitte eum surgere RW
135 Item A, Aliter RW
136 predictum A, supradictum RW
137 recludi A, precludi RW
138 sericum RW, A *om.*

pills. Afterwards, make him rise up again,[31] and give him some comforting food. Do this until he is healthy again.

There is yet another remedy for this illness, which although it is unclean, is certainly practical. The meat of the hawk should be soaked in the excrement of a young unweaned boy, and given to the hawk in this way.

<9> Nephew: After saying this, let us talk about fungi.

Adelard: Amongst fungi, some are dry, others moist. If he is suffering from dry fungi, take him and put him into a lamb's skin, open his mouth and scrape off the dry fungus, or cut it off on the outside.[32] Afterwards, apply the following kind of medication to the wound. Take very fine filings of copper, ground pepper and mistletoe, sift this through a cloth. Add salt and butter and apply it to the wound. But if he suffers from moist fungi, take mead and wine, a small cup of each, and mix them together. Afterwards, take a hot red tile,[33] pour part of this liquid onto it, open the hawk's beak, drip it into the beak through a linen cloth and then close the beak. After that, apply a reed to his nostrils and suck out the liquid together with the fungi, and do this often. Afterwards, let him rise up, and do this every three days.

Again, there are two little holes beside the outermost parts of the beak.[34] Make an incision in these little holes, pour in the liquid described above and then extract the illness by sucking through the wounds. If you see that the wound is liable to close up again, take a piece of silk cloth or a thread, scrape the area, wrap the scraped area (?)

medicamine involve et vulnera obtura ne claudantur, donec sic medicando sanus fiat.[139]

<10> [Si accipiter nares obturas habuerit][140]

<NEPOS:> Amodo de narium obturatione dicamus.[141]

<ADELARDUS:> Si accipiter nares obturatas[142] habuerit,[143] pone caput eius et nares usque ad oculos in aqua tepida. Postea in nervosis carnibus trahere fac et curabitur. Universalem[144] autem capitis medicinam hanc esse dico:[145] ut eum accipias et super nares ubi vene iunguntur coquas, pedibusque compedes addas ne cocturas attingat[146] et, cum hoc feceris, singulis noctibus ipsum[147] sub caldaria tepida pone, et sub caldaria lapides pone ne terram tangat et sic curabitur. Item,[148] ad caput eius[149] purgandum, salvia frica palatum eius et fac trahere.[150]

<11> [Si carnem vomuerit][151]

<NEPOS:> De vomitu sequitur.[152]

<ADELARDUS:> Si carnem suam accipiter vomuerit, mane ei aquam[153] porrige, et usque ad vesperam ieiunare fac[154] et tunc aut sorices aut passeres ei[155] prebe. Item,[156] carnem suam in bullientem aquam[157] mersam extrahe et ei prebe. Item, carnem suam siccam in iure trifolii[158] intinge.

139 fiat A *add.* a commodo
140 Si accipiter nares obturas (*for* obturatas) habuerit *rubric in* A, de narium obturatione *instructions for rubric in margin of* R, W *om.*
141 Amodo...dicamus W, a commodo A, R *om.*
142 obturatas R, obturas AW
143 habuerit A *add.* ut de narium obturatione dicamus
144 Universalem AW, Utilem R
145 hanc esse dico RW, esse credo vel dico hanc A
146 attingat A, tangat RW
147 ipsum A, eum W, R *om.*
148 Item A, Amplius RW
149 eius A, RW *om.*
150 RW *add.* et sanabitur
151 Si carnem vomuerit *rubric in* A, contra vomitum carnis *rubric in* R, W *om.*
152 De vomitu sequitur W (*cf. β-version*: Nepos. De vomitu quid faciam? Alardus), AR *om.*
153 aquam RW, carnem A
154 fac RW, permitte A
155 ei A, RW *om.*
156 Item A, Amplius RW
157 bullientem aquam A, bullienti aqua RW
158 trifolii A, cerefolii RW

with the above medicine and stop up the wounds to prevent them from closing, until the bird gets well, by being healed this way.

<10> Nephew: Now let us talk about the blocking up of the nostrils.

Adelard: If the hawk has his nostrils blocked up,[35] put his head and his nostrils up to his eyes in warm water. Afterwards, let him tear at some sinewy meat, and he will be cured. I say that a general remedy for the illnesses of the head is as follows: you take him and make a cauterization above the nostrils where the veins come together, and you put fetters on his feet to prevent him from reaching the cauterizations. When you have done this, put him each night under a warm cauldron and put stones under <the rim of> the cauldron, so that it does not touch the earth, and thus he will be cured.[36] Again, in order to purge his head, rub his palate with sage and let him tear <at some meat>.

<11> Nephew: Vomiting is the next topic.

Adelard: If the hawk vomits his meat, give him water[37] in the morning, let him fast until evening and then give him some shrew-mice or sparrows. Again, put his meat in boiling water, take it out and offer it to him. Again, dip his dry meat into the juice of trefoil.

<12> NEPOS: His dictis[159] circa caput expositis, de accidentibus pectoris subdamus, et in primis de pulsione quid laudas ?

ADELARDUS: Heris limaturam[160] da ei frequenter et eboris.[161] Ad hoc[162] item, butirum maii[163] et radices[164] in igne coctas ieiuno[165] dabis. Item, auripigmenti pulverem in offis ei dabis aut in corde columbe vel pulli et curabitur.

<13> NEPOS: De inflatione quid dices?

ADELARDUS: Herbam que anigilia[166] vocatur ei da et eum[167] ab inflatione et vento[168] purgabit.[169] Indefinite vero[170] dico, si de ignota aliqua egritudine de accipitre desperaveris, lac capre[171] ei porrige et, si hauserit, evadet.

<14> [De pediculis][172]

NEPOS: His visis,[173] de pediculis quid dices?

ADELARDUS: Si pediculos habuerit et pinguis fuerit, nichil obest. Si macer, de iure absinthii plumas infunde et ad solem sedere[174] fac. Item, adipe gallinaceo vel anserino filum laneum inunge[175] et circa collum liga, rostrum etiam eius et[176] pedes et alas intrinsecus unge. Vel[177] de arbore quam fusenum[178] vocant excorticata[179] perticam fac ei et morientur.

<15> NEPOS: His susceptis, de tineis quid dices?[180]

ADELARDUS: Ad solem sedentem melle perunge et exibunt. Deinde

159 dictis R, visis A, W *om.*
160 Heris limaturam A, De eris limatura RW
161 da ei frequenter et eboris R, dabis ieiunis exeuntibus dabis A, ieiunis frequenter dabis W
162 Ad hoc *scripsimus* (*cf. β-version*: 'ad idem'), adhuc A, RW *om.*
163 maii A, maiale R, W *om.*
164 radices A, raphani radiculas R, W *om.*
165 ieiuno R, -nis AW
166 anigilia W, anglica A, verbenam R, angalia *SdH*
167 et eum A, eumque RW
168 vento AW, ventositate R
169 purgabit A, curabis R, curabit W. *At this point* R *has* 'de ignota egritudine' *in the margin.*
170 vero A, autem RW
171 capre AW, caprinum R
172 De pediculis *rubric in* A, contra pediculos *rubric in* R, W *om.*
173 visis A, R *om.*, auditis W
174 sedere AW, siccare R
175 inunge RW, infunde A. R *add.* et staphisagria
176 etiam eius et *scripsimus*, etiam et eius A, eius et R, et W
177 Vel AR *a.c.*, Item R *p.c.*W
178 fusenum W, fuscum AR
179 excorticata A, -tam RW
180 dices A, doces W, R *om.* R *adds rubric* contra tineas

<12> Nephew: Now that these sayings concerning the head have been explained, we should go on to the illnesses affecting the breast. In the first place, what do you recommend for the short-winded?[38]

Adelard: You will give him filings of brass and ivory frequently.[39] Again, you will give him, when his stomach is empty, may-butter[40] and radishes[41] cooked in fire. Again, you will give him powder of orpiment in his meal, or in the heart of a pigeon or chicken, and he will be cured.

<13> Nephew: What would you say about inflation?[42]

Adelard: Give him the herb called *anigilia*,[43] and this will purge him from inflation and wind. But in general I say that if you despair of curing the hawk from a disease that is unknown, offer him goat's milk, and if he drinks it up he will be saved.

<14> Nephew: Now we have seen this, what would you say about lice?

Adelard: If he has lice and is fat, there is no trouble. If he is thin, douse his feathers with wormwood juice and make him sit in the sun. Again, douse a woollen thread with hen- or goose-fat,[44] tie it around his neck and anoint also his beak, his feet and the inside of the wings. Or make a perch for him from the tree they call a 'spindle-tree', having taken off the bark, and the lice will die.

<15> Nephew: Having dealt with these things, what would you say about mites?

Adelard: When he sits in the sun, smear him with honey,[45] and

pulverem atramenti et nitri et corticis fraxini[181] superasperge et morientur. Item, pulverem[182] absinthii et salis cum aceto bullire fac et unge locum ubi tinee assunt.[183] Item, dentem suspensi aut os tenue[184] quod maiori ossi brachii coniungitur[185] quod escinum[186] vocant, in pulverem redige et cum carne sua da ei et evadet. Item, cum eodem pulvere ossis equo pondere da ei de pulvere radicis rubee parelle.[187]

<16> [De ictericia][188]

<NEPOS:> His intellectis, de fellera quid dices ?

ADELARDUS: Quamplures istam egritudinem incurabilem esse iudicant,[189] set hii felleram a ruptura dividere nesciunt; ruptura quippe[190] incurabilis est; fellera vero, nisi fallor, potest curari. Prius intellige quia a felle fellera dicitur, eo quod colera rubea que in felle sedem habet tunc habundat et per corpus diffunditur, adeo ut rostrum[191] ac pedes discoloret.[192] Tunc vero id accidit ut more solito audacior fit, quia vis audacie omnium viventium[193] in colera rubea consistit. Sic autem mederi temptabis: herbam quandam que iuxta rivos[194] in aquosis locis crescit[195] accipies. Hanc poete 'electrum', phisici[196] vero †'boricam'†[197] vocant, Angli[198] suo ydiomate 'nigram herbam' vocant, cuius ramus[199] quadrangulus est ad modum †dite feris†[200] factus, in foliis vero magne urtice assimilatur;[201] in summitate vero ramorum[202] quasi racemi quidam nascuntur. Hanc, inquam, herbam cape

181 et nitri et corticis fraxini RW, sume et corticem fraxini et nitri et A
182 pulverem W, pulvis AR
183 assunt A, sunt RW
184 tenue W, illud A, R *om.*
185 coniungitur W, adheret vel A, R *om.*
186 escinum W, esuum A, R *om.*
187 parelle *scripsimus*, patelle A*SdH*, paralle W, R *om.* (radice rubee et parattele *DN*). W*SdH add.* et evadet.
188 De ictericia *rubric in* A, contra feleram R *i.m.*, W *om.*
189 esse iudicant A, dicunt RW
190 quippe W, quoque A, quidem R
191 R *add.* ceram, W *add.* ac ceram *post* pedes
192 discoloret RW, discolorat A
193 omnium viventium RW, A *om.*
194 rivos A, rivulos et RW
195 crescit A, nascitur RW
196 phisici W, philosophi A, pħi R
197 boricam W, brie A, boris R, boric *DN SdH*
198 Angli W, Anglici AR
199 R *add.* vel radix
200 dite feris *scripsimus* (*cf.* dite foris *SdH*), dite A, R *om.*, elice feris W
201 in foliis vero magne urtice assimilatur A, folia vero foliis magne urtice similia RW
202 ramorum RW, illorum A

they will go away. Afterwards, sprinkle powder of atrament and of natron and of the bark of an ash tree over him, and they will die. Again, bring to the boil powder of wormwood and salt with vinegar and anoint the place where the mites are present. Again, take the tooth of a hanged man, or that bone which is attached to the larger bone of the arm which they call the 'escinum', reduce it to powder and give it to him with his meat, and he will be saved.[46] Again, with the same bone-powder give him the same quantity of powder of the root of a red dock.

<16 Nephew>. Now that we have understood this, what would you say about *fellera?*[47] Adelard: Very many people judge that this illness cannot be cured, but they do not know the difference between *fellera* and *ruptura. Ruptura* is indeed incurable, but *fellera*, if I am not mistaken,[48] can be cured. In the first place, understand that *fellera* is derived from 'gall-bladder' (*fel*), because the red bile that has its seat in the gall-bladder is then abundant and spreads through the body to such an extent that it makes the beak and the feet[49] become discoloured. But it happens then that the hawk becomes bolder than normal, because in all animals boldness takes its force from the red bile. You will try to cure it in this way:[50] you should take a certain plant that grows on river banks in wetlands. This, the poets call 'electrum', while doctors[51] call it 'borica' (?) and the English call it in their own tongue the 'black herb'.[52] Its stem is rectangular in the same way as *dite feris*(?),[53] but in its leaves it resembles the big nettle; at the extremity of the branches[54] there grow some clusters of berries.

et in iure eius carnem intinge[203] et sic accipitri porrige et credo quod[204] non vituperabis me.

<17> [De febre][205]

NEPOS: De febre quid faciendum esse censes ?

ADELARDUS: Febrem quidem[206] patitur, cum supra perticam sedens,[207] oculos claudit, dormit, caput inflatur. Tu vero lenticulam fontis accipe et factum[208] inde pulverem cum carne sibi propone et curabitur.[209] Item, in iure ubi grus cocta fuerit[210] eum balneari facias[211] et sanabitur.[212]

<18> [De aculeis][213]

NEPOS:[214] De aculeis quos patitur quando[215] uvariam explumat, quid dicis?[216]

ADELARDUS: Carnem pulli cum satureie pulvere da ei et morientur. Item, ius herbe serpentarie in[217] budello galline vel[218] columbe utrinque ligato[219] include et quantum poterit capere in ore eius inicito.[220] Hoc fac tribus diebus et morientur. Vel limaturam eris ei cum carne dabis et morientur. Item, vermem qui dicitur ades in os eius inice et os firmiter claude, ut transglutiat et sanabitur.[221] Item, pro pediculis et aculeis simul, tolle ederam et fac de illa vasculum et lac caprinum infunde et claude, clausumque[222] super cineres[223]

203 intinge W, inunge AR
204 credo quod A, R *om.*, forsitan W
205 De febre *rubric in* A, contra febrem R *i.m.*, W *om.*
206 quidem R, quidam A, quid W
207 sedens R, ponitur A, W *om.*
208 factum RW, fac A
209 sibi propone et curabitur W, et ei da tunc sanabitur A, R *om.*
210 fuerit R, erit A, fuit W
211 balneari facias RW, balniaris A
212 sanabitur A, curabitur RW
213 De aculeis *rubric in* A, RW *om.*
214 Nepos W, AR *om.*
215 A *add.* in
216 De aculeis . . . dicis A, Si plume de aculeis cadant R, De aculeis quos patitur quod est cum uvariam explumat quod disseres W
217 in RW, cum A
218 galline vel A, RW *om.*
219 ligato R *p.c.*, ligando AR *a.c.*(?)W
220 inicito AR, mitte W
221 et morientur. Item, vermem qui dicitur ades (*followed by a lacuna and a wrongly-copied passage subsequently underscored*) ut transglutiat et sanabitur W, Item in os eius inice et os firmiter claude, ut transglutiat et sanabitur A, et sanabitur R
222 clausumque W, clausum vero A, et clausum R
223 AR, W *add.* calidos

So, I say, take this plant, soak the meat in the juice and give it to the hawk in this way, and I think you will not inveigh against me.

<17> Nephew: What do you think should be done about fever?

Adelard: A bird suffers from fever when on the perch he closes his eyes, sleeps, and has a swollen head. But as for you, take water-lentil, offer the powder made from it with his meat and he will be healed. Again, make him bathe in the liquid in which a crane has been cooked, and he will be healed.

<18> Nephew: What do you say about the worms (?),[55] which he suffers from when he loses the feathers around his belly?[56]

Adelard: Give him the meat of a chicken with powder of savory, and they will die. Again, enclose the juice of dragonwort in the gut of a hen or a pigeon, making a knot at both ends, and put into his beak as much as he can take.[57] Do this for three days, or give him brass filings with his meat, and they will die. Again, put the reptile called the 'ades'[58] into his beak and close his beak firmly so that he can swallow it, and he will be healed. Again, for lice and worms together, take ivy<-wood>[59] and make a little vessel from it, pour goat's milk into it, close it and make the bundle boil over ashes; then soak his

bullire fac et deinde carnes[224] eius in hoc lacte intingue.[225] Item, interuscum geneste[226] in budello pulli inclusum da ei. Probatum est.

<19> [De petra][227]

NEPOS: His satis expeditis,[228] de petra quid intelligas doce.

ADELARDUS: De grumillo[229] ius exprime[230] et ibi carnem intinctam[231] ei prebe. Item, de semine huius herbe[232] pulvere facto, cum carne offer. Item, carnem eius in iure petrosilini intinge.

<20> [De tenasmon][233]

NEPOS: Pro stricto budello quid faciendum censes?[234]

ADELARDUS: Magnos morsellos[235] <carnis>[236] vaccine in aqua tepefactos donec albescant ei appone.

<21> [De bistico][237]

NEPOS: De bistico[238] quid tibi videtur?

ADELARDUS: Bisticus[239] dicitur, cum pro fundamenti obturatione bis simul egerit. Coquas vero malvam et satureiam in aqua porcinoque addito adipe recenti,[240] et cum optime coctum fuerit, cum cocleari tepidum[241] in os eius infunde, et sic evadet. Item, pulverem lapidis unde ignis eicitur,[242] qui galesta[243] vocatur vel parites secundum alios[244] – pulverem, inquam, combustum[245] et putredinem corili quam[246]

224 carnes A, carnem RW
225 intingue W, inunge AR
226 interuscum geneste RW, succus intercusti A (*explained in the β-version as* 'corticem medianam')
227 De petra *rubric in* A, R *i.m.*, W *om.*
228 expeditis A, R *om.*, intellectis W
229 grumillo W, grumilio R, granillo A
230 ius exprime W, excipe ius A, ius accipe R
231 ibi carnem intinctam A, carnem intinctam R, in triticum carne W
232 huius herbe A, R *om.*, grumilli W
233 De cenasmon (*sic*) *rubric in* A, RW *om.*
234 censes A, esse iudicas W
235 morsellos W, vitellos A
236 carnis *addidimus*, ARW *om.*
237 De bistico *rubric in* A, RW *om.*
238 bistico A, R *om.*, bistuco W
239 bisticus A, strictura budelli R, bistucus W
240 recenti A, frisso R, frisco W *a.c.*, frisca W *p.c.*
241 tepidum R, tepicum A, W *om.*
242 eicitur AW, excutitur R
243 galesta A, galestra R, lapis galesta W
244 vel parites secundum alios A, RW *om.* (parites *sc.* pyrites)
245 combustum R, eius combusti dico A, combusti W
246 quam W, quod AR

meat in this milk. Again, put the inner bark of broom in a chicken's gut and give it to him. This is a proven remedy.

<19> Nephew: Having treated these things sufficiently, teach what you know about the stone.[60]

Adelard: Squeeze out the juice of gromwell and give him meat that has been soaked in it. Again, make powder of the seed of the same plant and present it with his meat. Again, soak his meat in the juice of parsley.

<20> Nephew: What do you think should be done for an obstructed gut?[61]

Adelard: Give him large pieces of cow's meat that have been warmed in water until they become white and give them to him.

<21> Nephew: What do you think about the *bisticus*?

Adelard: The bird is called 'bisticus' when he defecates twice at the same time because of a hardening of the bottom.[62] You should cook mallow and savory in water to which you have added fresh pig's fat and, when this is perfectly cooked, pour it with a spoon into his beak while it is still warm, and thus he will be saved. Again, take powder of the stone used to produce fire, called 'galesta' or flint[63] according to others – take this burnt powder, I say, and the mould from a hazel-

vermulturam[247] dicunt equali pondere per lineum pannum[248] cribra et cum butiro in quo radix flammule cocta fuerit misce et ei butirum comedendum[249] prebe.

<22> [De verrucis][250]

NEPOS: De verrucis quod concilium prebes?

ADELARDUS: Accipe radicem prunellifere[251] spine et filicem[252] arboream et stupas et simul diminuendo tere in aceto[253] et cum serico filo[254] verrucam stringe et, cum medicari volueris, nodum laxa et unge et restringe, et hoc fac donec excidat.

<23> [De podagra][255]

NEPOS: His intellectis, de podagra quid faciam?

ADELARDUS: Corticem fraxini et pomerii et corticem spine prunellifere et quercus novelle: hos quatuor cortices[256] in vase eneo usque ad similitudinem et spissitudinem incausti coque et refrigerari[257] permitte. Deinde veterem saponem adde et sic accipitrem super lapidem planum pone et pedes eius simul et lapidem unge, et sanabitur.

<24> [De mutatione][258]

NEPOS: Pertractatis[259] eorum egritudinibus magis frequentibus et suppositis eorum curationibus, de mutatione amodo est dicendum.

ADELARDUS: Cum accipitrem in mutationem[260] miseris, mutam ita prepara ut tribus horis diei solem habeat, nec super ipsam pluat; sufficienter carnes diversas adde. Quod si bene non mutaverit, dabis carnem soricinam.[261] Si non per hoc, iuvenes[262] ciconias dabis ei. Si non per illud, pulverem de †inmusia†[263] fac et cum carne ei prebe. Item, pulverem de favo mellis super carnem ei prebe et mutabitur. Item, serpentem coque in aqua et in iure illo triticum ebullias, et pullos quos ei dabis inde nutrias. Item, viridem lacertam in olla rudi

247 vermulturam *scripsimus*, inde vulturis A, vermulturas R, vermultam W
248 lineum pannum R, pannum A, lineam W
249 comedendum RW, comedere A
250 De verrucis *rubric in* A, RW *om.*
251 prunellifere RW, prunifere A
252 filicem RW, salicem A
253 in aceto A, et in aceto pone RW
254 filo W, AR *om.*
255 De podagra *rubric in* A, RW *om.*
256 hos quatuor cortices A, RW *om.*
257 refrigerari A, -re RW
258 De mutatione *rubric in* A, RW *om.*
259 Pertractatis A, tractatis W, R *om.*
260 mutationem A, mutam R, vitam W
261 carnem soricinam A, ei sorices RW
262 iuvenes W, invenies A, R *om.*
263 inmusia W, īmusia A, R *om.*, ramusia *SdH*

tree which they call 'vermultura',[64] an equal weight of both, and then sift it through a linen cloth and mix it with butter in which the root of spearwort has been cooked and give him the butter to eat.

<22> Nephew: What advice do you provide for warts?

Adelard: Take sloe-thorn root, oak fern and tow, grind them into small pieces in vinegar. Bind the wart with a silk thread and, when you want to apply the medicine, slacken the knot, anoint the wart and bind it again, and do this until it falls out.

<23> Nephew: This being understood, what should I do about bumblefoot?[65]

Adelard: Take the barks of an ash, an apple-tree, a sloe-thorn and a young oak, boil these four barks in a bronze vessel until the mixture acquires the likeness and thickness of ink, and let it get cold. Then add old soap, put the hawk on a flat stone in this way,[66] and anoint the feet of the hawk together with the stone, and he will be healed.

<24> Nephew: Since their most frequent illnesses have been fully dealt with and their medications added, we should now talk about mewing.

Adelard: When you put a hawk to mew, prepare the mew in such a way that it receives the sun for three hours a day and that it does not rain on it, and provide sufficient and varied meat. If he does not mew well, you will give him the meat of shrew-mice. If he does not mew with this, you will give him young storks. If he still does not mew with that, make powder from *inmusia* (?) and offer it to him with his meat. Again, give him powder from a honey-comb on his meat and he will mew. Again, cook a snake in water, boil wheat in that juice and feed with this wheat the chickens you intend to give him.[67] Again, you should burn a green lizard in a rough pot, and you will give the

comburas et eius pulverem in carne dabis. Item, sanguisugas frustatim[264] incisas cum carne ei prebe. Item, tolle araneas quas futuelas[265] vocant et cum carne ei dabis et mutabitur.

<25> [De grassicie et macie][266]

NEPOS: Restat nunc, ut michi videtur, quatinus exaginare sciam, sicque documentum tuum perfectum esse patebit.

ADELARDUS: Ut hoc perfecte scias, carnium natura dividenda est. Sunt autem carnes alie pinguedinem, alie macredinem inferentes.[267] Pinguedinem inferunt parvi ocelli[268] pingues, et precipue passeres, et sorices pingues et cattina caro[269] et gallinacea; porcina etiam, set superbiam prestat;[270] caprolina[271] et caprina. Ad hoc universalis valitudo est, ut frequenter ei carnes mutes. Macredinem inferunt pullina caro, vaccina et leporina.

<26> [Si volare neglexerit][272]

NEPOS: Adhuc dubito, si volare neglexerit quid agam?

ADELARDUS: Teneram carnem vacce in urinam positam ad vesperum modice prebe et in crastina de lingua porci[273] sumat parum, et sic volandi cupidus erit.[274] Plumatas in singulis septimanis prebe duas.

(MSS AW)

<27> [De salvis][275]

NEPOS: His satis visis,[276] de salvis faciendis audire desidero sicque propositum opusculum non erit semiplenum.[277]

ADELARDUS: Salva alia confortativa, que apud veteres Anglos vocatur 'milda', alia purgativa que apud eosdem[278] 'strica'[279] dicitur. Mildam sic facies: betonice, serpilli, malve, hastule regie quam Angli verdetenam veterem[280] vocant, plantaginis, millefolii, equales mensuras in butiro coque, deinde secundum arbitrium tuum mel liquidum

264 frustatim *scripsimus*, frustratim AW, R *om.*
265 futuelas W, farvellas A, R *om.*, fatuales *SdH* (ranae que vulgo futuecine *LUP*)
266 De grassicie et macie *rubric in* A, RW *om.*
267 A *add.* vel habentes
268 ocelli RW, porcelli et A
269 et cattina caro W, A *om.*, catina R
270 prestat W, prestant A, parat R
271 caprolina RW, capita similiter ovina A
272 Si volare neglexerit *rubric in* A, RW *om.*
273 lingua porci RW, porcellina A
274 *With this word* R *ends*
275 De salvis *rubric in* A, W *om.*
276 visis A, divisis W
277 non erit semiplenum A, estimabiliter non interruptum W
278 apud eosdem W, A *om.*
279 strica W, stricta A
280 verdetenam veterem A, Woerouam W

resulting powder to him on his meat. Again, give him leeches cut into pieces with his meat. Again, take the spiders they call 'futuele' and give them to him with his meat and he will mew.

<25> Nephew: There remains now, I think, that I should know how to enseame him,[68] and thus your lesson will prove to be complete.

Adelard: In order that you know this completely, the nature of different meats should be differentiated. Some meats bring about fatness, others thinness. The kinds that bring about fatness are little plump birds and especially sparrows, fat shrew-mice and cat and cockerel meat; pig meat also – but it causes the bird to be disdainful – and the meat of wild and domestic goats.[69] It is a general rule of health that you should often change the meat you serve him. Thinness is brought about by the meat of chicken, cow and hare.[70]

<26> Nephew: I am still in doubt about what to do if he refuses to fly.

Adelard: Give him in the evening a little tender cow's meat that you have put in urine and let him take a little pig's tongue on the following day, and in this way he will be eager to fly. Provide him with roughage twice a week.[71]

<27>[72] Nephew: Having seen these things sufficiently, I would like to hear about making salves, and thus the proposed little work will not be half-complete.

Adelard: One salve is comforting, which is called 'mild' by the old English; another is purgative, which is called 'strong' by the same people.[73] You will make the mild one in this way: cook together in butter equal measures of betony, thyme, mallow, *hastula regia* (which the English call 'old woodruff'[74]), plantain and milfoil; then add

admisce et usui reserva, et cum tibi videtur necessarium[281] cum carne propone.[282]

<28> [De †tiriaca†][283]

Stricam[284] sic compones: consolide, petrosilini, betonice, serpilli, malve equales mensuras in butiro in quo prius radix flammule cocta fuerit, coque; deinde vermulturam corili subtilissimam cum melle mixtam adde et commisce et usui reserva et, cum dederis, a carne abstinere fac[285] donec bene purgatum esse cognoscas.[286]

(MS A) <29> [Ut libentius ad manum veniat][287]

NEPOS: His liquido edoctis, doceri cupio quomodo eum possim facere redire a ludo.

ADELARDUS: Accipe lardum porcinum pinguem et recentem, et decoque in forti vino. Postea in sacculo lineo ponatur et sic suspendatur in aqua fluenti, ita quod saccus maneat in aqua per tres dies vel quatuor. Postea ponatur inter duos asseres et exprimatur bene, iacens ibi per biduum. De hoc da ei parum quolibet mane dum ieris lusum, vel per se vel cum alia carne, et redibit libentius ad te quotiens ipsum iactaveris. Si vero accipitri silvestri inde ter dederis, domesticus erit.

(MS R)[288] Ut libens iactatus redeat

Accipe lardum recens et decoque optime in vino forti. Postea in sacco lineo suspendatur in aqua fluente; ita saccus totus sit sub aqua tribus vel quatuor diebus. Postea ponatur inter duos asseres et exprimatur bene, iacens ibi per biduum. Et quando vis ire ad ocellandum cum ancipitre, da ei mane parum de eadem carne et redibit per se. Si silvatico ex eadem carne dederis statim fiet domesticus.

<30> Hec habui de cura accipitrum que dicerem. Ceterum tibi vel alicui si suam placet addere sententiam non invideo.

281 tibi videtur necessarium A, volueris simul W
282 propone W, propina A
283 De tiriaca *rubric in* A, W *om.*
284 Stricam *scripsimus*, Tyriacam A, Strictura W
285 abstinere fac W, abstineas A
286 esse cognoscas A, intelligas W
287 Ut libentius ad manum veniat *rubric in* A
288 *This chapter is placed before chapter 7 in R. It is missing in W*

according to your discretion some liquid honey and reserve it for use, and when you think it necessary give it with his meat.

<28> You will prepare the strong[75] salve in this way: cook equal quantities of consound,[76] parsley, betony, thyme and mallow in butter in which spearwort-root has previously been cooked. Afterwards, add very fine hazel-mould mixed with honey and blend them together and reserve for use; and when you give it to him, keep him from eating meat until you know he is well purged.

<29> (MS A) To make him return more willingly to the hand.

Nephew: Now these have been clearly explained, I would like to be taught how I can make him return from the flight.[77]

Adelard: Take fatty and fresh pig's bacon and cook it in strong wine. Then this should be put into a little linen sack and hung in running water in such a way that the sack stays in the water for three or four days. Afterwards, this must be put between two boards and squeezed well, staying there for two days. Give him a little of this each morning you go out to hunt, either on its own or with other meat, and he will come back to you willingly whenever you let him fly. If you give some of this three times to a wild hawk, he will become tame.

(MS R) So that he may return willing, when he has been thrown.[78]

Take fresh pig's bacon and cook it very well in strong wine. Then, this should be hung in a linen sack in running water; in this way the whole sack is under the water for three or four days. Afterwards, this must be put between two boards and squeezed well, staying there for two days. And when you wish to go out to hunt, give him a little of the same meat in the morning and he will come back on his own. If you give some of the same meat to a wild hawk, he will become tame.

<30> This was what I had to say about the care of hawks. But if you or anybody else wants to add his own judgement, I shall not begrudge that.

Notes

1 The mention of 'rerum causis' can be seen as a direct allusion to Adelard's *QN*, especially considering that the motto at the beginning of that text is 'Sic faciunt cause rerum.'

2 'Slow' or 'slack'; MS W gives the reading 'mollis' ('soft').

3 Although 'knowledge' ('scientia', as in MS W) might seem a more natural reading and receives the support of the Anglo-Norman version in British Library, Harley 978, fol. 116v, line 19 ('savoient'), 'judgement' ('sententia' as in MS A) is probably correct, since it takes up the verb 'sentire' earlier in the sentence, and is, in turn, taken up in the last sentence of the text.

4 Compare Guibert of Nogent's criticism of Gaudry, bishop of Laon, for his immoderate passion for hunting, which Guibert blames on the English: 'De rebus nempe militaribus, canibus et accipitribus loqui gratum habuerat, quod apud Anglos didicerat' (Guibert de Nogent, *Autobiographie*, ed. E. R. Labande, Paris, 1981, p. 295).

5 Adelard refers to the danger of being accused of ignorance and spite at the beginning of the *DED*.

6 The books, signalled for the first time by Charles Haskins ('King Harold's Books'), do not seem to have survived; using Adelard's material, Daude de Pradas refers to *un libre del rei Enric d'Anclaterra* (v.1930) while the Anglo-Norman fragment in British Library, Harley 978, fol. 116v (line 16; ed. G. Tilander, 'Fragment d'un traité de fauconnerie anglo-normand en vers', *Studier i Modern Språkvetenskap*, 15, Uppsala, 1943, pp. 26–44, see p. 33) refers to 'El livere al bon rei Edward' – presumably Edward the Confessor – as his source.

7 Note that Adelard speaks only of taking young birds (i.e., 'eyasses') from the nest and not of catching mature birds, called 'haggards' or 'passagers' in the language of falconers and hawkers.

8 In comparison to the advice in other treatises and to modern practice, this delay of seven days sounds very short: to capture birds at this very weak stage rather multiplies the risks in handling them.

9 I.e., wings and feet.

10 I.e., they will abandon the nesting site.

11 This may point to an accurate observation, as the parents often bring branches with green leaves during the nesting time, to renew the bedding of the nest.

12 This sentence must refer to their making a shade with their wings rather

than with twigs, unless Adelard thought that the provision of fresh leaves (preceding note) was intended to provide shade.

13 'Passim' in this sense is recorded in the *Novum Glossarium*, s.v. 2.

14 Present-day falconers confirm that a bird fed with low-vitamin food gets a bone-disease called rickets, in which the legs get bowed inwards. Raw meat lacks the necessary calcium.

15 'Pellis exterior' could be understood also as the shell, but this would be less suited to the context, and would make the rest of the sentence redundant. From the translation of Daude de Pradas, v.594 ('e de la sotil pel mundatz') it appears to be the outer skin or membrane, once the shell is removed.

16 These cooked preparations are paralleled in various treatises, e.g., Frederick II, *De arte venandi*, II, p. 142.

17 This is a very accurate observation. Zoologists have noted that very young birds of prey, when still unable to eat without help, are predisposed to seize anything that is red.

18 It is important not to hurry the process, because the feathers of young birds, when still growing, are filled with blood, and any damage can cause a lasting handicap.

19 This stage is reached at about ten weeks.

20 Jesses are leather strips used to prevent the bird from flying away. They are tied around the legs of the bird, and the free ends are held in the hand, or attached to the leash.

21 The leash is a longer strip of leather, used to keep the bird from straying once he has been put on the perch, or when taken on the hand for a trip. Its extremity is tied to the jesses, and the free part can be knotted around a perch, or twisted and held in the hand.

22 From Adelard's words, it appears that this kind of knowledge was transmitted by experience and direct apprenticeship. This apparently also applies to the actual training of the birds, which Adelard does not describe.

23 It is most appropriate to take a hawk by the feet, because, unlike the falcons, goshawks take their prey and kill it not with the beak, but with the feet, so the feet must be immobilized.

24 Adelard's text is the only Latin treatise to mention this mobile perch; it is mentioned, however, in a fifteenth-century German treatise about the care of goshawks: 'Ältere deutsche Habichtslehre', section 23, in K. Lindner, *Die deutsche Habichtslehre. Das Beizbüchlein und seine Quellen*, Berlin, 1964, p. 115.

25 The meaning of 'lucanar' in most of the few contexts in which it appears is unclear, though 'window' or 'louvre' is most clearly attested. Adelard, however, may have been using it in the sense of 'candle' (i.e., as equivalent to 'lucerna'), since the usual practice was to put the bird before a lit candle for the first few nights, in order to weaken his resistance: see e.g., *Dancus Rex*, c.16.

26 The reason for this practice, which can be recognized in many illustrations, is given in later vernacular texts (e.g., *Livre du Roy Modus*, c.91, line 25):

touching the feathers of the bird with a small stick was recommended, so as not to take away their natural shine. Modern falconers call this shine 'bloom'; it provides a water-proofing for the feathers.

27 The reading of MS W, 'morem' ('habit'), is equally plausible.

28 This is the classical *a capite ad calcem* order found in medieval medical texts.

29 MSS ARW agree on the reading 'equum' (the chapter is not in the β-version). It sounds surprising to kill a horse for this disease, and the differentiation between a domestic and a wild horse is also unexpected. It is possible that the original text had 'porcum': 'porcus silvestris' would make sense as the boar. However, later on pig's meat is not recommended (c. 25). Another possibility is that the original text had 'capram', the wild goat being the roedeer. This is Swaen's suggested emendation. Simon de Hembrad gives 'qualiam' (= quail), a bird which would seem to be too small for the procedure required.

30 This would correspond to the size of a casting (the undigested fur and feathers regurgitated by the bird), and hence provides a good measure for what the bird can swallow. MS A states that four pills are to be made, presumably on the assumption that a separate pill would be used for clearing each nostril.

31 One should suppose, although Adelard does not say it, that during the whole operation the bird was held flat on some surface by a servant.

32 From the name and the curing of this illness, one can assume that this chapter concerns what the falconers call the 'frounce', a stomatitis which is signalled by a cheese-like plaque (*caseum*) in the mouth.

33 Simon de Hembrad explains this: 'accipe tegulam unam et calefac quousque rubefiat' ('take one tile and heat it until it becomes red').

34 What is meant here is not certain. The only visible holes in the region of the beak are the nostrils, adequately called 'nares' in the preceding chapter. They cannot be meant here. But at the extremity of the beak of birds of prey, there is, although invisible, an open place in the skull, and this could be intended. Abel Boyer still advised in 1948 cauterisation in the palate of the bird with the aid of silver nitrate: see A. Boyer and M. Planiol, *Traité de fauconnerie et d'autourserie*, Paris, 1948, p. 13.

35 This symptom could be related to a form of sinusitis, a frequent disease for captive raptors.

36 The Latin text is somewhat cryptic and the scene is difficult to reconstruct. The insertion of 'the rim of' provides an imaginable situation: the hawk is underneath a cauldron which has been turned upside down and rests on stones. This interpretation is also suggested by the reading in the *Liber de universis passionibus* of MS Cambridge University Library, Ff.6.13, fol. 71r, which, in copying the chapter of Adelard, instructs the reader to 'put stones under the rim of the cauldron so that it is raised half a foot from the ground' ('sub ora caldarie mitte petras ut sit a terra dimidio pedis').

37 MS A is isolated in giving 'carnem', which gives an awkward recipe: 'if the hawk vomits his meat, give him meat . . .'. The reading 'aquam' of MSS

RW is supported by the Latin compilations; however, birds of prey drink little.

38 Asthmatic breathing occurs as a symptom of one of the most frequent diseases of birds of prey, namely aspergillosis.

39 Note that the β-version has 'filings of iron' ('ferri'). MS A's text implies that one should give the birds this medication when they are going out ('exeuntibus'). MSS A and W both include the words 'dabis ieiunis' ('you will give it when they are hungry'), but the presence of the same words in the next sentence makes one suspicious. Only MS R retains the reference to a single hawk.

40 May-butter is unsalted butter put aside in May for medicinal use.

41 MS R reads 'raphani radiculas', i.e., little roots of a black radish.

42 The meaning of 'inflatio' is not clear: one would like to translate it as 'flatulence', but according to falconers this does not occur in birds of prey.

43 'Anigilia' (MS W) and 'anglica' (MS A) both suggest the Latin plant-name 'anagallis' or 'anagallicum' (= *Linaeus* anagallis). This identification is confirmed by the reading of MS R in which the name 'verbena' is substituted. Verbena is Albertus Magnus's name for 'anagallis'; see Albertus Magnus, *De vegetabilibus*, VI.471, ed. C. Jessen, Berlin, 1867, p. 579.

44 To these ingredients, MS R adds 'staphisagria' (i.e., stavesacre).

45 The reason for this procedure is explained in the *Liber de universis passionibus* as being that the sweetness of the honey entices the mites to the surface ('mellis dulcedine tinee exibunt foras'). Smearing the bird with a sticky substance would however damage the feathers.

46 This recipe is totally isolated in Latin falconry medicine, where the use of human bones does not occur. It is probably a transposition from another tradition, but equivalents are not known. 'Escinum' is from Anglo-Norman 'eskignon', and is used in the same sense as here by Robert of Torigni: 'os brachii quod eschinum vocant' (see *DMLBS*, s.v. eschinum).

47 In the heading to this chapter MS A gives the usual word for 'jaundice' ('ictericia'). The word 'fellera' would seem to be formed from 'fel' ('gall-bladder') on the analogy of 'cholera' from 'cholē' (= yellow/red bile; see *DMLBS*, s.v.). Jaundice, however, does not occur in birds; the term has been transposed from human to bird medicine; see Van den Abeele, *La Fauconnerie au moyen âge*, pp. 202–3.

48 Note the pun with *fellera* and *fallor*.

49 MSS R and W add 'and the cere' (i.e., the membrane at the base of the beak).

50 On the following very detailed plant description, containing some puzzling names, see Evans, 'Difficulties in the Botanical Lexis'. According to his analysis, the plant must be the Water Betony.

51 'Doctors' ('phisici', with MS W; MS R's abbreviation is ambiguous) is the term expected here, rather than 'philosophers' ('philosophi', MS A). The two words are easily confused, especially when abbreviated, but it is clearly to the pharmaceutical works of contemporary doctors that Adelard is

referring, even though Adelard's term has not been found in these works. The reading of MS W ('borica') is close to the form 'boric', which is found in the *Doctrina nutritionis*, Simon de Hembrad, Daude de Pradas (v. 2967, reading 'e boric') and the unpublished French translation in MS Lyon, Bibliothèque municipale 765, fol. 179v ('borich'). This plant would not seem to be the same, however, as the Biblical 'borith' = 'soapwort' (AS 'lēaþorwyrt'); see *MlatWb*, s.v. borith.

52 The Latin is clearly a calque on the Anglo-Saxon 'brūnwyrt'.

53 Swaen's suggestion that this is 'wild ivy' ('helix fera'), made on the basis that, in Greek, 'helix' (literally 'helix' or 'spiral') can mean a tendril of ivy, is tenuous. For an illustration of 'herba nigra' see T. G. Leporace, *Un inedito erbario*, Florence, 1952, p. 37.

54 MS A implies 'leaves' rather than 'branches'.

55 Although 'aculeus' usually means 'sting', some sort of vermin must be referred to here since the 'aculei' are said to 'die'. The prescription that Adelard uses here is very similar to that used by Gerard the Falconer (hence Albertus Magnus, *De animalibus*, XXIII, c. 19, para. 86, ed. Stadler, p. 1479) for 'anguille' = 'worms', and the terms 'stings' ('acus') and 'worms' ('angille') are interchanged in the falconry treatise of Alexander Medicus: see Van den Abeele, *La Fauconnerie au moyen âge*, pp. 186–7.

56 'Uvaria' means the bird's 'belly', from the fact that this is the place where the eggs ('ova') are formed.

57 This method, which appears for the first time in Adelard's text, reflects a point of practical experience. Birds of prey drink very little, so it will not help to present a liquid medicine in a cup. The expedient of a gut was a very adequate method to ensure that the liquid was effectively swallowed, and its action was thus delayed until the gut was dissolved in the stomach. For a detailed parallel description, see *Gerardus Falconarius*, c.8. Present-day falconers use a syringe with a plastic tube.

58 As Swaen suggested (p. 26, n. 204), this must be the Anglo-Saxon word for lizard or newt: 'âðexe' (Bosworth and Toller, s.v.).

59 'Wood' has to be supplied for reasons of coherence, because leaves or stems of ivy will not be adequate for the prescribed procedure. The addition is supported by the translation of Daude de Pradas, vv. 3435–6: 'en un fust d'edra vert e gros / per costa faretz far un cros.'

60 'Stone', not explained here, is probably what falconers used to call 'cray', a form of constipation.

61 In the heading to this chapter MS A names this disease 'tenasmon', i.e., what is now known as 'tenesmus' (from Greek τεινεσμός) and is defined as 'an ineffective painful straining to empty the bowels or bladder' (Collins English Dictionary, s.v.).

62 'Bisticus' is an unidentified term, which elsewhere occurs only in the *Epistola ad Ptolomeum*, c. 15. The symptom 'defecates twice' could refer to scattered instead of concentrated droppings, which is regarded by falconers

as a symptom of illness; see Van den Abeele, *La Fauconnerie au moyen âge*, pp. 190–1.

63 The flint-stone 'pyrites' is described in Pliny, *Natural History*, 34, 135 and 36, 137.

64 Cf. the word 'vermoulure', attested in French from 1283 onwards (*Dictionnaire historique de la langue française*, p. 2231).

65 'Podagra' ('gout') in human medicine corresponds in treatises on falconry to 'bumblefoot', a very frequent problem. It is an inflamed and infected wound on the sole of the foot.

66 To put the bird on a stone instead of the usual perch is certainly effective, as it changes the position of the feet.

67 This treatment for mewing is the most widely diffused recipe in medieval falconry. With slight variants, it occurs in at least five Latin texts, three French, one Provençal and one German, up to well into the sixteenth century.

68 To 'enseame' is to rid a bird of his surplus fat. The control of the correct body weight is one of the most important aspects of daily care. The order of the chapters makes good sense, because enseaming is necessary after the period of mewing.

69 The reading of MS A, probably a result of scribal errors, implies 'heads of sheep and goats'.

70 These instructions do not seem to be consistent: while chicken is not rich and beef is average, hare is a very rich food. The preceding rule of changing the meat often is certainly effective.

71 'Roughage' is in the language of the falconers some rough substance such as a hare's foot, some feathers or a bit of cotton, given to the bird. This undigested material cleans the stomach and is regurgitated in the form of round 'castings'. In captivity, especially when fed with meat, the hawks do not get enough rough substance to form castings.

72 Chapters 27–9 have no equivalent in the β-version, and chapters 27 and 28 are also lacking in MS R. They all concern certain preparations for special use, and are perhaps miscellaneous chapters added to the end of the original treatise.

73 The two Old English words transcribed here are 'milde' and 'stræc/strec' ('mild' and 'strong'; see Bosworth and Toller, s.v.), and it is clear that it is the second word that has been corrupted into 'tiriaca' (MS A) and 'strictura' (MS W) respectively, in chapter 28. The only text deriving from the *De avibus tractatus* which includes chapters 27 and 28 is the unpublished French translation in MS Lyon, Bibliothèque municipale, 765, fol. 180v, where the terms are given as 'milda' and 'stritura'. Swaen suggests that these chapters come from a different source than the rest of the text.

74 The reading of MS W, 'Woerouam' instead of 'verdetenam veterem', represents the Old English form for woodruff, 'woderove'; cf. the thirteenth-

century gloss in MS Oxford, Corpus Christi 135, fol. 37r: 'haste regie id est woderove' (Hunt, *Teaching and Learning Latin*, I, p. 50).

75 The scribe of MS A has apparently misread the word 'strica' as 'tiriaca/tyriaca', i.e., 'theriac', a preparation made from a large number of ingredients, intended originally as an antidote against poison, which is found both in human medicine and in falconry treatises.

76 'Consound' ('Consolida') is the medieval term for the whole daisy family, including comfrey and may weed.

77 Literally, 'game'; cf. Middle High German 'Spil'; see D. Dalby, *Lexicon of the Medieval German Hunt*, Berlin, 1965, s.v.

78 In MS R this chapter is inserted before the beginning of the description of birds' illnesses and their cure (i.e., after the end of chapter 6). It does not occur in MS W.

Abbreviations and Select Bibliography

This bibliography includes all the abbreviations used in the Introduction and Notes, and primary and secondary sources referred to more than once, or of particular relevance to the texts edited here. The classical and patristic sources have been quoted according to the standard editions, and are not included here unless the page numbers of a particular edition have been used.

Reference Works and Titles of Series

Bosworth and Toller: J. Bosworth and Y. Northcote Toller, *An Anglo-Saxon Dictionary*, Oxford, 1898.

CCCM: Corpus Christianorum Continuatio Mediaevalis.

Collins English Dictionary, 2nd edn, London and Glasgow, 1986.

Dictionnaire historique de la langue française, Paris, 1992.

DMLBS: E. Latham and D. Howlett, *Dictionary of Medieval Latin from British Sources*, Oxford, 1975–

Hunt: T. Hunt, *Teaching and Learning Latin in 13th Century England*, London, 1991.

LLNMA: J. W. Fuchs ad O. Weijers, *Lexicon latinitatis Nederlandicae medii aevi*, Leiden, 1970–

MlatWb: *Mittellateinisches Wörterbuch*, Munich, 1959–

Novum Glossarium mediae latinitatis, Copenhagen, 1957–

PL: Patrologia Latina, ed. J. P. Migne, 221 vols, Paris, 1844–64.

TLL: *Thesaurus linguae latinae*, Leipzig, 1900–

Tobler-Lommatzsch: A. Tobler and E. Lommatzsch, *Altfranzösisches Wörterbuch*, Stuttgart, 1915–

v. Wartburg: W. v. Wartburg, *Französisches Etymologisches Wörterbuch*, Tübingen, 1950–

Walther: H. Walther, *Proverbia sententiaeque latinitatis medii aevi*, 9 vols, Göttingen, 1963–86.

The Works of Adelard of Bath

The works edited in this volume, and their previous editions

DAT: De avibus tractatus

Swaen: *De cura accipitrum: a Mediaeval Latin Treatise by Adelard of Bath*, ed. A. E. H. Swaen, Groningen, 1937.

DED: De eodem et diverso

Willner: *Des Adelard von Bath Traktat De eodem et diverso*, ed. H. Willner, Beiträge zur Geschichte der Philosophie des Mittelalters, IV.1, Münster, 1903; for Hebrew versions see Gollancz.

QN: Questiones naturales

Müller: *Die Quaestiones naturales des Adelardus von Bath*, ed. M. Müller, Beiträge zur Geschichte der Philosophie und Theologie des Mittelalters, XXXI.2, Münster, 1934.

"Quaestiones naturales" di Adelardo di Bath, eds S. Balossi, A. di Giovanni and B. Ferrari, Scientia Veterum, collana di studi di storia della medicina diretta e curata da G. Del Guerra, Rapallo, 1965.

Other works by Adelard

Original works:

De opere astrolapsus, edited in Dickey, 'Adelard of Bath', pp. 112–229.

Regule abaci, edited in B. Boncompagni, 'Intorno ad uno scritto inedito di Adelardo di Bath intitolato "Regule abaci"', *Bulletino di bibliografia e di storia delle scienze matematiche e fisiche*, 14, 1881, pp. 1–134.

Translations:

Abū Maʿšar (Iafar), *Ysagoga minor*, edited in Abū Maʿšar, *The Abbreviation of the Introduction to Astrology Together with the Medieval Translation of Adelard of Bath*, eds C. Burnett, K. Yamamoto and M. Yano, Leiden, 1994.

Euclid, *Elements*. Adelard's translation edited in H. L. L. Busard, *The First Latin Translation of Euclid's Elements Commonly Ascribed to Adelard of Bath*, Studies and Texts 64, Toronto, Pontifical Institute of Mediaeval Studies, 1983; for the *commentum*-version see Robert of Chester (?), *Redaction*.

al-Khwārizmī, Astronomical Tables (*Ezig*), edited in *Die astronomischen Tafeln des Muḥammed ibn Mūsā al-Khwārizmī in der Bearbeitung des Maslama ibn Aḥmed al-Madjrīṭī und der lateinische Übersetzung des Athelhard von Bath*, eds H. Suter, A. Bjørnbo and R. Bestborn, Copenhagen, 1914.

Other Primary Sources

Alexander, Pseudo-, *Problemata*, ed. in *Problemata Aristotelis . . .*, Venice, 1505.

Beringerius of Tours, *Rescriptum contra Lanfrannum* (*De sacra coena*), I, ed. R. B. C. Huygens, CCCM, Turnhout, 1988.

Bernardus Silvestris, *Cosmographia*, ed. P. Dronke, Leiden, 1978.

Boethius, *De arithmetica* and *De musica*, ed. G. Friedlein, Leipzig, 1867.

Cicero, *De natura deorum*, ed. and trans. H. Rackham, Loeb Classical Library, London and New York, 1933.

Constantine the African, *Pantegni*, in *Opera omnia Isaac*, Lyons, 1515.
Dancus Rex: see Tilander, *Dancus Rex*.
Daude de Pradas, *Dels Auzels Cassadors*, ed. H. Schutz, Columbus, Ohio, 1945.
Frederick II, *De arte venandi cum avibus*, ed. C. A. Willemsen, Leipzig, 1942.
Geometria incerti auctoris, in N. Bubnov, *Gerberti opera mathematica*, Berlin, 1899, pp. 310–65.
Gerard the Falconer: see Tilander, *Dancus Rex*.
Honorius Augustodunensis, *Clavis physicae*, ed. P. Lucentini, Rome, 1974.
Elucidarium, ed. Y. Lefèvre, *L'Elucidarius et les lucidaires*, Paris, 1954.
John of Salisbury, *Metalogicon*, ed. J. B. Hall, CCCM, Turnhout, 1991.
John Scotus Eriugena, *Periphyseon*, ed. and trans. I. P. Sheldon-Williams, L. Bieler, E. A. Jeauneau and J. J. O'Meara, 4 vols, Dublin, 1969–95.
Les livres du Roy Modus et de la Royne Ratio, ed. G. Tilander, Paris, 1932.
Martial, *Epigrams*, ed. D. R. Shackleton Bailey, 3 vols, Cambridge Mass., 1993.
Nemesius, *De natura hominis*, trans. Burgundio of Pisa, eds G. Verbeke and J. R. Moncho, Leiden, 1975.
Premnon physicon, trans. Alfano, ed. C. Burkhard, Leipzig, 1917.
Problemata physica arabica, ed. L. S. Filius, Culemborg, 1989.
The Prose Salernitan Questions, ed. B. Lawn, London, 1979.
Robert of Chester (?), *Redaction of Euclid's* Elements, *the so-called Adelard II Version*, eds H.L.L.Busard and M. Folkerts, 2 vols, Basel-Boston-Berlin, 1992.
Roger Bacon, *Opus maius*, ed. R. Bridges, 2 vols, Oxford, 1897.
Vincent of Beauvais, *Speculum naturale*, Douai, 1624.
William of Conches, *Dragmaticon*, ed. I. Ronca, CCCM, Turnhout, 1997.

Other Works

Bliemetzrieder, F., *Athelhard von Bath: Blätter aus dem Leben eines englischen Naturphilosophen des 12. Jahrhunderts und Bahnbreckers einer Wiedererweckung der griechischen Antike*, Munich, 1935.
Bubnov, N., *Gerberti Opera Mathematica*, Berlin, 1899.
Burnett, C., 'Adelard of Bath and the Arabs', in *Rencontres de cultures dans la philosophie médiévale*, eds M. Fattori and J. Hamesse, Louvain-la-Neuve and Cassino, 1990, pp. 89–108.
Adelard of Bath: An English Scientist and Arabist of the Early Twelfth Century, Warburg Institute Surveys and Texts, 14, London, 1987.
'Adelard of Bath's Doctrine on Universals and the *Consolatio Philosophiae* of Boethius', *Didascalia*, 1, Tokyo, 1995, pp. 1–14.
'Adelard, Music and the Quadrivium', in Burnett (ed.), *Adelard of Bath*, pp. 69–86.
'*Algorismi vel helcep decentior est diligentia*: the Arithmetic of Adelard of Bath and his Circle', in *Mathematische Probleme im Mittelalter: Der*

lateinische und arabische Sprachbereich, ed. M. Folkerts, Wolfenbüttel, 1996, pp. 221–331.

'The Latin and Arabic Influences on the Vocabulary Concerning Demonstrative Argument in the Versions of Euclid's *Elements* Associated with Adelard of Bath', in *Aux origines du lexique philosophique européen*, ed. J. Hamesse, Louvain-la-Neuve, 1997, pp. 117–35.

'Master Theodore, Frederick II's Philosopher', in *Federico II e le nuove culture*, Atti del XXXI Convegno storico internazionale, Todi, 9–12 ottobre 1994, Spoleto, 1995, pp. 225–85.

'Ocreatus', in *Vestigia Mathematica: Studies in Medieval and Early Modern Mathematics in Honour of H. L. L. Busard*, eds M. Folkerts and J. P. Hogendijk, Amsterdam, 1993, pp. 69–77.

'*Omnibus convenit Platonicis*: an Appendix to Adelard of Bath's *Quaestiones naturales?*', in *From Athens to Chartres, Neoplatonism and Medieval Thought: Studies in Honour of E. Jeauneau*, ed. H. J. Westra, Leiden, 1994, pp. 259–81.

'Talismans: Magic as Science? Necromancy among the Seven Liberal Arts', article I in Burnett, *Magic and Divination in the Middle Ages: Texts and Techniques in the Islamic and Christian Worlds*, Aldershot, 1996.

'The Works of Petrus Alfonsi: Questions of Authenticity', *Medium Ævum*, 66, 1997, pp. 42–79.

Burnett and I. Ronca, 'The Transmission of Adelard's *Questiones naturales*', to be published in *Revue d'histoire des textes*.

Clagett, M., 'The Medieval Latin Translations from the Arabic of the *Elements* of Euclid', *Isis*, 44, 1953, pp. 16–42.

Cochrane, L., *Adelard of Bath: The First English Scientist*, London, 1994.

Courcelle, P., 'Adelard de Bath et la *Consolation* de Boèce', *Kyriakon: Festschrift Johannes Quasten*, 2 vols, Münster, 1970, II, pp. 572–5.

Daiber, H., 'Masā'il wa-adjwiba'', *Encyclopedia of Islam*, 2nd edn., VI, Leiden, 1991, pp. 636–9.

Dickey, B., 'Adelard of Bath: An Examination Based on Heretofore Unexamined Manuscripts', Ph. D. dissertation, University of Toronto, 1982.

Drew, A., 'The *De eodem et diverso*', in Burnett, C. (ed.), *Adelard of Bath: An English Scientist and Arabist of the Early Twelfth Century*, pp. 16–24.

Evans, D., 'Adelard on Falconry', in Burnett, C. (ed.), *Adelard of Bath: An English Scientist and Arabist of the Early Twelfth Century*, pp. 25–7.

'Difficulties in the Botanical Lexis of the Old Provençal Falconry Treatise: Daude de Pradas and Adelard of Bath', in *Studia Occitanica in memoriam Paul Remy*, II, Kalamazoo, 1986, pp. 281–9.

Fredborg, K. M., 'The Commentaries on Cicero's de Inventione and Rhetorica ad Herennium by William of Champeaux', *Cahiers de l'Institut du moyen-âge grec et latin*, 17, 1976, pp. 1–39.

Geymonat, F., ed. and comm., *Le Questioni filosofiche*, Pisa, 1999.

Gibson, M., 'Adelard of Bath' in Burnett, C. (ed.), *Adelard of Bath: An English Scientist and Arabist of the Early Twelfth Century*, pp. 7–16.

Gollancz, H., *Dodi ve-Nechdi of Berachya Hanakdan*, London, 1920.

Haskins, C. H., 'King Harold's Books', in *English Historical Review*, 37, 1922, pp. 398–400.

Studies in the History of Mediaeval Science, 2nd edn, Cambridge, Mass., 1927.

Hauréau, B., *Histoire de la philosophie scolastique*, Paris, 1872.

Jolivet, J., 'Adélard de Bath et l'amour des choses', in *Metaphysique, histoire de la philosophie. Recueil d'études offert à Fernand Brunner*, Neuchâtel, 1981, pp. 77–84.

'Les "Quaestiones naturales" d'Adélard de Bath ou la nature sans le Livre', in *Études de civilisation médiévale. Mélanges E.-R. Labande*, Poitiers, 1974, pp. 437–46.

Lawn, B., *The Salernitan Questions*, Oxford, 1963; Italian translation by A. Spagnuolo (with corrections and additions), Naples, 1969. (All the references in this volume are to the English version.)

Maurach, G., 'Adelhard von Bath - Ein Naturdenker des 12.Jahrhunderts', *Abhandlungen der Braunschweigischen Wissenschaftlichen Gesellschaft*, 34, 1982, pp. 173–97.

Mews, C. J., 'Nominalism and Theology before Abaelard: New Light on Roscelin of Compiègne', *Vivarium*, 30, 1992, pp. 4–33.

North, J., 'Some Norman Horoscopes', in Burnett, C. (ed.), *Adelard of Bath: An English Scientist and Arabist of the Early Twelfth Century*, pp. 147–61.

Pabst, B., *Prosimetrum*, 2 vols, Cologne et al., 1994.

Reynolds, L. D., *Texts and Transmission*, Oxford, 1983.

Silverstein, T., 'Adelard, Aristotle and the *De natura deorum*', *Classical Philology*, 47, 1952, pp. 82–6.

Southern, R. W., *Robert Grosseteste: The Growth of an English Mind in Medieval Europe*, 2nd edn, Oxford, 1992.

Speer, A., *Die entdeckte Natur*, Leiden, 1995.

Thorndike, L., *A History of Magic and Experimental Science*, II, New York, 1923.

Tilander, G., *Dancus Rex, Guillelmus Falconarius, Gerardus Falconarius: Les plus anciens traités de fauconnerie de l'Occident*, Lund, 1963.

'Fragment d'un traité de fauconnerie anglo-normand en vers', *Studier i Modern Språkvetenskap*, 15, 1944, pp. 26–44.

Van den Abeele, *La Fauconnerie au moyen âge. Connaissance, affaitage et médecine des oiseaux de chasse d'après les traités latins*, Paris, 1994.

'Les Traités de fauconnerie latins du xiie s. Manuscrits et perspectives', *Scriptorium*, 44, 1990, pp. 276–86.

Index

This index provides a guide to the subject matter, proper names, leitmotivs, and terms left in their original languages, in the English translations of the *De eodem et diverso, Questiones naturales* and *De avibus tractatus*.